Engineering Sustainability Goals

Also of Interest

Advanced Energy Materials.
Natural Polymers, Renewable Energy, Sustainable Development
Verma, Sonika, Sharma, Rajput (Eds.), 2025
ISBN 978-3-11-128890-1, e-ISBN 978-3-11-128907-6

Nanocellulose-Reinforced Thermoplastic Starch Composites.
Sustainable Materials for Packaging
Ilyas, Sapuan, Norrrahim (Eds.), 2023
ISBN 978-3-11-077356-9, e-ISBN 978-3-11-077360-6

Science, Engineering, and Sustainable Development.
Cases in Planning, Health, Agriculture, and the Environment
Integrated Global STEM, Vol 1
Krueger, Telliel, Soboyejo (Eds.), 2024
ISBN 978-3-11-075749-1, e-ISBN 978-3-11-075750-7

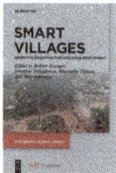

Smart Villages.
Generative Innovation for Livelihood Development
Integrated Global STEM, Vol 2
Krueger, Vedogbeton, Fofana, Soboyejo (Eds.), 2024
ISBN 978-3-11-078621-7, e-ISBN 978-3-11-078623-1

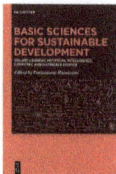

Basic Sciences for Sustainable Development.
Energy, Artificial intelligence, Chemistry, and Materials Science
Ramasami (Ed.), 2023
ISBN 978-3-11-099097-3, e-ISBN 978-3-11-091336-1

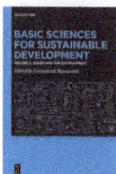

Basic Sciences for Sustainable Development.
Vol 2: Water and the Environment
Ramasami (Ed.), 2023
ISBN 978-3-11-107089-6, e-ISBN 978-3-11-107120-6

Engineering Sustainability Goals

UNSDG 12: Responsible Consumption and Production

Edited by
David S-K. Ting and Jacqueline A. Stagner

DE GRUYTER

Editors
David S-K. Ting
Turbulence & Energy Laboratory
University of Windsor
Windsor, ON
Canada N9B 3P4

Jacqueline A. Stagner
Turbulence & Energy Laboratory
University of Windsor
Windsor, ON
Canada N9B 3P4

ISBN 978-3-11-156293-3
e-ISBN (PDF) 978-3-11-156304-6
e-ISBN (EPUB) 978-3-11-156324-4

Library of Congress Cataloging-in-Publication Data
A CIP catalog record for this book has been applied for at the Library of Congress.

Bibliographic information published by the Deutsche Nationalbibliothek
The Deutsche Nationalbibliothek lists this publication in the Deutsche Nationalbibliografie;
detailed bibliographic data are available on the Internet at http://dnb.dnb.de.

© 2026 Walter de Gruyter GmbH, Berlin/Boston, Genthiner Straße 13, 10785 Berlin
Cover image: Petmal/iStock/Getty Images Plus
Typesetting: Integra Software Services Pvt. Ltd.

www.degruyterbrill.com
Questions about General Product Safety Regulation:
productsafety@degruyterbrill.com

To everyone who aspires to achieve sustainability goals.

Preface

"Keep in existence" is a synonym for "sustain." Cradle-to-cradle mode of operation is intrinsic to keeping the human species in existence. This naturally encompasses, among other elements, ecofriendly energy and renewables, along with conservational building energy management. Along these lines, we strive to contribute to the engineering of the United Nations Sustainable Development Goals (SDGs).

This volume commences from the wellspring that sustains everyone, that is, food. Food is inextricably linked to SDG 1: No Poverty, SDG 2: Zero Hunger, and SDG 3: Good Health and Well-being. Sustainable food production begins with sustainable farming. Encouraging farmers to use sustainable practices can be challenging because a change from their accustomed practice is required. **Savari**, **Rezai**, **Shokati Amghani**, and **Mojtahedi** enlighten us with a way to overcome this challenge in Chapter 1. The idea is to alter farmers' behavior. They synergize eight predominant theories that assess environmentally friendly behaviors. The strengths and variables of these theories are revealed, paving the way to promote environmentally friendly and safe behaviors in farming.

Intertwined with food are water and energy. On that account, **Farmandeh**, **Choobchian**, and **Karami** explore the challenges associated with the water-energy-food nexus in Chapter 2. Specifically, they identify and analyze the essential elements of the water-energy-food nexus in agricultural development. Within the predominant water nexus, total water usage, water consumption for energy production, and water efficiency are dynamically coupled. They recommend six policy implications: (1) Prioritization of water management in policy frameworks; (2) integration of energy efficiency in agricultural policies; (3) focused investment in sustainable food production; (5) comprehensive nexus-based planning; and (6) localized policy adaptation.

Medicinal and aromatic plants may not come to mind when we consider the 17 United Nations Sustainable Development Goals. These plants play a critical role in traditional medicine and culture, and hence, SDG 3: Good Health and Well-being. **Bahadori** and **Sadighi** disclose these realities of medicinal and aromatic plants in Chapter 3. Medicinal and aromatic plants are natural resources that contribute significantly to biodiversity, ecological stability, and cultural heritage. They find that environmental and legislation challenges are the most prevalent barriers to the conservation of these precious greens.

A sustainable tomorrow necessitates green architecture and engineering globally. **Awazi** brings us up to date with the situation in the Congo Basin in Chapter 4. The Congo Basin faces numerous challenges, including limited use of renewable energy, energy efficiency, water conservation techniques, and limited implementation of forest conservation strategies. Governance, weak institutional and policy frameworks, high costs, green grabbing, lack of standardized certification systems, poor urban and rural planning, shortage of utility data, and lack of awareness and knowledge are

https://doi.org/10.1515/9783111563046-202

obstacles to be overcome. Certain subregional champions are making inroads in green architecture and engineering, specifically, SGD 7, SDG 9, SDG 11, and SDG 13.

The progress in greening Africa is the subject of Chapter 5. In this chapter, **Awazi** conveys the latest status, rationale, and national transition to green growth in Africa. The continent has been striving for the Millennium Development Goals since the early 2000s and the United Nations SDGs since 2015. They are confronted by land degradation, deforestation, water scarcity, air pollution, and energy consumption and waste management associated with rapid urbanization and industrialization. Despite many advancements in greening of several nations in Africa, inadequate financing, weak institutional frameworks, and limited technological capacity hinder the widespread implementation of green growth strategies.

Plastic waste is a serious hindrance against our striving toward sustainability. **Jain, Jain, Kushwah, Solanki, Malviya**, and **Singh** present a positive update in confronting plastic waste in Chapter 6. Why do we love plastics in the first place? Because they are strong, durable, waterproof, lightweight, and flexible. The three Rs, Reduce, Reuse, and Recycle, are in order. Other than reducing plastic production and usage, the next best thing is to reuse, and this is closely tied to recycling. They found that incorporating plastic waste into construction materials substantially enhances environmental sustainability and offers economic benefits. One highlight is the significant improvement in the compressive strength of concrete with the optimal plastic waste inclusion.

Engineering infrastructures consume a massive amount of energy and materials. These constructions can make use of recycled matter, biocrete, and other advanced materials to enhance sustainability. **Sharma, Dehalwar, Kumar, Yadav**, and **Verma** present the utilization of advanced materials for permeable paving, biocrete, and piezoelectric materials in walkways to transit stations in Chapter 7. Permeable paving improves stormwater management and reclamation, promoting pedestrian safety and environmental resilience. Self-healing and eco-friendly biocrete can lower maintenance costs and enhance structural longevity. Traffic kinetic energy can be harnessed to power lighting and sensors. Cost is one hurdle to overcome.

Building energy management is an integral part of engineering sustainability goals. Approximately 40% of global energy is used in heating, ventilation, and air conditioning (HVAC) alone. **Wang, Carriveau**, and **Ting** furnish regarding this in Chapter 8. Integrating solar technology into HVAC systems offers a promising pathway toward net-zero energy buildings. To hasten this expedition, solar-assisted HVAC should be complemented by heat pumps, energy storage systems, and auxiliary renewable sources in a synergistic manner. In parallel, advanced energy management strategies should be implemented to bring about potent efficiency, reliability, and sustainability.

It would be amiss if engineering education is not brought up to speed to appropriately enlighten and prepare young minds in the realization of a sustainable tomorrow. In the final chapter, **Schuelke-Leech** contends that there are three distinct visions for engineering reform: Vision 1 – industry-servicing engineering; Vision 2 – technology-responsive engineering; and Vision 3 – socially responsive engineering.

A Hebrew proverb teaches, "And do not force your children into your learning, for they were born in another time." The chapter strives to spur discussion about these three competing visions and how to better prepare the future engineer to effectively engineer sustainability.

This discourse aims to foster cradle-to-cradle engineering, the only mode of operation for a sustainable tomorrow. Progresses associated with behavior and culture, politics and policies, technologies, and economics are expounded. Many outstanding hurdles have been revealed, and possible strategies to overcome them are proposed.

Acknowledgments

This book was only realized because of the scrupulous experts who compiled the chapters and the anonymous reviewers who furthered the quality of the chapters through rigorous review. The editors truly enjoyed collaborating with the amazing publishing team who epitomized simplicity and eased our work while boosting our efficiency. Providence carried this dream from inception to consummation.

David S-K. Ting and Jacqueline A. Stagner
Turbulence & Energy Laboratory
University of Windsor
Canada

https://doi.org/10.1515/9783111563046-203

Contents

List of Contributing Authors

Moslem Savari
Department of Agricultural Extension and
Education
Agricultural Sciences and Natural Resources
University of Khuzestan
Mollasani
Iran
Savari@asnrukh.ac.ir

Marzieh Rezai
Department of Natural Resources Engineering
University of Hormozgan
Bandar Abbas
Iran
M.Rezai@Hormozgan.ac.ir

Mohammad Shokati Amghani
Department of Agricultural Extension and
Education
College of Agriculture
Tarbiat Modares University (TMU)
Tehran
Iran
m.shokati@modares.ac.ir

Mehrdad Mojtahedi
Department of Agricultural Management and
Development
Faculty of Agriculture
University of Tehran
Karaj
Iran
mehrdad.mojtahedi@ut.ac.ir

Ebrahim Farmandeh
Department of Agricultural Extension and
Education
College of Agriculture
Tarbiat Modares University (TMU)
Tehran
Iran

Shahla Choobchian
Department of Agricultural Extension and
Education
College of Agriculture
Tarbiat Modares University (TMU)
Tehran
Iran
shchoobchian@modares.ac.ir

Shobeir Karami
Persian Gulf Research Institute
Persian Gulf University
Bushehr
Iran

Mitra Bahadori
Department of Rural Development
College of Agriculture
Isfahan University of Technology
Isfahan
Iran
m.bahadori@alumni.iut.ac.ir
ORCID: 0009-0009-3940-8103

Hassan Sadighi
Department of Agricultural Extension and
Education
College of Agriculture
Tarbiat Modares University
Tehran
Iran
sadigh_h@modares.ac.ir
https://www.scopus.com/authid/detail.uri?au
thorId=22941821100

Nyong Princely Awazi
Department of Forestry and Wildlife Technology
College of Technology (COLTECH)
The University of Bamenda
Bamenda
Cameroon
and
FOKABS INC.
2500 St. Laurent Blvd
Ottawa, ON
Canada K1H 1B1

https://doi.org/10.1515/9783111563046-205

and
Department of Forestry
Faculty of Agronomy and Agricultural Sciences
The University of Dschang
Dschang
Cameroon
awazinyong@uniba.cm, pnyong@fokabs.com,
nyongprincely@gmail.com
ORCID: 0000-0002-0801-0719

Devansh Jain
L. N. Malviya Infra Projects Pvt. Ltd.
Bhopal 462023
Madhya Pradesh
India

Harshita Jain
Jai Narain College of Technology
Bhopal 462038
Madhya Pradesh
India

Suresh Singh Kushwah
Rajiv Gandhi Proudyogiki Vishwavidyalaya
Bhopal 462033
Madhya Pradesh
India

Vijay Singh Solanki
L. N. Malviya Infra Projects Pvt. Ltd.
Bhopal 462023
Madhya Pradesh
India

Laxmi Narayan Malviya
L. N. Malviya Infra Projects Pvt. Ltd.
Bhopal 462023
Madhya Pradesh
India

Dungar Singh
L. N. Malviya Infra Projects Pvt. Ltd.
Bhopal 462023
Madhya Pradesh
India
dsdudi97@gmail.com

Shashikant Nishant Sharma
Maulana Azad National Institute of Technology
Bhopal
Madhya Pradesh
India
urp2025@gmail.com

Kavita Dehalwar
Maulana Azad National Institute of Technology
Bhopal
Madhya Pradesh
India

Gopal Kumar
Maulana Azad National Institute of Technology
Bhopal
Madhya Pradesh
India

Krishna Yadav
Maulana Azad National Institute of Technology
Bhopal
Madhya Pradesh
India

Devraj Verma
Maulana Azad National Institute of Technology
Bhopal
Madhya Pradesh
India

Xi Wang
Turbulence and Energy Laboratory
University of Windsor
401 Sunset Avenue
Windsor, ON
Canada
Wang1st@uwindsor.ca

Rupp Carriveau
Turbulence and Energy Laboratory
University of Windsor
401 Sunset Avenue
Windsor, ON
Canada

David S-K. Ting
Turbulence and Energy Laboratory
University of Windsor
401 Sunset Avenue
Windsor, ON
Canada

Beth-Anne Schuelke-Leech
University of Windsor
Windsor, ON
Canada
basl@uwindsor.ca

Moslem Savari, Marzieh Rezai, Mohammad Shokati Amghani*, and
Mehrdad Mojtahedi

Dominant Socio-psychological Theories in Assessing Eco-friendly Behaviors of Farmers

Abstract: Nowadays, food demand for the world's growing population has increased.
Accordingly, despite the global restrictions on water extractions and arable land quality, the indiscriminate use of chemicals to produce sufficient food to meet the needs
of the growing population has been increased in developing countries. Today, environmental preservation is one of humanity's significant challenges. Accordingly, the
production of healthy agricultural products in accordance with global standards, syn-
chronized with the preservation of the environment, has received great attention in
the realization of sustainable agriculture. Among the most significant problems aris-
ing from excessive exploitation of the natural environment are biodiversity loss,
rangeland degradation, depletion of water resources, increased dust storms, deforestation, excessive use of chemical pesticides, soil erosion, and land use change. Studies
indicate that many environmental issues stem from irresponsible human behavior toward the environment. Therefore, understanding and changing individuals' behaviors
are recognized as essential prerequisites for environmental management programs.
Understanding people's thought processes and their comprehension of natural resources, as well as their willingness to take necessary actions to protect the environment,
are crucial for resolving environmental crises. Hence, researchers and policymakers
argue that environmental problems can be mitigated through the adoption of environmentally friendly behaviors by farmers. In recent years, various theories have
been employed to assess environmentally friendly behaviors in different studies. This
chapter examined eight predominant theories utilized by various researchers and
outlined all their components and relationships. This research attempts to discuss the
strengths and variables of each theory to serve as a guide for researchers and policy-

*Corresponding author: Mohammad Shokati Amghani**, Department of Agricultural Extension and
Education, College of Agriculture, Tarbiat Modares University (TMU), Tehran, Iran,
e-mail: m.shokati@modares.ac.ir
Moslem Savari, Department of Agricultural Extension and Education, Agricultural Sciences and Natural
Resources University of Khuzestan, Mollasani, Iran, e-mail: Savari@asnrukh.ac.ir
Marzieh Rezai, Department of Natural Resources Engineering, University of Hormozgan, Bandar Abbas,
Iran, e-mail: M.Rezai@Hormozgan.ac.ir
Mehrdad Mojtahedi, Department of Agricultural Management and Development, Faculty of
Agriculture, University of Tehran, Karaj, Iran, e-mail: mehrdad.mojtahedi@ut.ac.ir

https://doi.org/10.1515/9783111563046-001

makers seeking to promote environmentally friendly and safe behaviors in the environment.

Keywords: Safe behaviors in the environment, social psychological theories, healthy food, sustainable development, environmentally friendly behaviors

1 Introduction

In recent decades, special measures have been implemented to intensify the utilization of chemical fertilizers in response to the escalating demand for food attributed to population growth across various countries (Xiang et al., 2020). The widespread use of chemical fertilizers to enhance soil fertility expanded in the 1950s and 1960s, leading to the Green Revolution. Although this phenomenon contributed to increased food production worldwide, it had adverse effects on the environment (Agegnehu et al., 2016). Conversely, traditional and conventional agricultural practices often rely on the excessive application of chemical inputs to achieve higher yields (Zheng et al., 2022). Hence, conventional agricultural practices reliant on heavy usage of chemical fertilizers present a significant threat to both human health and the environment (Duan et al., 2021). At the same time, the ongoing rise in food production and the enhancement of food security to some extents have been facilitated by the adoption of chemical fertilizers, pesticides, and novel technologies (Bahrulolum et al., 2021), as chemical fertilizers play a significant role in improving soil efficiency and ensuring agricultural product availability (Wang et al., 2023). However, these practices also limit the quality and sustainability of agricultural development (Moya et al., 2019). Consequently, the escalating use of chemical fertilizers and pesticides has led to environmental issues such as soil degradation, greenhouse gas emissions, and water pollution (Huang et al., 2020).

On the other hand, in Iran, as one of the leading countries in the Middle East, there is an increased urgency to safeguard the environment compared to other nations (Maleksaeidi and Keshavarz, 2019). Since, Iran has the highest soil erosion rate among Asian countries, with over 94% of its agricultural lands being impacted by erosion (FAO, 2017). This erosion is occurring at an annual rate of 6.16 tons per hectare, indicating a concerning upward trend (Ataei et al., 2021). Additionally, Iran's forest area has dwindled from 18.4 million hectares to 11 million hectares over the past 50 years, with the remaining forests in suboptimal condition (Savari et al., 2022; Savari and Khaleghi, 2023). Despite the growth in agricultural production in recent decades, there is growing concern about unsustainable agricultural practices and their detrimental impacts on natural resources and the environment (Cao et al., 2022; Helferich et al., 2023; Savari, 2023; Savari and Maymand, 2013). Evidence indicates that extensive use of chemical fertilizers diminishes soil fertility in the long run (Lu et al., 2021).

In light of global concerns regarding farmers' behaviors, numerous studies have investigated environmentally friendly behaviors (Wang et al., 2018), documenting a discrepancy: while many individuals identify as environmentalists, their actions often fail to align with conservation efforts, as evidenced by statistics and data (Englis & Philips, 2013). Many environmental challenges stem from irresponsible human conduct (Vicente-Molina et al., 2013), as humans frequently prioritize exploitation and self-interest over sustainable resource management (Liu, 2015). Consequently, experts assert that the fundamental issue underlying environmental problems lies within the psychological framework of human behavior (Bijani et al., 2017; Shafiei & Maleksaeidi, 2020; Savari et al., 2021). In other words, environmental issues fundamentally arise from human behavioral patterns (Kim et al., 2015; Chen, 2015). As a result, researchers and policymakers contend that by promoting environmentally friendly behaviors, a substantial portion of these challenges can be mitigated (Shin et al., 2018). Proenvironmental behavior includes actions that benefit the environment and entail a form of social behavior that involves a conflict between personal interests and decisions that serve to preserve the environment (Shafiei and Maleksaeidi, 2020), often at a significant individual cost (Lange et al., 2020).

Internal motivations play a fundamental role in fostering environmentally friendly behaviors in real social environments. A growing body of psychological-social studies indicates how individuals may become restricted (or unmotivated) in adopting environmentally supportive behaviors. Psychological research has proposed and evaluated various strategies aimed at enhancing individual environmental behavior (Price and Leviston, 2014). These strategies encompass education, feedback mechanisms, reconsideration requests, financial incentives, and social comparisons (Van Valkengoed et al., 2022). Measures based on socio-psychological studies to promote environmentally conscious behavior among individuals are particularly promising (Bergquist et al., 2019; Cialdini and Jacobson, 2021; Vesely et al., 2022). Moreover, the practical importance of social-psychological interventions has been demonstrated in addressing various environmental issues such as sustainable transportation, sustainable consumption (food), and energy conservation (de Kort et al., 2008; Jaeger and Schultz, 2017). Interventions based on social norms exploit a fundamental aspect of human motivation – the tendency to seek social approval or to obtain socially acceptable information by observing the behavior of others in certain situations (Cialdini et al., 2021). However, the differential impact of interventions based on social norms can be attributed to the different structures of social norms on which they are based (Bergquist et al., 2023). Norm-based interventions are highly popular because they are perceived as a convenient, inexpensive, and easy way to manage behavior change interventions (Anderson et al., 2017). Therefore, various models have been proposed to explain ecological behaviors, among which the most important ones are presented below.

2 Theory of Planned Behavior (TPB)

The theory of planned behavior (TPB) was introduced by Ajzen (1991), representing an evolved version of the theory of reasoned action (TRA) proposed by Fishbein and Ajzen in 1975 (Koen et al., 2012). This theory offers a foundational framework for understanding the underlying factors driving individual behaviors (Lubran, 2010). By incorporating the interplay of individual, social, and environmental factors, the TPB stands out as a valuable and robust instrument for forecasting inclinations and behavioral intentions toward environmentally friendly actions (Rahmanian et al., 2012; Savari et al., 2023a). Widely acclaimed in psychology, this theory holds significant potential for promoting environmentally supportive behaviors (Goh et al., 2017; Savari and Khaleghi, 2023). Previous studies consistently demonstrated that its constructs account for a considerable portion of the variability in environmental behaviors (Goh et al., 2017). Moreover, it has found application across diverse domains, spanning biodiversity conservation, forest preservation, adoption of green and organic products, water conservation efforts, and the utilization of environmentally friendly transportation methods (Savari and Khaleghi, 2023).

TPB is a highly popular conceptual framework used in human behavior investigation (Tseng et al., 2022). It is a social-psychological theory that suggests intentions or behavioral inclinations can predict actual behavior (Popa et al., 2019). The theory states that the intention to perform a certain action in the environment reliably predicts the probability that this behavior will be performed since most human behaviors are within one's control (Savari, 2023). "Inclination" refers to an internal state directing an individual's focus and motivation toward a particular behavior (Ajzen, 1991b).

Essentially, it reflects one's readiness, motivation, and willingness to embrace a certain action. Therefore, assessing individuals' behavioral inclinations is crucial in analyzing and predicting behaviors (Eldredge et al., 2016). This theory outlines three primary predictors of behavioral inclinations, namely: (a) attitude toward the behavior, (b) subjective norms, and (c) perceived behavioral control (Ababneh et al., 2022). Attitude toward the behavior refers to an individual's positive or negative evaluation of engaging in the desired behavior (Karimi and Saghaleini, 2021). Cultivating a positive attitude toward environmentally friendly behaviors is crucial for promoting them effectively (Holt et al., 2021; Ullah et al., 2021). Numerous studies have underscored the pivotal role of attitude as one of the primary predictors of intentions (Trihadmojo et al., 2020). Essentially, attitude reflects individuals' evaluations of engaging in a behavior, and it profoundly influences their behavioral intentions (Ullah et al., 2021; Sanchez et al., 2018). Therefore, creating conducive conditions to enhance people's beliefs regarding environmental conservation actions is paramount (Empidi and Emang, 2021).

Another variable in this theory is perceived social norms or mental norms for encountering or not encountering a behavior (Wauters et al., 2010). Mental norms are considered as social pressure or influence individuals face when confronted with a behavioral choice (Sanchez et al., 2018) and comprise a combination of an individual's perception of social pressures to act and their motivation for environmentally supportive behavior (Holt et al., 2021). Essentially, mental norms refer to a type of deliberate behavior based on individuals' beliefs and normative pressures regarding a subject, its dos and don'ts, and how they perceive it (Lu et al., 2014). Mental norms represent an individual's perception of social pressures influencing their adoption or avoidance of environmentally friendly behavior (Savari and Khaleghi, 2023). These norms stem from the influence of behaviors and speech of significant individuals in a person's life (Goh et al., 2017).

Another variable in this theory is perceived behavioral control, as the third determinant of behavioral intentions, which pertains to an individual's perception of the ease or difficulty of engaging in environmentally supportive behaviors (Wauters et al., 2010). In other words, this content refers to the perceived ease or difficulty and ultimately the performance of a specific behavior (Sanchez et al., 2018). This theory encompasses two aspects: the degree of control an individual has over the desired behavior and the individual's self-confidence in their ability to engage in or refrain from environmentally friendly behavior. Perceived behavioral control refers to individuals' perception of the ease or difficulty of performing the desired behavior (Ullah et al., 2021). The previous version of the TPB is the TRA proposed by Ajzen (1991) and A distinguishing feature of the TPB from its previous version (i.e., the TRA) is the addition of perceived behavioral control as another key predictor of behavioral intention (Beck and Ajzen, 1991).

2.1 Limitations of the Model

– The model of planned behavior does not work well in collectivist countries.
– The model of planned behavior has limited variables and therefore has low explanatory power in some fields.
– In this theory, the consequences of inappropriate behavior in the environment are not seen (Figure 1).

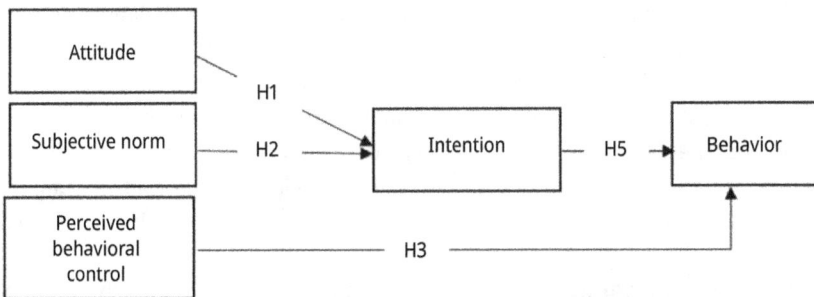

Figure 1: Theory of planned behavior (TPB).

3 Norm Activation Model (NAM)

The norm activation model (NAM) was introduced by Schwartz (1977), and aims to examine and develop environmentally friendly behavioral goals (Kim and Hwang, 2020; Han, 2014; Onwezen, Antonides, and Bartels, 2013). The NAM primarily pertains to personal concerns for the welfare of others and is rooted in behaviors of a prosocial nature. It also serves as an influential model explaining how normative considerations affect environmental behavior. This model has been extensively utilized to predict people's altruistic and prosocial behaviors (De Groot and Steg, 2009; Shin et al., 2018). Conservation behavior refers to an individual's actions aimed at aiding others, encompassing a wide spectrum of prosocial behaviors (Zhang et al., 2013). This theory closely relates to personal ethics, as individual ethical intensity determines the level of prosocial behavior (Schwartz, 1977). Conservation behavior is often regarded as a prosocial one because it entails positive consequences for others (Steg and De Groot, 2010). Prosocial behavior refers to actions that mitigate the adverse effects of an individual's actions on the environment (Kollmuss and Agyeman, 2002; Shin et al., 2018). NAM aims to demonstrate a shift from personal interests to the interests of others, as ethical considerations play a crucial role in this model (Han et al., 2019). Therefore, the core element in NAM is the personal norm variable (Han et al., 2019; Setiawan et al., 2014). In other words, personal norm delineates an individual's sense of ethical commitment toward performing or refraining from an action (Lopes et al., 2019). It aims to assess desirable sentiments of personal beliefs and values (Arvola et al., 2008) and pertains to an individual's beliefs regarding the rightness or wrongness of engaging in a behavior (Bakhtiyari et al., 2017).

According to this theory in Figure 2, the activation of personal norms requires two fundamental stages: the first stage is awareness, wherein individuals become cognizant of their behaviors potentially having adverse effects on others (awareness of consequences). In the second stage, individuals acknowledge responsibility for their actions having negative impacts on others (situational responsibility) (Bamberg, 2013). However, if individuals fail to activate personal norms, no appropriate action will be recognized, and no suitable action will be taken in this regard (Turaga et al., 2010). Additionally, the personal norm is activated by other variables. Firstly, awareness of need refers to an individual's awareness of the necessity of assistance and engaging in prosocial and protective behaviors (Klockner, 2013), wherein individuals realize the need to attend to others (Lopes et al., 2019). Secondly, outcome efficacy is about identifying actions that meet the needs of others (Steg and De Groot, 2010), whereby individuals, after recognizing the negative consequences of their behavior and the importance of assisting others, seek alternative actions to replace their previous behavior (Shin et al., 2018).

The third important variable in this model is ability, which indicates the level of perceived ease or difficulty in performing a specific behavior. According to this definition, individuals with higher levels of ability are more likely to engage in a specific behavior (De Leeuw et al., 2015). As ability encompasses a wide range of emotions, inclinations, and readiness to behave regarding environmental protection (Habibi and Mostafizadeh, 2017), individuals with higher levels of ability consistently employ safer behaviors in environmental protection (Damalas and Koutroubas, 2018). The fourth important variable in this model is denial of responsibility, which refers to an individual's unwillingness to take responsibility for the negative consequences of their behavior (Lopes et al., 2019). In this stage, individuals are unwilling to accept responsibility for their behavior. Accordingly, they will freely behave in the environment without considering others because they do not believe in their responsibility for destructive behaviors (Harland et al., 2007). In NAM, it is assumed that all variables influence emotions (guilt and pride) and individual behavior through personal norms (Rezaei et al., 2019; Harland et al., 2007). Guilt emotion implies that an individual faces a negative consequence of their behavior and believes that the action has occurred because of their behavior. This feeling is typically painful and unpleasant for the individual because the behavior contradicts their norms (Han, 2014). In contrast, pride emotion is gratifying for individuals and arises when individuals confront positive consequences of their behavior because the behavior is consistent with individuals' norms (Setiawan et al., 2014).

NAM has been widely used to elucidate behavioral objectives in various domains and plays a fundamental role in predicting socio-psychological factors to a large extent. However, different researchers have also attempted to enhance the model's robustness by combining various models or adding additional variables to NAM (Onwezen et al., 2013; Shin et al., 2018; Rezaei et al., 2019; Wang et al., 2019; Song et al., 2019; Kim and Hwang, 2020). Thus, considering that NAM is designed to study prosocial behavior and proenvironmental behavior, one of its major weaknesses is the limited inclusion of environmental aspects in this model (Song et al., 2019; Chen and Tung, 2014). Consequently, with the intensification of global environmental problems, a growing number of researchers have been intensively engaged in studies on environmentally friendly behaviors and environmental concerns (De Groot and Steg, 2007).

3.1 Limitations of the Model

- Limited environmental aspects (such as environmental concern) have not been seen in the model.
- Human behavior does not always directly follow social norms, but has a complex process.

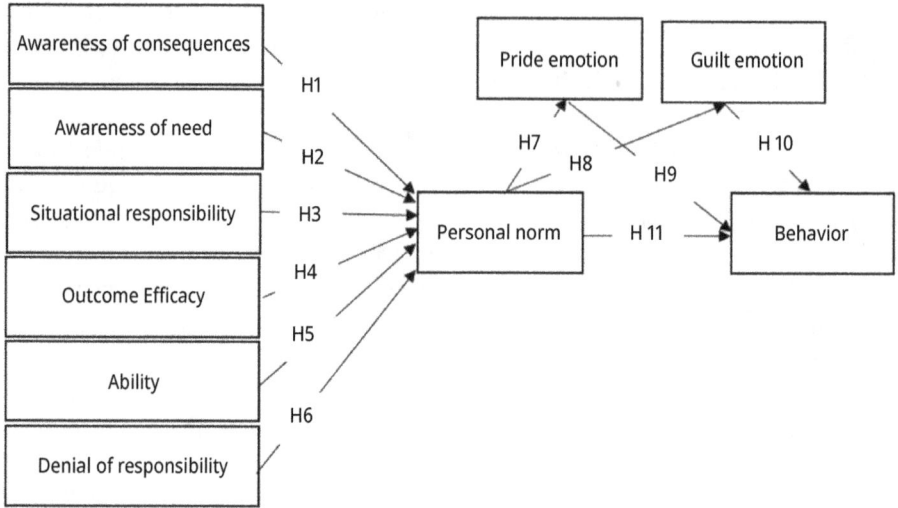

Figure 2: Norm activation model (NAM).

4 Health Belief Model (HBM)

The health belief model (HBM) was introduced by Strecher and Rosenstock (1950) (Raheli et al, 2020) and is considered one of the oldest and most widely used models in health psychology. It represents one of the first comprehensive attempts to explore health behavior based on principles with promising value (Jeong and Ham, 2018). This framework focuses its attention on changes in beliefs and assumes that changes in beliefs lead to changes in behavior (Yazdanpanah et al., 2015a). However, earlier research has underscored the central role of the HBM as one of the most important theories for explaining and predicting preventive behavior toward human hazards (Martin et al., 2010). Health beliefs have been employed for various objectives concerning environmentally friendly behaviors, among which pivotal examples include water conservation (Raheli et al, 2020), agricultural pattern alterations (Boazar et al., 2020), food safety practices (Rezaei and Mianaji, 2019), adoption of organic foodstuffs (Yazdanpanah et al., 2015a), utilization of renewable energy sources (Yazdanpanah et al., 2015b), safe pesticide application (Yazdanpanah et al., 2016), and inclination toward biofuels (Bakhtiyari et al., 2017).

According to this model, people are more likely to adopt health-promoting behaviors if they have a disposition to maintain well-being and believe that proenvironmental behaviors will lead to an improvement and increase in their health status (Rezaei Mianaji, 2019). Health education creates the necessary impetus for behavioral change by raising individual awareness and teaching health-related attitudes and inclinations (Hanson et al., 2002). Researchers argue that the motivation to initiate a healthy behavior (such as proenvironmental behavior) is crucial and considered fundamental (Groenewold et al., 2012). This model is based on the assumption that a person accepts a health-related measure if they believe that it will protect them from illness. Under this model, the individual has the positive expectation that health and disease prevention will be achieved by following recommendations. Consequently, they expect that by following the recommendations they will avoid becoming ill. This fosters the belief and confidence that the achievement of goals is feasible through adherence to recommendations (Sheppard and Thomas, 2020). The HBM focuses on two aspects of health behavior perceived threat, which pertains to the individual's perception of the issue, and behavior assessment, which involves evaluating the balance between benefits and barriers (Vassallo et al., 2009; Yazdanpanah et al., 2015b).

The perceived threat itself encompasses two subsidiary components: perceived severity and perceived susceptibility, collectively indicating an individual's perception of the threat posed by a stressful situation. These two factors prompt individuals to look for risk-reduction strategies (Boazar et al., 2020; Raheli et al., 2020; Huang et al., 2020b; Sheppard and Thomas, 2020; Rezaei Mianaji, 2019; Jeong and Ham, 2018; Yazdanpanah et al., 2015b). Perceived susceptibility indicates individuals' belief in the existence of a problem (such as environmental degradation), acceptance of its reality, and feeling of vulnerability to its impact on their health (Huang et al., 2020b). In addition, the perceived severity indicates that the individual perceives the identified problem as a serious health issue and understands the consequences for the various dimensions of physical, social, psychological and economic health (Boazar et al., 2020). However, behavior assessment comprises two subsidiary components: perceived benefits and perceived barriers. These factors significantly influence an individual's attitude toward employing risk-reduction strategies (Boazar et al., 2020; Raheli et al., 2020; Huang et al., 2020b; Sheppard and Thomas, 2020; Rezaei Mianaji, 2019; Jeong and Ham, 2018; Yazdanpanah et al., 2015b). Perceived benefits denote individuals' belief and understanding regarding the usefulness of taking action to reduce risk or the benefits derived from health-related actions (such as adopting environmentally friendly behaviors) (Boazar et al., 2020; Ejeta et al. 2016; Vassallo et al., 2009). Thus, perceived vulnerability and perceived benefits can generate impetus toward behavior emergence but cannot predict specific actions unless their practicality and usefulness are acknowledged (Ejeta et al., 2016). Perceived barriers signify obstacles to adopting a health behavior (such as environmentally friendly behaviors). For instance, the cost of implementing a protective behavior may outweigh its benefits (Azadi et al., 2019). In other words, individuals engage in health behavior when they perceive it to be less costly than its benefits (Vassallo et al., 2009). Moreover, subsequent research in the application of this theory has included other cognitive or motivational components to enhance or predict behavior change. These include two motivational structures: self-efficacy and cue to action (Vassallo et al., 2009; Orji et al., 2012; Akey et al., 2013; Yazdanpanah et al., 2015b; Raheli et al, 2020; Boazar et al., 2020).

Cue to action refers to internal or external stimuli that prompt an individual's behavior toward acting. In other words, internal stimuli such as inclinations and needs affect individuals from within, while external stimuli such as mass communication tools and interpersonal relationships impact individuals from the outside (Boazar et al., 2020). Therefore, guided by the environment, individuals discern the extent of benefits derived from the behavior they engage in (Devitt et al., 2016; Straub and Leahy 2014). Perceived self-efficacy has been added as a predictor of health behavior to the model. This concept refers to an individual's beliefs regarding their abilities to accept recommended behaviors to perform necessary actions, accompanied by achieving desired outcomes. In other words, self-efficacy refers to an individual's beliefs (self-confidence) about their abilities to motivate themselves, cognitive resources, and necessary action sequences required for successfully executing a specific task (Bakhtiyari et al., 2017; Yazdanpanah et al., 2015a). Accordingly, Buglar et al., (2010) argued that self-efficacy enhances the predictive power of behavior in health belief behavior. Perceived self-efficacy reflects individuals' beliefs and judgments about their abilities to perform tasks and responsibilities (Yadav and Pathak, 2016). According to this definition, people with a higher level of self-efficacy are more likely to behave in an environmentally friendly way (Keesstra et al., 2018).

4.1 Limitations of the Model

–   It does not appear to be suitable for long-term behavioral change.
–   Other factors besides health beliefs, such as cultural factors, socioeconomic status, and past experiences, also influence behavior formation, which are not considered in the model.
–   The predictive power of the HBM is relatively weak compared to theories of rational action and planned behavior. This necessitates emphasizing the need for expanding predictive factors of the HBM. However, this issue has been somewhat addressed by adding the construct of self-efficacy construct.

5 Protection Motivation Theory (PMT)

Protection motivation theory (PMT) in Figure 3 is commonly used to explain risk reduction behaviors or the intention to engage in protective behaviors in the environment. This theory encompasses individual and social factors in the cognitive decision-making process (Salehi et al., 2020; Savari et al., 2022). Thus, PMT posits that the perception of threat and the desire to prevent harm motivate individuals to change their behavior or protect themselves (Mitter et al., 2019). Introduced by Rogers (1975) to elucidate preventive behaviors (e.g., environmental degradation) (Bockarjova and Steg, 2014), PMT

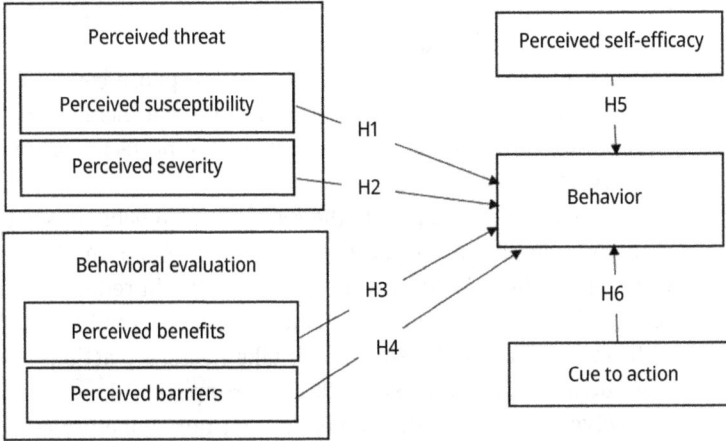

Figure 3: Health belief model (HBM).

describes how fear of a future hazardous event influences individuals' behaviors and their motivation to protect themselves (Poelma, 2018). Therefore, this theory combines individual and social structures to understand the cognitive decision-making process (Rainear and Christensen, 2017). PMT assumes that individuals' decisions to engage in preventive behaviors are based on their motivation to protect themselves against threats (Janmaimool, 2017; Savari et al., 2023b).

PMT is one of the significant cognitive-social theories used to examine factors influencing individuals' motivation and ultimately their protective behaviors. On one hand, PMT serves as a general decision-making model when facing various threats (Le Dang et al., 2014; Kristoffersen et al., 2017) and provides valuable tools to describe decision-making regarding environmental concerns (Karrer, 2012). On the other hand, it is an important socio-psychological model (Haer et al., 2016), clearly associated with economic, social, and psychological factors (Van Duinen et al., 2011). Furthermore, this theory has been applied in research related to water conservation (Kuruppu & Liverman, 2011; Tapsuwan & Rongrongmuang, 2015), forest protection (Savari et al., 2022), and environmentally friendly behaviors (Nelson et al., 2011; Rainear and Christensen, 2017; Shafiei & Maleksaeidi, 2020; Janmaimool, 2017).

In PMT (Figure 4), individuals facing potential threats may have two evaluations (Truelove et al., 2015), namely threat appraisal and coping appraisal (Wang et al., 2019a). The first factor includes two variables: perceived severity and perceived vulnerability (Chen, 2020b). Perceived vulnerability indicates an individual's sensitivity to existing threats (Truelove et al., 2015). If the threat is perceived as high-risk, individuals become motivated to protect themselves against it (Truelove et al., 2015). Whereas, perceived severity refers to the supposed seriousness of potential harms understood by an individual (Janmaimool 2017). Perceived vulnerability refers to individuals' mental perception of the likelihood of a threat, representing their defensive ability to cope with the threat

(Liao et al., 2020). Thus, the mentioned content indicates an individual's vulnerability to an existing threat (Keshavarz and Karami 2016).

Coping appraisal consists of variables such as self-efficacy, response costs (RC), and response efficacy (Chen, 2020a). This process evaluates factors that increase self-efficacy and response efficacy, or decrease RC, and the likelihood of protective action (Rongrongmuang and Tapsuwan, 2015).

Response efficacy refers to a person's belief that the recommended behaviors will effectively reduce or eliminate the threat (Cismaru et al., 2011; Kuruppu and Liverman., 2011). In other words, it predicts the effectiveness of an action in reducing the risk (Truelove et al., 2015). Self-efficacy, on the other hand, refers to an individual's perceived ability to perform recommended behaviors to take necessary actions accompanied by desirable outcomes (Truelove et al., 2015; Kuruppu and Liverman., 2011). RC refer to the perceived costs, including monetary and non-monetary costs such as time, effort, discomfort, inconvenience, and hardship associated with engaging in protective actions (Cismaru et al., 2011).

PMT operates through two cognitive processes: threat appraisal and coping appraisal, without considering social and environmental background factors (Mitter et al., 2019). Therefore, it was attempted to include additional variables in this framework and improve the model over time (Boss et al., 2015; Oakley et al., 2020; van Valkengoed & Steg, 2019; Liao et al., 2020; Vance et al., 2012).

This theory assumes that coping appraisal has a stronger effect on individuals' behavioral intentions than threat appraisal (Zhao et al., 2016). In this theory, only the variable of RC has a negative effect on the inclination toward protective behavior, while the remaining variables have a positive effect in this regard (Bockarjova and Steg, 2014). Additionally, among the PMT variables, studies have shown that self-efficacy has a greater influence on environmentally friendly behaviors (Prasetyo et al., 2020).

5.1 Limitations of the Model

- Some argue that this model primarily emphasizes the collection of individual behavior variables, yet they contend that these factors alone may not inherently induce behavioral change.
- Perceived threat is a construct closely related to behavior enactment, with its performance stemming from perceived severity and vulnerability. However, the relationship between threat and severity in the formation of the perceived threat is not always clear.
- This model is grounded in cognitive foundations and does not take into account the emotional aspects of behavior.

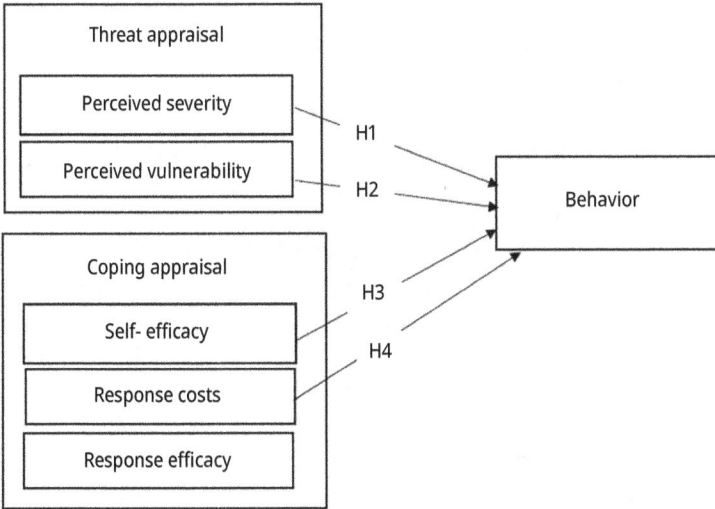

Figure 4: Protection motivation theory (PMT).

6 Theory of Value-Belief-Norm (VBN)

The value-belief-norm (VBN) was initially developed by Stern (2000), originating from a broader version of the NAM. This model places special emphasis on individuals' values and ethical norms in shaping behavior (Cao et al., 2022). The VBN theory integrates the NAM (especially altruistic values) to explain individuals' behaviors toward the environment and society holistically (Stern, 2000). VBN encompasses values, beliefs, and personal norms, linking them causally to the new ecological paradigm, awareness of consequences, the ascription of responsibility, personal norms, and behaviors (Nordlund and Garvill, 2002).

This underscores the importance of achieving harmony between humans and nature, demonstrating that awareness of negative consequences and having a sense of social responsibility can influence environmental protection behaviors (Han et al., 2017). Values, beliefs, norms, and responsible behaviors are all encompassed within the VBN model, which assumes a hierarchical relationship between these socio-psychological factors and their potential impacts on each other (Liu et al., 2018).

In new ecological paradigm model, individuals classify themselves into three subgroups based on their values: egoistic, altruistic, and biospheric. Those with egoistic values evaluate environmental aspects based on how they affect them personally, potentially resisting adopting protective behaviors if they perceive a high cost involved (Savari, 2023). Conversely, individuals with altruistic values typically assess environmental aspects based on the benefits and costs to human groups (Savari, 2023). Those with biospheric values (BVs) judge the environment based on benefits and costs to the ecosystem, likely to prevent threatening ecosystem conditions due to their value for natural resources and the environment (Wang et al., 2019a). Studies have demonstrated that BVs and AVs positively affect the New Ecological Paradigm, while egoistic values have a negative impact (De Groot & Steg, 2008).

Therefore, individuals' values regarding environmental conservation influence the formation of self-ascribed responsibility concerning the consequences they seek for themselves, others, and other ecosystems (Wensing et al., 2019). According to VBN, beliefs are directly related to personal norms. In other words, individuals with common beliefs about environmental well-being are more likely to establish personal norms for proenvironmental behavior (Savari, 2023). Personal norms become active when individuals realize that not performing the target behavior will have undesirable consequences and they must be responsible for their unfavorable consequences (Schwartz, 1977).

The first stage is awareness, meaning individuals become aware of their behaviors that may have adverse effects on others, termed awareness of consequences. In the second stage, individuals accept responsibility for this action, acknowledging that their behavior negatively affects others, termed ascribed responsibility (Bamberg, 2013). In the context of the VBN model (Figure 5), the variable awareness of consequences can thus activate the personal norms of the individual, that is, the individual becomes aware of the undesirable consequences of not behaving prosocially toward others. The stronger this sense of consequence, the stronger the moral commitment, increasing the likelihood of activating personal norms for engaging in prosocial behavior (Nguyen & Nguyen, 2020). Awareness of how one's actions impact the environment involves understanding actions that can mitigate the adverse effects of one's behavior on the environment. This awareness is associated with a sense of responsibility and an increase in personal norms (Ghazali et al., 2019; Landon et al., 2018). Self-expectation of performing a specific behavior under certain conditions is referred to as personal norms. Violating personal norms leads to guilt while adhering to them fosters pride (Cao et al., 2022). These norms signify an individual's moral commitment to performing or refraining from an action (Lopes et al., 2019). This content aims to assess desirable emotions of belief and value within the individual (Arvola et al., 2008) and pertains to the individual's beliefs about the rightness or wrongness of performing a behavior (Bakhtiyari et al., 2017). Ultimately, the relationships between variables based on the VBN model (Stern, 2000) are presented as follows.

6.1 Limitations of the Model

-   In this theory, too much emphasis has been placed on the value beliefs of people, while behavior is most affected by people's awareness.
-   This theory argues that the gradual destruction of the "natural environment" followed by increased human intervention. This is despite the fact that in addition to human resources, environmental factors are also influential.

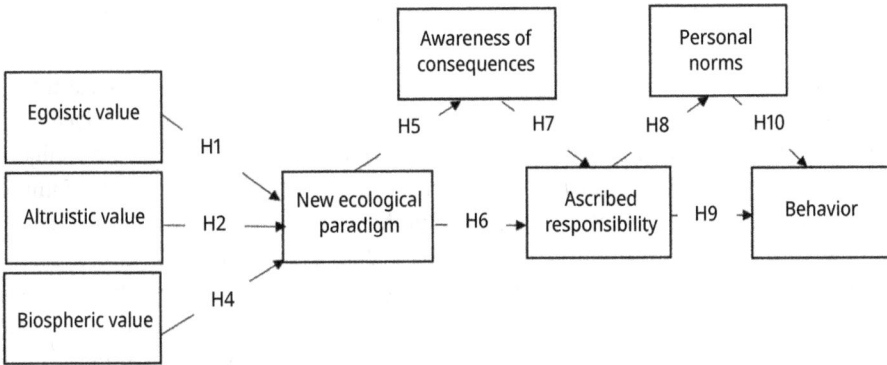

Figure 5: Theory of value-belief-norm (VBN).

7 Technology Acceptance Model (TAM)

The technology acceptance model (TAM) is one of the models used to predict the level of stakeholders' acceptance of a new technology (PUTRI et al., 2023). TAM was introduced by Davis et al. (1989) and has been widely utilized in studies. This model is an adaptation of the theory developed by Fishbein, known as the TRA, and assumes that a person's reaction and perception determine that person's attitude and behavior. TAM resulted in the addition of two main structures to TRA (Peng & Xu, 2023). Over the past few decades, several models have been presented to understand the characteristics of technology acceptance among users. However, TAM remains the most important and fundamental basis for technology acceptance (Rho et al., 2014; Kamal et al., 2020; Savari et al., 2024a) and has been utilized as the most common model in various research fields such as sociology, psychology, and agriculture (Cacciamani et al., 2018; Gokcearslan, 2017; Ifenthaler & Schweinbenz, 2016; Kim & Jang, 2015; Sharifzadeh et al., 2017; Bagheri et al., 2021). The main objective of TAM is to predict the factors influencing technology acceptance among users and identify issues with a new technology among the public (Nikou & Economides, 2017; Sung et al., 2019). This theory has been applied in the context of environmentally friendly behaviors (Savari et al., 2021), biological pest control (Rezaie et al., 2021), the use of ecological inputs (Bagheri et al., 2021; Savari et al., 2024b), and others.

This model assumes that the adoption of an innovation or new activity is directly influenced by individuals' behavioral intentions (Zheng & Li, 2020). It elucidates how beliefs and attitudes toward the use of "things" drive behavior. Accordingly, an individual's attitude toward engaging in a specific action is determined by their beliefs about the outcomes of that action and their evaluation of these outcomes (Clarke & Abbott, 2016; Dündar & Akçayir, 2014; Ferguson, 2017). In this framework, users' inclinations play a pivotal role in the successful implementation and utilization of technology (Aggelidis et al., 2009).

Davis et al. (1989) proposed that attitude toward use influences behavioral intention, with attitude toward use primarily determined by two structures: perceived usefulness and perceived ease of use. Consequently, Davis et al. (1989) eliminated subjective norms from the TRA model to develop the TAM. Essentially, perceived usefulness and perceived ease of use are crucial factors in the acceptance of new behaviors (such as environmentally friendly behaviors) (Aung & San, 2021; Chen & Aklikokou, 2020). However, Davis also argued that the mentioned factors are influenced by other external variables (Davis et al., 1989). Therefore, numerous studies have demonstrated that TAM, when expanded to include external variables, offers a more comprehensive explanation of continuous intention for use (Wu and Chen, 2017). Furthermore, perceived ease of use notably impacts perceived usefulness, subsequently influencing continuous intention to use and resulting in actual acceptance of the new behavior (Dai et al., 2020; Hao et al., 2017). Considering that TAM (Figure 6), as proposed by Davis et al., (1989), has been widely employed to elucidate environmentally friendly behaviors among individuals in numerous studies on the acceptance of new technologies, it has proven to be highly applicable for explaining proenvironmental behaviors. Therefore, the main components of Davis's Technology Acceptance Model include perceived ease of use, perceived usefulness, attitude toward use, and behavioral intention (Mohr & Kühl, 2021). Perceived ease of use refers to an individual's belief regarding the difficulty of using an innovation or the complexity of using it Zhong et al., 2019). In other words, this content signifies the extent to which a user expects that using the intended system will require no effort (Ullah et al., 2021). Perceived usefulness indicates that an individual believes that using a particular system enhances their performance (Kim et al., 2008).

Attitude toward a behavior reflects a framework in which an individual evaluates the desirability or undesirability of that behavior (Liu & Bridget, 2020; Ullah et al., 2021). Furthermore, the intention variable serves as a very good predictor for actual environmental behaviors (Marcos et al., 2021; Empidi & Emang, 2021). Intention reflects the motivation or plan to engage in an action (Zhong et al., 2019; Sánchez et al., 2018) and also reflects the level of motivation, readiness, and willingness of an individual to adopt a behavior (Eldredge et al., 2016).

7.1 Limitations of the Model

– Failure to consider social norms in accepting new behaviors.
– Consider the limited aspects of technology adoption.

Perceived ease of us

H1

H4

H3

Attitude — H5 → Intention — H6 → Behavior

Perceived usefulness — H2

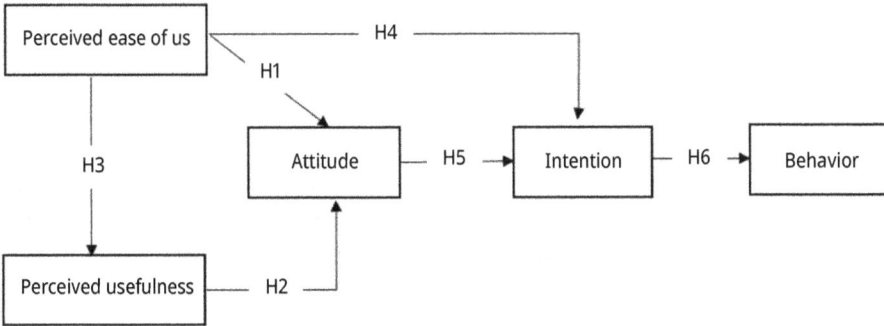

Figure 6: Technology acceptance model (Davis et al. 1989; Venkatesh and Davis 1996).

8 Social Cognitive Theory (SCT)

The social cognitive theory (SCT) provides a framework for understanding the psychosocial mechanisms influencing human thoughts, emotions, and behaviors (Bandura, 1986). SCT is one of the most common behavior change theories (Stallones et al., 2009; Shahangian et al., 2021). The fundamental assumption of this theory is reciprocal determinism. In this concept, the individual, environment, and behavior continuously interact (Valois et al., 2017). One may exert predominant influence over the others at different times. This allows various stimuli to influence human thoughts, emotions, and behaviors in different ways (Young et al., 2005). SCT explains how people acquire and maintain behavioral patterns (Murphy et al., 2001). The main constructs in SCT include outcome expectancies, perception of others' behavior, socio-structural factors, self-efficacy, and behavioral intentions (Bandura, 1986; Yazdanpanah et al., 2015b; Shahangian et al., 2021; Savari et al., 2022). In this theory, these variables influence individuals' behaviors (Bandura, 1986; Yazdanpanah et al., 2015b). Behavioral factors include primary and long-term goals, with environmental factors encompassing both barriers and facilitators (Ramirez et al., 2012). Social support is associated with how and to what extent others' assistance influences the performance of specific behaviors and acts as a facilitator (Ramirez et al., 2012). Socio-structural barriers encompass personal, social, and structural barriers that are directly perceived as impediments to behavior change. Notably, the greater the number of barriers, the less likely engagement in behavior change will occur (Ramirez et al., 2012).

Outcome expectancies are another key construct of SCT, which affects intention and behavior. These expectations can be found in both positive and negative forms (Thøgersen and Grønhøj, 2010). When the expectancy is more positive, there is a higher likelihood of engagement in the behavior (Ramirez et al., 2012). Outcome expectancies refer to an individual's judgment regarding the likelihood of outcomes resulting from performing or not performing a specific action (Savari et al., 2022). This theory states that individuals act in ways they believe will produce positive and valuable outcomes, while actively avoiding behaviors they expect to have undesirable consequences (Young et al., 2014). Another construct of SCT is the perception of others' behavior, which means that individuals learn not only from their personal experiences but also from observing the behavior and outcomes of others (Thøgersen and Grønhøj, 2010). It also refers to an individual's intention regarding how much others (important individuals in life) engage in that behavior (Thøgersen and Grønhøj, 2010). According to Bandura (2002), self-efficacy is a central construct in SCT, which has a direct influence on behavior as well as an indirect influence via all other components of the model. Perceived self-efficacy refers to an individual's belief in their abilities to adopt recommended behaviors to perform necessary actions, accompanied by achieving desirable outcomes (Truelove et al., 2015; Kuruppu & Liverman., 2011).

Finally, the last variable in this theory is intention, which stands as a robust predictor of actual behavior (Marcos et al., 2021). Intention reflects the motivation or plan to engage in an action (Zhong et al., 2019; Sánchez et al., 2018). It also reflects an individual's level of motivation, readiness, and willingness to adopt a behavior (Eldredge et al., 2016). This theory has been applied in various environmental-friendly behaviors, including water conservation (Shahangian et al., 2021), soil conservation (Savari et al., 2022), employing irrigation technologies for optimal agricultural water use (Yazdanapanah et al., 2015), and the use of renewable energies (Bakhtiyari et al., 2017). Accordingly, based on the explanations provided within the framework of SCT, it is presented in Figure 7.

8.1 Limitations of the Model

– According to this theory, learning is done through observation, imitation, and modeling, and there is no need for the reinforcement factor—which is emphasized by Thorndike and Skinner.
– Great emphasis on the interactive relationship between humans and the environment.

9 Nature Connectedness via Values Model

The nature connectedness model, proposed by Schwartz (1977), presented a basis for understanding how the environment provides a context for nurturing behavior. This theory suggested that stronger ethical commitments are more translatable to proenvironmental behaviors, involving an orientation toward the environment when personal attachment is activated (Cock et al, 1978; cited in Hoot and Friedman, 2011). Studies have shown that environmental values are strongly associated with environmental. Environmental values are associated with a wide range of environmental preferences and actions, including accepting climate change policies, sustainable consumption, environmental activities, and proenvironmental behaviors (Steg et al 2011; Thogersen and Olan-

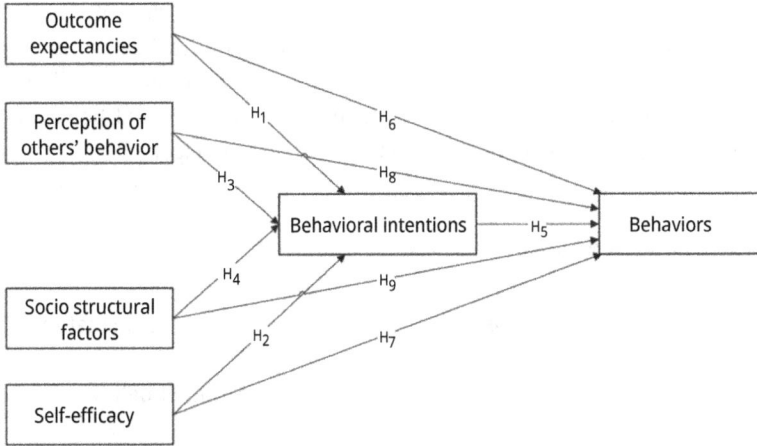

Figure 7: Social cognitive theory (SCT).

der, 2002; Schultz and Zelezny, 1998). The nature connectedness variable in relation to responsible environmental behaviors has been highlighted in Wilson's theory. Accordingly, Wilson et al (1984) proposed the biophilia hypothesis, suggesting that individuals' emotional connection with nature may motivate them to protect the environment. Therefore, individuals' emotional relationships with nature have been studied as determinants of interaction in environmental behaviors by connecting with nature (Pereira and Forster, 2015). Humans have an inherent need for belonging and connection with other living beings. According to this theory, humans instinctively strive to build meaningful emotional relationships with other living beings, laying the foundation for our motivation to care for and protect the environment. Humans' empathy and fondness for living things is evident in their fascination with nature, appreciation of natural landscapes, cultivation of indoor plants, and compassionate treatment of animals.

Moreover, various studies have confirmed the role of nature connectedness in proenvironmental behaviors. Kals et al. (1999) demonstrated that nature connectedness predicts 25–39% of environmental behaviors. Mayer and Frantz (2004) also elucidated a positive correlation between nature connectedness and proenvironmental behaviors. Dutcher et al. (2007) concluded that nature connectedness explains 10% of the variance in environmental behaviors. Similarly, Clayton (2003) also demonstrated that environmental identity indices have a positive correlation with responsible environmental behaviors such as cooperation with environmental organizations and institutions (Badri Gargari et al., 2011).

Researchers argue that environmental values play a significant role in the relationship between connectedness to nature and environmental behaviors (Arnocky et al, 2007; Gosling and Williams, 2010). According to the biophilia hypothesis, placing a higher value on nature leads to a stronger connection with nature and a higher likelihood of engaging in environmental behaviors (Wilson, 1984; Kellert, 1993). Arnocky et al. (2007) found that environmental values serve as an intermediary component in the relationships between proenvironmental behaviors and personal beliefs; this resembles a construct of connectedness with nature, which is assumed as an empathic connection to all beings. Gasling and Williams (2010) concluded in their study that environmental concern acts as an intermediary variable between connectedness to nature and proenvironmental behaviors (Pereira and Forster, 2015). This framework is presented as follows in Figure 8:

9.1 Limitations of the Model

– Failure to pay attention to the norms and beliefs of people in the formation of behavior
– Failure to pay attention to external motivations in the formation of behavior

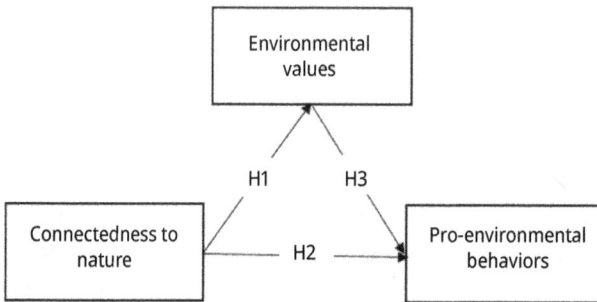

Figure 8: Nature connectedness via values model.

10 Conclusion

In the 1950s and 1960s, the excessive use of chemical fertilizers to improve soil fertility rose steeply, catalyzing the Green Revolution. While this initiative significantly increased global food production, it also had detrimental consequences for the environment. Indeed, conventional agricultural practices emphasizing the use of chemical fertilizers pose threats to both human health and the environment. The introduction of chemical fertilizers, pesticides, and new food production technologies has somewhat improved food security. However, the escalated consumption of chemical fertilizers and pesticides has gradually resulted in environmental problems such as soil degradation, greenhouse gas emissions, and water pollution. Additionally, pesticide residues and chemicals are considered significant threats to environmental pollution and human health.

In the 1960s, the excessive use of chemical fertilizers and the emergence of their effects posed a serious challenge to humanity's survival. However, in 1962, following the publication of Rachel Carson's "silent spring," awareness of the adverse effects of excessive fertilizer and chemical toxin use on humans and the environment surged.

Promoting and developing environmentally compatible agriculture became an inevitable necessity, leading to a growing interest in sustainable agriculture and environmentally friendly behaviors. Proenvironmental behaviors play a crucial role in preserving the environment and maintaining ecological balance. This research aimed to discuss and examine eight prominent theories in the study of environmentally friendly behaviors to foster safer environmental practices. The scrutiny of these theories reveals that while some sociologists emphasize external factors such as social and economic characteristics, psychologists highlight internal factors namely values, beliefs, attitudes, motivations, perceptions, knowledge, personality traits, and worldview. However, it's evident among researchers that incentives and economic factors alone cannot reliably predict proenvironmental behaviors. When incentives are removed, behavior tends to revert to its initial state. Psychologists assert that normative behaviors are vital for societal cohesion, and non-compliance may lead to social breakdown and isolation. Therefore, social and psychological variables play crucial roles in facilitating sustainable behavior. Considering their significance in human behavior, considerable attention should be directed toward understanding these theories and their variables. They hold the potential to initiate fundamental changes in human behavior toward a more sustainable future.

Active participation in environmental conservation can be identified through psychological variables such as values and social norms. In elucidating and predicting behaviors that have consequences beyond the individual level, such as environmental behaviors, norms play a crucial role and have been employed in numerous environmental behavior studies. Norms morally compel individuals to act in a manner that aligns with societal values. Norms serve as ethical commitments to perform or refrain from specific actions that lead to environmentally supportive behaviors. Therefore, social norms describe desirable and acceptable ways of living.

This research had two limitations in spite of its in-depth investigation in the field of prevailing psychological-social theories in the field of safe behaviors in the environment: (1) in this research, only eight theories were examined, it is better to collect other theories in this field and add them to this section in future research, and (2) in future research, it is better to examine the power of theories in different countries to determine their importance in all countries.

Accordingly, some general recommendations for developing environmentally supportive behaviors are presented as follows:
- Utilizing the capacities of rural women and children as nature allies to promote protective norms
- Promoting an environmentally friendly culture, such as collecting litter in nature, proper disposal of food waste, and respecting the rights of other creatures
- Promoting environmentally friendly behaviors in the community by individuals with high social status and farmers who value their actions and professions
- Utilizing media and provincial networks to develop and promote environmentally friendly behaviors

- Conducting educational and promotional classes by agricultural experts in the Jihad organization on how to implement environmentally friendly behaviors
- Developing non-governmental organizations and environmental conservation co-operatives and encouraging individuals to join these organizations to strengthen their environmentally friendly behaviors in natural environments
- Encouraging and reinforcing the motivation of community members to adopt protective and environmentally friendly behaviors through economic and social incentives

References

Ababneh, K. I., Ahmed, K., & Dedousis, E. (2022). Predictors of cheating in online exams among business students during the Covid pandemic: Testing the theory of planned behavior. *The International Journal of Management Education, 20*(3), 100713. doi: 10.1016/j.ijme.2022.100713.

Agegnehu, G., Nelson, P. N., & Bird, M. I. (2016). Crop yield, plant nutrient uptake and soil physicochemical properties under organic soil amendments and nitrogen fertilization on Nitisols. *Soil and Tillage Research, 160*, 1–13. doi: 10.1016/j.still.2016.02.003

Aggelidis, V. P. & Chatzoglou, P. D. (2009). Using a modified technology acceptance model in hospitals. *International Journal of Medical Informatics, 78*(2), 115–126.

Ajzen, I. (1991). The theory of planned behavior. *Organizational Behavior and Human Decision Processes, 50*(2), 179–211.

Ajzen, I., & Fishbein, M. (1980). Understanding attitudes and predictiing social behavior. Englewood cliffs. N, J. Prentice-Hall.

Akey, J. E., Rintamaki, L. S., & Kane, T. L. (2013). Health Belief Model deterrents of social support seeking among people coping with eating disorders. *Journal of Affective Disorders, 145*(2), 246–252.

Anderson, K., Song, K., Lee, S., Krupka, E., Lee, H., & Park, M. (2017). Longitudinal analysis of normative energy use feedback on dormitory occupants. *Applied Energy, 189*, 623–639.

Anh, H. Q., Le, T. P. Q., Da Le, N., Lu, X. X., Duong, T. T., Garnier, J., . . . Nguyen, T. A. H. (2021). Antibiotics in surface water of East and Southeast Asian countries: A focused review on contamination status, pollution sources, potential risks, and future perspectives. *Science of the Total Environment, 764*, 142865. doi: 10.1016/j.apenergy.2016.12.086

Arnocky, S., Stroink, M., & DeCicco, T. (2007). Self-construal predicts environmental concern, cooperation, and conservation. Journal of Environmental Psychology, 27(4), 255–264.

Arvola, A., Vassallo, M., Dean, M., Lampila, P., Saba, A., Lähteenmäki, L., & Shepherd, R. (2008). Predicting intentions to purchase organic food: The role of affective and moral attitudes in the Theory of Planned Behaviour. *Appetite, 50*(2–3), 443–454.

Ataei, P., Gholamrezai, S., Movahedi, R., & Aliabadi, V. (2021). An analysis of farmers' intention to use green pesticides: The application of the extended theory of planned behavior and health belief model. Journal of Rural Studies, 81, 374–384.

Aung, M. & San, K. (2021). Usefulness of Google Classroom for Management Students of a Thai Private University. *APHEIT International Journal, 10*(2), 1–10.

Azadi, Y., Yazdanpanah, M., & Mahmoudi, H. (2019). Understanding smallholder farmers' adaptation behaviors through climate change beliefs, risk perception, trust, and psychological distance: Evidence from wheat growers in Iran. *Journal of Environmental Management, 250*, 109456.

Badri Gargari, R., H. Lotfinia, and R. Mohamadnajad. (2011). "Staff Burnout in Tabriz hospitals: The role of behavior types and coping strategies." Journal of Urmia Nursing & Midwifery Faculty (2228-6411) 9, no. 4.

Bagheri, A., Bondori, A., Allahyari, M. S., & Surujlal, J. (2021). Use of biologic inputs among cereal farmers: application of technology acceptance model. Environment, Development and Sustainability, 23(4), 5165–5181.

Bahrulolum, H., Nooraei, S., Javanshir, N., Tarrahimofrad, H., Mirbagheri, V. S., Easton, A. J., & Ahmadian, G. (2021). Green synthesis of metal nanoparticles using microorganisms and their application in the agrifood sector. *Journal of Nanobiotechnology, 19*(1), 1–26. doi: 10.1186/s12951-021-00834-3.

Bakhtiyari, Z., Yazdanpanah, M., Forouzani, M., & Kazemi, N. (2017). Intention of agricultural professionals toward biofuels in Iran: Implications for energy security, society, and policy. *Renewable and Sustainable Energy Reviews, 69*, 341–349.

Bamberg, S. (2013). Changing environmentally harmful behaviors: A stage model of self-regulated behavioral change. *Journal of Environmental Psychology, 34*, 151–159.

Bandura, A. (1986). *Social Foundations of Thought and Action: A Social Cognitive Theory.* Englewood Cliffs, NJ: Prentice-Hall.

Bandura, A. (2002). Social cognitive theory in cultural context. Applied psychology, 51(2), 269–290.

Bergquist, M., Nilsson, A., & Schultz, W. P. (2019). A meta-analysis of field-experiments using social norms to promote pro-environmental behaviors. Global Environmental Change, 59, 101941.

Bergquist, M., Thiel, M., Goldberg, M. H., & van der Linden, S. (2023). Field interventions for climate change mitigation behaviors: A second-order meta-analysis. *Proceedings of the National Academy of Sciences, 120*(13), e2214851120. doi: 10.1073/pnas.2214851120.

Bijani, M., Ghazani, E., Valizadeh, N., & Haghighi, N. F. (2017). Pro-environmental analysis of farmers' concerns and behaviors towards soil conservation in central district of Sari County, Iran. *International Soil and Water Conservation Research, 5*(1), 43–49. doi: 10.1016/j.iswcr.2017.03.001.

Boazar, M., Abdeshahi, A., & Yazdanpanah, M. (2020). Changing rice cropping patterns among farmers as a preventive policy to protect water resources. *Journal of Environmental Planning and Management,* 1–17.

Bockarjova, M. & Steg, L. (2014). Can Protection Motivation Theory predict pro-environmental behavior? Explaining the adoption of electric vehicles in the Netherlands. *Global Environmental Change, 28*, 276–288.

Boss, S., Galletta, D., Lowry, P. B., Moody, G. D., & Polak, P. (2015). What do systems users have to fear? Using fear appeals to engender threats and fear that motivate protective security behaviors. *MIS Quarterly (MISQ), 39*(4), 837–864.

Buglar, M. E., White, K. M., & Robinson, N. G. (2010). The role of self-efficacy in dental patients' brushing and flossing: testing an extended Health Belief Model. Patient education and counseling, 78(2), 269–272.

Cacciamani, S., Villani, D., Bonanomi, A., Carissoli, C., Olivari, M. G., Morganti, L., & Confalonieri, E. (2018). Factors affecting students' acceptance of tablet PCs: A study in Italian high schools. *Journal of Research on Technology in Education, 50*(2), 120–133.

Cao, H., Li, F., Zhao, K., Qian, C., & Xiang, T. (2022). From value perception to behavioural intention: Study of Chinese smallholders' pro-environmental agricultural practices. *Journal of Environmental Management, 315*, 115179.

Chen, L. & Aklikokou, A. K. (2020). Determinants of E-government adoption: Testing the mediating effects of perceived usefulness and perceived ease of use. *International Journal of Public Administration, 43*(10), 850–865.

Chen, M. F. (2020a). Moral extension of the protection motivation theory model to predict climate change mitigation behavioral intentions in Taiwan. *Environmental Science and Pollution Research, 27*(12), 13714–13725.

Chen, M. F. (2020b). Moral extension of the protection motivation theory model to predict climate change mitigation behavioral intentions in Taiwan. *Environmental Science and Pollution Research,* 1–12.

Chen, M. F., & Tung, P. J. (2014). Developing an extended theory of planned behavior model to predict consumers' intention to visit green hotels. International journal of hospitality management, 36, 221–230.

Cialdini, R. B. & Jacobson, R. P. (2021). Influences of social norms on climate change-related behaviors. *Current Opinion in Behavioral Sciences, 42*, 1–8. doi: 10.1016/j.cobeha.2021.01.005

Cismaru, M., Cismaru, R., Ono, T., & Nelson, K. (2011). "Act on climate change": An application of protection motivation theory. *Social Marketing Quarterly, 17*(3), 62–84.

Clarke, L. & Abbott, L. (2016). Young pupils', their teacher's and classroom assistants' experiences of i P ads in a N orthern I reland school: "Four and five years old, who would have thought they could do that?". *British Journal of Educational Technology, 47*(6), 1051–1064.

Clayton, S. (2003). Environmental identity: A conceptual and an operational definition. Identity and the natural environment: The psychological significance of nature, 45–65.

Cock, M. J. W. (1978). The assessment of preference. The Journal of Animal Ecology, 805–816.

Dai, H. M., Teo, T., Rappa, N. A., & Huang, F. (2020). Explaining Chinese university students' continuance learning intention in the MOOC setting: A modified expectation confirmation model perspective. *Computers & Education, 150*, 103850.

Damalas, C. A., & Koutroubas, S. D. (2018). Current status and recent developments in biopesticide use. Agriculture, 8(1), 13.

Davis, F. D. (1989). Perceived usefulness, perceived ease of use, and user acceptance of information technology. *Management Information Systems Quarterly*, 319–340.

Davis, F. D., Bagozzi, R. P., & Warshaw, P. R. (1989). User acceptance of computer technology: A comparison of two theoretical models. *Management Science, 35*(8), 982–1003.

De Groot, J. I., & Steg, L. (2007). Value orientations and environmental beliefs in five countries: Validity of an instrument to measure egoistic, altruistic and biospheric value orientations. Journal of cross-cultural psychology, 38(3), 318–332.

De Groot, J. I. & Steg, L. (2008). Value orientations to explain beliefs related to environmental significant behavior: How to measure egoistic, altruistic, and biospheric value orientations. *Environment and Behavior, 40*(3), 330e354. https://doi.org/10.1016/j.tourman.2016.06.018.

De Groot, J. I., & Steg, L. (2009). Mean or green: which values can promote stable pro-environmental behavior? Conservation Letters, 2(2), 61–66.

De Kort, Y. A., McCalley, L. T., & Midden, C. J. (2008). Persuasive trash cans: Activation of littering norms by design. *Environment and Behavior, 40*(6), 870–891. doi: 10.1177/0013916507311035.

De Leeuw, A., Valois, P., Ajzen, I., & Schmidt, P. (2015). Using the theory of planned behavior to identify key beliefs underlying pro-environmental behavior in high-school students: Implications for educational interventions. *Journal of Environmental Psychology, 42*, 128–138.

Devitt, C., O'Neill, E., & Waldron, R. (2016). Drivers and barriers among householders to managing domestic wastewater treatment systems in the Republic of Ireland; implications for risk prevention behaviour. *Journal of Hydrology, 535*, 534–546.

Duan, W., Peng, L., Zhang, H., Han, L., & Li, Y. (2021). Microbial biofertilizers increase fruit aroma content of Fragaria× ananassa by improving photosynthetic efficiency. *Alexandria Engineering Journal, 60*(6), 5323–5330. doi: 10.1016/j.aej.2021.04.014.

van Duijn, S., Nabuurs, R. J., van Rooden, S., Maat-Schieman, M. L., van Duinen, S. G., van Buchem, M. A., . . . & Natté, R. (2011). MRI artifacts in human brain tissue after prolonged formalin storage. Magnetic resonance in medicine, 65(6), 1750–1758.

Dündar, H. & Akçayır, M. (2014). Implementing tablet PCs in schools: Students' attitudes and opinions. *Computers in Human Behavior, 32*, 40–46.

Dutcher, D. D., Finley, J. C., Luloff, A. E., & Johnson, J. B. (2007). Connectivity with nature as a measure of environmental values. Environment and behavior, 39(4), 474–493.

Ejeta, L. T., Ardalan, A., Paton, D., & Yaseri, M. (2016). Predictors of community preparedness for flood in Dire-Dawa town, Eastern Ethiopia: Applying adapted version of Health Belief Model. *International Journal of Disaster Risk Reduction, 19*, 341–354.

Eldredge, L. K. B., Markham, C. M., Ruiter, R. A., Fernández, M. E., Kok, G., & Parcel, G. S. (2016). *Planning Health Promotion Programs: An Intervention Mapping Approach*. John Wiley & Sons.

Empidi, A. V. A. & Emang, D. (2021). Understanding Public Intentions to Participate in Protection Initiatives for Forested Watershed Areas Using the Theory of Planned Behavior: A Case Study of Cameron Highlands in Pahang, Malaysia. *Sustainability, 13*(8), 4399. doi: 10.3390/su13084399.

Englis, B. G. & Phillips, D. M. (2013). Does innovativeness drive environmentally conscious consumer behavior? *Psychology & Marketing, 30*(2), 160–172. doi: 10.1002/mar.20595.

Ferguson, J. M. (2017). Middle school students' reactions to a 1: 1 iPad initiative and a paperless curriculum. *Education and Information Technologies, 22*(3), 1149–1162.

Ghazali, E. M., Nguyen, B., Mutum, D. S., & Yap, S. F. (2019). Pro-environmental behaviours and Value-Belief-Norm theory: Assessing unobserved heterogeneity of two ethnic groups. *Sustainability, 11*(12), 3237.

Goh, E., Ritchie, B., & Wang, J. (2017). Non-compliance in national parks: An extension of the theory of planned behaviour model with pro-environmental values. *Tourism Management, 59*, 123–127. doi: 10.1016/j.tourman.2016.07.004

Gokcearslan, S. (2017). Perspectives of Students on Acceptance of Tablets and Self-Directed Learning with Technology. *Contemporary Educational Technology, 8*(1), 40–55.

Gosling, E., & Williams, K. J. (2010). Connectedness to nature, place attachment and conservation behaviour: Testing connectedness theory among farmers. *Journal of environmental psychology, 30*(3), 298–304.

Groenewold, G., De Bruijn, B., & Bilsborrow, R. (2012). Psychosocial factors of migration: Adaptation and application of the health belief model. *International Migration, 50*(6), 211–231.

Habibi, F. & Mostafizadeh, S. (2017). Investigating the environmental behaviors of Lake Merrivan Lake. *Geography and Development Quarterly, 47*, 163–184.

Haer, T., Botzen, W. W., & Aerts, J. C. (2016). The effectiveness of flood risk communication strategies and the influence of social networks Insights from an agent-based model. *Environmental Science and Policy, 60*, 44–52.

Han, H. (2014). The norm activation model and theory-broadening: Individuals' decision-making on environmentally-responsible convention attendance. *Journal of Environmental Psychology, 40*, 462–471.

Han, H., Hwang, J., & Lee, M. J. (2017). The value–belief–emotion–norm model: Investigating customers' eco-friendly behavior. *Journal of Travel & Tourism Marketing, 34*(5), 590–607.

Han, H., Hwang, J., Lee, M. J., & Kim, J. (2019). Word-of-mouth, buying, and sacrifice intentions for eco-cruises: Exploring the function of norm activation and value-attitude-behavior. Tourism Management, 70, 430–443.

Hanson, J. A. & Benedict, J. A. (2002). Use of the Health Belief Model to examine older adults' food-handling behaviors. *Journal of Nutrition Education and Behavior, 34*, S25–S30.

Hao, S., Dennen, V. P., & Mei, L. (2017). Influential factors for mobile learning acceptance among Chinese users. *Educational Technology Research and Development, 65*, 101–123.

Harland, P., Staats, H., & Wilke, H. A. (2007). Situational and personality factors as direct or personal norm mediated predictors of pro-environmental behavior: Questions derived from norm-activation theory. *Basic and applied social psychology, 29*(4), 323–334.

Helferich, M., Thøgersen, J., & Bergquist, M. (2023). Direct and mediated impacts of social norms on pro-environmental behavior. *Global Environmental Change, 80*(102680). doi: 10.1016/j.gloenvcha.2023.102680.

Holt, J. R., Butler, B. J., Borsuk, M. E., Markowski-Lindsay, M., MacLean, M. G., & Thompson, J. R. (2021). Using the Theory of Planned Behavior to Understand Family Forest Owners' Intended Responses to Invasive Forest Insects. *Society & Natural Resources, 34*(8), 1001–1018. doi: 10.1080/08941920.2021.1924330.

Hong, Y., Al Mamun, A., Masukujjaman, M., & Yang, Q. (2023). Significance of the environmental value-belief-norm model and its relationship to green consumption among Chinese youth. *Asia Pacific Management Review.*

Hoot, R. E., & Friedman, H. (2011). Connectedness and environmental behavior: Sense of interconnectedness and pro-environmental behavior. International Journal of Transpersonal Studies, 30(1), 10.

Huang, X., Dai, S., & Xu, H. (2020a). Predicting tourists' health risk preventative behaviour and travelling satisfaction in Tibet: Combining the theory of planned behavior and health belief model. *Tourism Management Perspectives, 33,* 100589.

Huang, Y., Luo, X., Tang, L., & Yu, W. (2020b). The power of habit: Does production experience lead to pesticide overuse? *Environmental Science and Pollution Research, 27*(20), 25287–25296. doi: 10.1007/s11356-020-08961-4.

Ifenthaler, D. & Schweinbenz, V. (2016). Students' acceptance of tablet PCs in the classroom. *Journal of Research on Technology in Education, 48*(4), 306–321.

Jaeger, C. M. & Schultz, P. W. (2017). Coupling social norms and commitments: Testing the underdetected nature of social influence. *Journal of Environmental Psychology, 51,* 199–208. doi: 1016/j.jenvp.2017.03.015

Janmaimool, P. (2017). Application of Protection Motivation Theory to Investigate Sustainable Waste Management Behaviors. *Sustainability, 9*(1079), 1–16.

Jeong, J. Y. & Ham, S. (2018). Application of the Health belief model to customers' use of menu labels in restaurants. *Appetite, 123,* 208–215.

Kals, E., Schumacher, D., & Montada, L. (1999). Emotional affinity toward nature as a motivational basis to protect nature. Environment and behavior, 31(2), 178–202.

Kamal, S. A., Shafiq, M., & Kakria, P. (2020). Investigating acceptance of telemedicine services through an extended technology acceptance model (TAM). *Technology in Society, 60,* 101212.

Karimi, S. & Saghaleini, A. (2021). Factors influencing ranchers' intentions to conserve rangelands through an extended theory of planned behavior. *Global Ecology and Conservation, 26,* e01513. doi: 10.1016/j.gecco.2021.e01513

Karrer, S. L. (2012). Swiss farmers' perception of and response to climate change. Ph.D. Thesis.

Koen, J., Klehe, U., & Van Vianen, A. (2012). Employability among the long-term unemployed: A futile quest or worth the effort? *Journal of Vocational Behavior, 82,* 37–48.

Keesstra, S., Mol, G., De Leeuw, J., Okx, J., Molenaar, C., De Cleen, M., & Visser, S. (2018). Soil-related sustainable development goals: Four concepts to make land degradation neutrality and restoration work. *Land, 7*(4), 133.

Kellert, S. R. (1993). Values and perceptions of invertebrates. Conservation biology, 7(4), 845–855.

Keshavarz, M. & Karami, E. (2016). Farmers' pro-environmental behavior under drought: Application of protection motivation theory. *Journal of Arid Environments, 127,* 128–136.

Kim, D. J., Ferrin, D. L., & Rao, H. R. (2008). A trust-based consumer decision-making model in electronic commerce: The role of trust, perceived risk, and their antecedents. *Decision Support Systems, 44*(2), 544–564.

Kim, H. J. & Jang, H. Y. (2015). Factors influencing students' beliefs about the future in the context of tablet-based interactive classrooms. *Computers & Education, 89,* 1–15.

Kim, H., Lee, S. H., & Yang, K. (2015). The heuristic-systemic model of sustainability stewardship: Facilitating sustainability values, beliefs and practices with corporate social responsibility drives and eco-labels/indices. *International Journal of Consumer Studies, 39*(3), 249–260. doi: 10.1111/ijcs.12173.

Kim, J. J. & Hwang, J. (2020). Merging the norm activation model and the theory of planned behaviour in the context of drone food delivery services: Does the level of product knowledge really matter? *Journal of Hospitality and Tourism Management, 42,* 1–11.

Klöckner, C. A. (2013). A comprehensive model of the psychology of environmental behaviour—A meta-analysis. Global environmental change, 23(5), 1028–1038.

Kollmuss, A. & Agyeman, J. (2002). Mind the gap: Why do people act environmentally and what are the barriers to pro-environmental behavior? *Environmental Education Research, 8*(3), 239–260.

Kristoffersen, A. E., Sirois, F. M., Stub, T., & Hansen, A. H. (2017). Prevalence and predictors of complementary and alternative medicine use among people with coronary heart disease or at risk for this in the sixth Tromsø study: A comparative analysis using protection motivation theory. *BMC Complementary and Alternative Medicine, 17*(1), 324.

Kuruppu, N. & Liverman, D. (2011). Mental preparation for climate adaptation: The role of cognition and culture in enhancing adaptive capacity of water management in Kiribati. *Global Environmental Change, 21*(2), 657–669.

Landon, A. C., Woosnam, K. M., & Boley, B. B. (2018). Modeling the psychological antecedents to tourists' pro-sustainable behaviors: An application of the value-belief-norm model. *Journal of Sustainable Tourism, 26*(6), 957–972.

Lange, F., Brick, C., & Dewitte, S. (2020). Green when seen? No support for an effect of observability on environmental conservation in the laboratory: A registered report. *Royal Society Open Science, 7*(4), 190189. doi: 10.1098/rsos.190189.

Le Dang, H., Li, E., Nuberg, I., & Bruwer, J. (2014). Understanding farmers' adaptation intention to climate change: A structural equation modelling study in the Mekong Delta, Vietnam. Environmental science & policy, 41, 11–22.

Liu, J. & Bridget, R. (2020). Food-energy-water nexus for multi-scale sustainable development. *Resources, Conservation and Recycling, 154*(C).

Liao, C., Yu, H., & Zhu, W. (2020). Perceived Knowledge, Coping Efficacy and Consumer Consumption Changes in Response to Food Recall. *Sustainability, 12*(7), 2696.

Lim, J. S. & Zhang, J. (2022). Adoption of AI-driven personalization in digital news platforms: An integrative model of technology acceptance and perceived contingency. *Technology in Society, 69,* 101965.

Liu, W., Shao, X. F., Wu, C. H., & Qiao, P. (2021). A systematic literature review on applications of information and communication technologies and blockchain technologies for precision agriculture development. *Journal of Cleaner Production, 298,* 126763. doi: 10.1016/j.jclepro.2021.126763

Liu, X., Zhou, S., Qi, S., Yang, B., Chen, Y., Huang, R., & Du, P. (2015). Zoning of rural water conservation in China: A case study at Ashihe River Basin. *International Soil and Water Conservation Research, 3*(2), 130–140. doi: 10.1016/j.iswcr.2015.04.003.

Liu, X., Zou, Y., & Wu, J. (2018). Factors influencing public-sphere pro-environmental behavior among Mongolian college students: A test of value–belief–norm theory. *Sustainability, 10*(5), 1384.

Lopes, J. R. N., de Araújo Kalid, R., Rodríguez, J. L. M., & Ávila Filho, S. (2019). A new model for assessing industrial worker behavior regarding energy saving considering the theory of planned behavior, norm activation model and human reliability. *Resources, Conservation and Recycling, 145,* 268–278.

Lubran, M. B. (2010). Factors influencing Maryland farmers' on-farm processing license application behavior. University of Maryland, College Park.

Lu, J., Hu, T., Zhang, B., Wang, L., Yang, S., Fan, J., . . . & Zhang, F. (2021). Nitrogen fertilizer management effects on soil nitrate leaching, grain yield and economic benefit of summer maize in Northwest China. Agricultural Water Management, 247, 106739.

Lu, Q. C., Zhang, J., Peng, Z. R., & Rahman, A. S. (2014). Inter-city travel behaviour adaptation to extreme weather events. Journal of transport geography, 41, 148–153.

Maleksaeidi, H., & Keshavarz, M. (2019). What influences farmers' intentions to conserve on-farm biodiversity? An application of the theory of planned behavior in fars province, Iran. Global Ecology and Conservation, 20, e00698.

Marcos, K. J., Moersidik, S. S., & Soesilo, T. E. (2021, March). Extended theory of planned behavior on utilizing domestic rainwater harvesting in Bekasi, West Java, Indonesia. In *IOP Conference Series: Earth and Environmental Science* (Vol. 716, No. 1, p. 012054). IOP Publishing.

Martin, F. P. J., Sprenger, N., Montoliu, I., Rezzi, S., Kochhar, S., & Nicholson, J. K. (2010). Dietary modulation of gut functional ecology studied by fecal metabonomics. Journal of proteome research, 9(10), 5284–5295.

Martin, S. M. & Lorenzen, K. (2016). Livelihood diversification in rural Laos. *World Development, 83,* 231–243.

Mayer, F. S., & Frantz, C. M. (2004). The connectedness to nature scale: A measure of individuals' feeling in community with nature. Journal of environmental psychology, 24(4), 503–515.

Mitter, H., Larcher, M., Schönhart, M., Stöttinger, M., & Schmid, E. (2019). Exploring farmers' climate change perceptions and adaptation intentions: Empirical evidence from Austria. *Environmental Management, 63*(6), 804–821.

Moglia, M., Cook, S., & Tapsuwan, S. (2018). Promoting water conservation: Where to from here? *Water, 10*(11), 1510.

Mohr, S. & Kühl, R. (2021). Acceptance of artificial intelligence in German agriculture: An application of the technology acceptance model and the theory of planned behavior. *Precision Agriculture, 22*(6), 1816–1844.

Moya, B., Parker, A., Sakrabani, R., & Mesa, B. (2019). Evaluating the efficacy of fertilisers derived from human excreta in agriculture and their perception in Antananarivo, Madagascar. Waste and Biomass Valorization, 10(4), 941–952.

Murphy, D. A., Stein, J. A., Schlenger, W., & Maibach, E. (2001). Conceptualizing the multidimensional nature of self-efficacy: Assessment of situational context and level of behavioral challenge to maintain safer sex. *Health Psychology, 20*(4), 281.

Nguyen, Y. T. H. & Nguyen, H. V. (2020). An alternative view of the millennial green product purchase: The roles of online product review and self-image congruence. *Asia Pacific Journal of Marketing and Logistics, 33*(1), 231–249.

Nigbur, D., Lyons, E., & Uzzell, D. (2010). Attitudes, norms, identity and environmental behaviour: Using an expanded theory of planned behaviour to predict participation in a kerbside recycling programme. *British Journal of Social Psychology, 49*(2), 259–284.

Nikou, S. A., & Economides, A. A. (2017). Mobile-Based Assessment: Integrating acceptance and motivational factors into a combined model of Self-Determination Theory and Technology Acceptance. Computers in Human Behavior, 68, 83–95.

Nikou, S. A. & Economides, A. A. (2018). Mobile-Based micro-Learning and Assessment: Impact on learning performance and motivation of high school students. *Journal of Computer Assisted Learning, 34*(3), 269–278.

Nordlund, A. M., & Garvill, J. (2002). Value structures behind proenvironmental behavior. Environment and behavior, 34(6), 740–756.

Oakley, M., Mohun Himmelweit, S., Leinster, P., & Casado, M. R. (2020). Protection Motivation Theory: A Proposed Theoretical Extension and Moving beyond Rationality – The Case of Flooding. *Water, 12*(7), 1848.

Onwezen, M. C., Antonides, G., & Bartels, J. (2013). The Norm Activation Model: An exploration of the functions of anticipated pride and guilt in pro-environmental behaviour. *Journal of Economic Psychology, 39*, 141–153.

Orji, R., Vassileva, J., & Mandryk, R. (2012). Towards an effective health interventions design: An extension of the health belief model. *Online Journal of Public Health Informatics, 4*(3).

Peng, M. Y. P. & Xu, Y. (2023). Enhancing Students' English Language Learning Via M-Learning: Integrating Technology Acceptance Model and SOR Model. *Available at SSRN 4247604.*

Pennock, D. (2019). Soil erosion: The greatest challenge for sustainable soil management. FAO.

Pereira, M., & Forster, P. (2015). The relationship between connectedness to nature, environmental values, and pro-environmental behaviours. Reinvention: An international journal of undergraduate research, 8(2).

Poelma, T. F. (2018). Transitioning to rice-shrimp farming in Kien Giang, Vietnam Determining rural household resilience to changing climatic conditions (Master's thesis).

Popa, B., Niță, M. D., & Hălălișan, A. F. (2019). Intentions to engage in forest law enforcement in Romania: An application of the theory of planned behavior. *Forest Policy and Economics, 100*, 33–43. doi: 10.1016/j.forpol.2018.11.005

Prasetyo, Y. T., Castillo, A. M., Salonga, L. J., Sia, J. A., & Seneta, J. A. (2020). Factors affecting perceived effectiveness of COVID-19 prevention measures among Filipinos during enhanced community quarantine in Luzon, Philippines: Integrating Protection Motivation Theory and extended Theory of Planned Behavior. *International Journal of Infectious Diseases, 99*, 312–323.

Price, J. C. & Leviston, Z. (2014). Predicting pro-environmental agricultural practices: The social, psychological and contextual influences on land management. *Journal of Rural Studies, 34*, 65–78. doi: 10.1016/j.jrurstud.2013.10.001

PUTRI, G. A., Widagdo, A. K., & Setiawan, D. (2023). Analysis Of Financial Technology Acceptance of Peer-to-Peer Lending (P2p Lending) Using Extended Technology Acceptance Model (Tam). *Journal of Open Innovation: Technology, Market, and Complexity*, 100027.

Raheli, H., Zarifian, S., & Yazdanpanah, M. (2020). The power of the health belief model (HBM) to predict water demand management: A case study of farmers' water conservation in Iran. *Journal of Environmental Management, 263*, 110388.

Rainear, A. M. & Christensen, J. L. (2017). Protection motivation theory as an explanatory framework for proenvironmental behavioral intentions. *Communication Research Reports, 34*(3), 239–248.

Ramirez, E., Kulinna, P. H., & Cothran, D. (2012). Constructs of physical activity behaviour in children: The usefulness of Social Cognitive Theory. *Psychology of Sport and Exercise, 13*(3), 303–310.

Rahmanian Koushkaki, M., Chizari, M., & Havasi, A. (2012). Study of factors affecting agricultural students' entrepreneurial Intentions University of Ilam. *Entrepreneurship Development, 4*(15), 125–144.

Rezaei, R., Safa, L., & Ganjkhanloo, M. M. (2020). Understanding farmers' ecological conservation behavior regarding the use of integrated pest management-an application of the technology acceptance model. *Global Ecology and Conservation, 22*, e00941.

Rezaei, R. & Mianaji, S. (2019). Using the Health Belief Model to Understand Farmers' Intentions to Engage in the On-Farm Food Safety Practices in Iran. *Journal of Agricultural Science and Technology, 21*(3), 561–574.

Rezaei, R., Safa, L., Damalas, C. A., & Ganjkhanloo, M. M. (2019). Drivers of farmers' intention to use integrated pest management: Integrating theory of planned behavior and norm activation model. *Journal of Environmental Management, 236*, 328–339.

Rho, M. J., Young Choi, I., & Lee, J. (2014). Predictive factors of telemedicine service acceptance and behavioral intention of physicians. *International Journal of Medical Informatics, 83*(8), 559–571.

Salehi, S. & Ebrahimi, H. (2020). Analysis of the status of students' knowledge and behavior towards water (with emphasis on the Danab project in Mazandaran). *Quarterly Journal of Environmental Sciences, 18*(2), 41–58. In person.

Sánchez, M., López-Mosquera, N., Lera-López, F., & Faulin, J. (2018). An extended planned behavior model to explain the willingness to pay to reduce noise pollution in road transportation. *Journal of Cleaner Production, 177*, 144–154.

Savari, M. (2023). Explaining the ranchers' behavior of rangeland conservation in western Iran. *Frontiers in Psychology, 13*, 1090723. doi: 10.3389/fpsyg.2022.1090723

Savari, M., Yazdanpanah, M., & Rouzaneh, D. (2022). Factors affecting the implementation of soil conservation practices among Iranian farmers. *Scientific Reports, 12*(1).

Savari, M. & Gharechaee, H. (2020). Utilizing the theory of planned behavior to predict Iranian farmers' intention for safe use of chemical fertilizers. *Journal of Cleaner Production*, 121512.

Savari, M. & Khaleghi, B. (2023). Application of the extended theory of planned behavior in predicting the behavioral intentions of Iranian local communities toward forest conservation. *Frontiers in Psychology, 14*, 1121396. doi: 10.3389/fpsyg.2023.1121396

Savari, M., Zhoolideh, M., & Khosravipour, B. (2021). Explaining pro-environmental behavior of farmers: A case of rural Iran. *Current Psychology*, 1–19.

Schwartz, S. H. (1977). Normative influences on altruism. *Advances in Experimental Social Psychology, 10*(1), 221–279.

Savari, M., Damaneh, H. E., & Damaneh, H. E. (2024a). Managing the effects of drought through the use of risk reduction strategy in the agricultural sector of Iran. *Climate Risk Management*, 100619.

Savari, M., Zhoolideh, M., & Limuie, M. (2024b). Factors affecting the use of climate information services for agriculture: Evidence from Iran. *Climate Services, 33*(100438).

Savari, M., Sheheytavi, A., & Amghani, M. S. (2023a). Factors underpinning Iranian farmers' intention to conserve biodiversity at the farm level. *Journal for Nature Conservation, 73*, 126419.

Savari, M., Sheheytavi, A., & Amghani, M. S. (2023b). Promotion of adopting preventive behavioral intention toward biodiversity degradation among Iranian farmers. *Global Ecology and Conservation, 43*, e02450.

Savari, M., Ebrahimi-Maymand, R., & Mohammadi-Kanigolzar, F. (2013). The Factors influencing the application of organic farming operations by farmers in Iran. *Agris On-line Papers in Economics and Informatics, 5*(4), 179–187.

Schultz, P. W., & Zelezny, L. C. (1998). Values and proenvironmental behavior: A five-country survey. Journal of cross-cultural psychology, 29(4), 540–558.

Setiawan, R., Santosa, W., & Sjafruddin, A. (2014). Integration of theory of planned behavior and norm activation model on student behavior model using cars for traveling to campus. *Civil Engineering Dimension, 16*(2), 117–122.

Shafiei, A. & Maleksaeidi, H. (2020). Pro-environmental behavior of university students: Application of protection motivation theory. *Global Ecology and Conservation*, e00908. doi: 10.1016/j.gecco.2020.e00908.

Shahangian, S. A., Tabesh, M., & Yazdanpanah, M. (2021). Psychosocial determinants of household adoption of water-efficiency behaviors in Tehran capital, Iran: Application of the social cognitive theory. *Urban Climate, 39*, 100935.

Sharifzadeh, M. S., Damalas, C. A., Abdollahzadeh, G., & Ahmadi-Gorgi, H. (2017). Predicting adoption of biological control among Iranian rice farmers: An application of the extended technology acceptance model (TAM2). *Crop Protection, 96*, 88–96.

Sheppard, J. & Thomas, C. B. (2020). Community pharmacists and communication in the time of COVID-19: Applying the health belief model. *Research in Social and Administrative Pharmacy*.

Shin, Y. H., Im, J., Jung, S. E., & Severt, K. (2018). The theory of planned behavior and the norm activation model approach to consumer behavior regarding organic menus. *International Journal of Hospitality Management, 69*, 21–29.

Song, S. B., Park, J. S., Chung, G. J., Lee, I. H., & Hwang, E. S. (2019). Diverse therapeutic efficacies and more diverse mechanisms of nicotinamide. Metabolomics, 15(10), 137.

Stallones, L., Acosta, M. S. V., Sample, P., Bigelow, P., & Rosales, M. (2009). Perspectives on safety and health among migrant and seasonal farmworkers in the United States and México: A qualitative field study. *The Journal of Rural Health, 25*(2), 219–225.

Steg, L., & De Groot, J. (2010). Explaining prosocial intentions: Testing causal relationships in the norm activation model. British journal of social psychology, 49(4), 725–743.

Steg, L. & Nordlund, A. (2018). Theories to explain environmental behaviour. *Environmental Psychology: An Introduction*, 217–227.

Stern, P. C. (2000). New environmental theories: toward a coherent theory of environmentally significant behavior. Journal of social issues, 56(3), 407–424.

Stern, P. C., Dietz, T., Abel, T., Guagnano, G. A., & Kalof, L. (1999). A value-belief-norm theory of support for social movements: The case of environmentalism. *Human Ecology Review*, 81–97.

Straub, C. L. & Leahy, J. E. (2014). Application of a modified health belief model to the pro-environmental behavior of private well water testing. *JAWRA Journal of the American Water Resources Association, 50*(6), 1515–1526.

Strecher, V. J., & Rosenstock, I. M. (1997). The health belief model. Cambridge handbook of psychology, health and medicine, 113, 117.

Sung, H. Y., Hwang, G. J., Chen, C. Y., & Liu, W. X. (2019). A contextual learning model for developing interactive e-books to improve students' performances of learning the Analects of Confucius. *Interactive Learning Environments*, 1–14.

Tapsuwan, S. & Rongrongmuang, W. (2015). Climate change perception of the dive tourism industry in Koh Tao Island, Thailand. *Journal of Outdoor Recreation and Tourism, 11*, 58–63.

Thøgersen, J., & Ölander, F. (2002). Human values and the emergence of a sustainable consumption pattern: A panel study. Journal of economic psychology, 23(5), 605–630.

Thøgersen, J., & Grønhøj, A. (2010). Electricity saving in households—A social cognitive approach. Energy policy, 38(12), 7732–7743.

Trihadmojo, B., Jones, C. R., Prasastyoga, B., Walton, C., & Sulaiman, A. (2020). Toward a nuanced and targeted forest and peat fires prevention policy: Insight from psychology. *Forest Policy and Economics, 120*, 102293. doi: 10.1016/j.forpol.2020.102293

Truelove, H. B., Carrico, A. R., & Thabrew, L. (2015). A socio-psychological model for analyzing climate change adaptation: A case study of Sri Lankan paddy farmers. *Global Environmental Change, 31*, 85–97.

Tseng, T. H., Wang, Y. M., Lin, H. H., Lin, S. J., Wang, Y. S., & Tsai, T. H. (2022). Relationships between locus of control, theory of planned behavior, and cyber entrepreneurial intention: The moderating role of cyber entrepreneurship education. *The International Journal of Management Education, 20*(3), 100682. doi: 10.1016/j.ijme.2022.100682.

Turaga, R. M. R., Howarth, R. B., & Borsuk, M. E. (2010). Pro-environmental behavior: Rational choice meets moral motivation. *Annals of the New York Academy of Sciences, 1185*(1), 211–224.

Ullah, S., Abid, A., Aslam, W., Noor, R. S., Waqas, M. M., & Gang, T. (2021). Predicting behavioral intention of rural inhabitants toward economic incentive for deforestation in Gilgit-Baltistan, Pakistan. *Sustainability, 13*(2), 617. doi: 10.3390/su13020617.

Valois, R. F., Zullig, K. J., & Revels, A. A. (2017). aggressive and violent behavior and emotional self-efficacy: is there a relationship for adolescents? *Journal of School Health, 87*(4), 269–277.

Van Valkengoed, A. M., Abrahamse, W., & Steg, L. (2022). To select effective interventions for pro-environmental behaviour change, we need to consider determinants of behaviour. *Nature Human Behaviour*, 1–11. doi: 10.1038/s41562-022-01473-w.

van Valkengoed, A. & Steg, L. (2019). Meta-analyses of factors motivating climate change adaptation behaviour. *Nature Climate Change*.

Vance, A., Siponen, M., & Pahnila, S. (2012). Motivating IS security compliance: Insights from habit and protection motivation theory. *Information & Management, 49*(3–4), 190–198.

Vassallo, M., Saba, A., Arvola, A., Dean, M., Messina, F., Winkelmann, M., . . . Shepherd, R. (2009). Willingness to use functional breads. Applying the Health Belief Model across four European countries. *Appetite, 52*(2), 452–460.

Venkatesh, V., & Davis, F. D. (1996). A model of the antecedents of perceived ease of use: Development and test. Decision sciences, 27(3), 451–481.

Vesely, S., Klöckner, C. A., Carrus, G., Tiberio, L., Caffaro, F., Biresselioglu, M. E., . . . Sinea, A. C. (2022). Norms, prices, and commitment: A comprehensive overview of field experiments in the energy domain and treatment effect moderators. *Frontiers in Psychology, 13*. doi: 10.3389/fpsyg.2022.967318.

Vicente-Molina, M. A., Fernández-Sáinz, A., & Izagirre-Olaizola, J. (2013). Environmental knowledge and other variables affecting pro-environmental behaviour: Comparison of university students from emerging and advanced countries. *Journal of Cleaner Production, 61*, 130–138. doi: 10.1016/j.jclepro.2013.05.015

Wang, S., Wang, J., Zhao, S., & Yang, S. (2019a). Information publicity and resident's waste separation behavior: An empirical study based on the norm activation model. *Waste Management, 87*, 33–42.

Wang, Y., Liang, J., Yang, J., Ma, X., Li, X., Wu, J., . . . Feng, Y. (2019b). Analysis of the environmental behavior of farmers for non-point source pollution control and management: An integration of the theory of planned behavior and the protection motivation theory. *Journal of Environmental Management, 237*, 15–23.

Wang, Y., Yang, J., Liang, J., Qiang, Y., Fang, S., Gao, M., . . . Feng, Y. (2018). Analysis of the environmental behavior of farmers for non-point source pollution control and management in a water source protection area in China. *Science of the Total Environment, 633*, 1126–1135. doi: 10.1016/j.scitotenv.2018.03.273

Wang, Y., Zhou, S., & Jiang, G. (2023). Can the application of environmentally friendly fertilisers reduce agricultural labour input? Empirical evidence from peanut farmers in China. *Sustainability, 15*(4), 2989. doi: 10.3390/su15042989.

Wauters, E., Bielders, C., Poesen, J., Govers, G., & Mathijs, E. (2010). Adoption of soil conservation practices in Belgium: An examination of the theory of planned behaviour in the agri-environmental domain. *Land Use Policy, 27*(1), 86–94. doi: 10.1016/j.landusepol.2009.02.009.

Wensing, J., Carraresi, L., & Bröring, S. (2019). Do pro-environmental values, beliefs and norms drive farmers' interest in novel practices fostering the Bioeconomy? Journal of environmental management, 232, 858–867.

Wilson, T. D., Dunn, D. S., Bybee, J. A., Hyman, D. B., & Rotondo, J. A. (1984). Effects of analyzing reasons on attitude–behavior consistency. Journal of Personality and Social Psychology, 47(1), 5.

Wu, B., & Chen, X. (2017). Continuance intention to use MOOCs: Integrating the technology acceptance model (TAM) and task technology fit (TTF) model. Computers in human behavior, 67, 221–232.

Xiang, X., Liu, J., Zhang, J., Li, D., Xu, C., & Kuzyakov, Y. (2020). Divergence in fungal abundance and community structure between soils under long-term mineral and organic fertilization. Soil and Tillage Research, 196, 104491.

Yadav, R., & Pathak, G. S. (2016). Young consumers' intention towards buying green products in a developing nation: Extending the theory of planned behavior. Journal of cleaner production, 135, 732–739.

Yazdanpanah, M., Komendantova, N., & Ardestani, R. S. (2015a). Governance of energy transition in Iran: Investigating public acceptance and willingness to use renewable energy sources through socio-psychological model. *Renewable and Sustainable Energy Reviews, 45*, 565–573.

Yazdanpanah, M., Komendantova, N., Shirazi, Z. N., & Linnerooth-Bayer, J. (2015b). Green or in between? Examining youth perceptions of renewable energy in Iran. *Energy Research & Social Science, 8*, 78–85.

Yazdanpanah, M., Forouzani, M., Abdeshahi, A., & Jafari, A. (2016). Investigating the effect of moral norm and self-identity on the intention toward water conservation among Iranian young adults. Water Policy, 18(1), 73–90.

Young, H. N., Lipowski, E. E., & Cline, R. J. (2005). Using social cognitive theory to explain consumers' behavioral intentions in response to direct-to-consumer prescription drug advertising. *Research in Social and Administrative Pharmacy, 1*(2), 270–288.

Young, M. D., Plotnikoff, R. C., Collins, C. E., Callister, R., & Morgan, P. J. (2014). Social cognitive theory and physical activity: A systematic review and meta-analysis. *Obesity Reviews, 15*(12), 983–995.

Zhang, Y., Wang, Z., & Zhou, G. (2013). Antecedents of employee electricity saving behavior in organizations: An empirical study based on norm activation model. *Energy Policy, 62*, 1120–1127.

Zhao, G., Cavusgil, E., & Zhao, Y. (2016). A protection motivation explanation of base-of-pyramid consumers' environmental sustainability. *Journal of Environmental Psychology, 45*, 116–126.

Zheng, J. & Li, S. (2020). What drives students' intention to use tablet computers: An extended technology acceptance model. *International Journal of Educational Research, 102*, 101612.

Zheng, S., Yin, K., & Yu, L. (2022). Factors influencing the farmer's chemical fertilizer reduction behavior from the perspective of farmer differentiation. Heliyon, 8(12).

Zhong, F., Li, L., Guo, A., Song, X., Cheng, Q., Zhang, Y., & Ding, X. (2019). Quantifying the influence path of water conservation awareness on water-saving irrigation behavior based on the Theory of Planned Behavior and structural equation modeling: A case study from Northwest China. *Sustainability, 11*(18), 4967. https://doi.org/10.3390/su11184967-.

Ebrahim Farmandeh, Shahla Choobchian*, and Shobeir Karami

Fields of Flourish: Exploring the Interwoven Water-Energy-Food Nexus in Agricultural Communities Through Content Analysis

Abstract: The water-energy-food nexus (WEFN) is a key framework for developing agricultural societies and managing the complex interactions that significantly impact agricultural progress. However, challenges remain, including a lack of comprehensive understanding and knowledge gaps about the key components of this nexus, which can impede effective decision-making and policy formulation in the agricultural sector.

This chapter addresses these challenges by thoroughly identifying and analyzing the essential elements of the WEFN in agricultural development. Using a qualitative content analysis method, which included manual analysis and Sankey 5.1 software, the study reviewed 295 relevant articles published in international journals from 2013 to 2023, ultimately selecting 110 for detailed examinations. The findings highlight that the concepts within the WEFN are crucial for shaping agricultural community development. These concepts were organized into three main themes, 29 subthemes, and 384 distinct concepts. The primary themes include the water, energy, and food nexus, with the most frequently researched subthemes being water consumption, energy consumption and fuel, and food production.

By identifying these key concepts, the research offers valuable insights for planners and policymakers, enabling the development of practical solutions and strategies tailored to the evolving needs of agricultural communities. The study's policy recommendations include integrated resource management, promoting renewable energy to reduce reliance on fossil fuels, adopting sustainable agricultural practices to improve water and energy efficiency, and implementing community engagement and education initiatives to enhance stakeholders' understanding of the nexus. These recommendations aim to support informed policy assessment and evaluation, ultimately fostering sustainable development in agricultural societies.

Keywords: Community development, interwoven interactions, policy formulation

*Corresponding author: Shahla Choobchian, Department of Agricultural Extension and Education, College of Agriculture, Tarbiat Modares University (TMU), Tehran, Iran,
e-mail: shchoobchian@modares.ac.ir
Ebrahim Farmandeh, Department of Agricultural Extension and Education, College of Agriculture, Tarbiat Modares University (TMU), Tehran, Iran
Shobeir Karami, Persian Gulf Research Institute, Persian Gulf University, Bushehr, Iran

https://doi.org/10.1515/9783111563046-002

1 Introduction

Today's world faces a myriad of natural and human-made risks, which have been exacerbated by the global aftermath of the COVID-19 pandemic. This crisis has highlighted that traditional survival strategies are no longer viable. Inflation, rising living costs, overt and covert conflicts over resource access, and widespread social discontent have further compounded these challenges. Additionally, the erosion of established social and political structures, coupled with escalating environmental crises, threatens the very foundation of development and its sustainability. At the same time, the evolution of societies, particularly in rural areas, faces a range of risks. These include social fragmentation, economic upheaval disrupting resource production and supply chains, and environmental degradation leading to the depletion of natural resources. For example, global resource projections for the next decade indicate that only 9% of resources are sustainably recoverable, with over 20% at risk of loss due to environmental and human-induced hazards. Economically, the period leading up to 2022 has seen an average inflation increase of over 30% in the production of food and drinking water worldwide. Furthermore, the discourse on climate change highlights the world's struggle to meet specified adaptation goals, as noted by the World Economic Forum (2023).

According to the FAO (2022), from 2000 to 2022, the total arable land for agriculture worldwide has approached five billion hectares, of which only 33% is used for agricultural production; the remainder is allocated to pastures and grasslands. From this arable land, the production of key agricultural products like wheat and corn has exceeded 3.9 billion tons. During this period, the global population of the hungry has increased by more than 150 million people. Additionally, 134 million hectares of agricultural land have been removed from production due to destructive and damaging uses. Focusing on Iran, the country's agricultural lands are estimated at around 7 million hectares, accounting for 2.4% of its total land area. It is noteworthy that 3–5% of Iran's population suffers from malnutrition, with nearly 10 million individuals affected by 2022, including 10% children under 5 years old. Iran faces one of the highest levels of water stress globally (75–100%). Given the current state of food production, the agricultural sector consumes over 90% of the country's water resources, necessitating the import of approximately 1.181 million tons of various grains to meet the growing food demand (FAO, 2022).

The preceding discussion highlights the critical challenges facing agricultural communities worldwide, with a particular focus on Iran. The WEFN, referred to as the "nexus" here, illustrates the complex interconnections among water, energy, and food sources, including agricultural production and consumable goods (Ibrahim et al., 2019). This nexus provides a dynamic and integrated approach crucial for achieving sustainable development goals. It functions as an interdisciplinary strategy that examines the interactions among water, energy, and food resources and their inherent characteristics (Saray et al., 2022).

The nexus framework is well-suited for addressing sustainable development challenges due to its notable attributes. It helps in understanding the intricate and interdependent relationships among production resources, considering both temporal and spatial dimensions. The focus broadens to include all relevant sectors, emphasizing not only the effectiveness of the system but also production, productivity, and wider sectoral considerations. By prioritizing resource interactions, the framework facilitates intersectoral planning to optimize resource use and improve resource availability (Cheng et al., 2023; Haghjoo et al., 2022; Li et al., 2019b; Purwanto et al., 2019; Zhang et al., 2021a). Additionally, it aims for equitable resource distribution and fosters inclusive decision-making processes that benefit all members of society.

The nexus framework also incorporates diverse decision-making models and advanced planning strategies, based on comprehensive assessments of current and future resource statuses (Sadeghi et al., 2020; Simpson et al., 2022). Its ultimate goal is to ensure triple nexus security by addressing climate change and promoting robust growth, sustainable development, and ecosystem health (Enayati et al., 2021).

Globally, the water-energy-food (WEF) organization has developed a comprehensive indicator known as the WEF indicator. Published annually for all countries, this indicator provides an aggregate score across the nexus dimensions. Table 1, from the Global Nexus Index Organization's latest 2022 report, shows the top and bottom five countries. Iran is ranked 94th with a score of 56.21. The WEF indicator includes three primary dimensions – water nexus, energy nexus, and food nexus – each with two key subindicators: access and availability. It also uses 21 subindicators, such as average food production value, cereal yield per hectare, obesity prevalence among adults over 18, and net energy usage, to evaluate nexus rankings and scores for individual countries (Figure 1) (WEF Nexus, 2022).

Table 1: The status of world countries in terms of the Global Nexus Index in 2022.

Country	Rank	Water pillar	Energy pillar	Food pillar	WEF Nexus Index score
Iceland	1	78.41	94.04	69.88	80.77
Canada	2	77.84	80.98	84.66	77.84
Norway	3	77.26	73.79	94.34	77.26
USA	4	75.3	77.70	73.76	75.30
New Zealand	5	74.75	77.74	74.47	74.75
IRAN	94	46.13	61.21	61.66	56.21
Republic of Yemen	157	29.36	56.97	27.99	38.11
Haiti	158	41.07	45.62	27.55	38.08
Central African Republic	159	44.39	N/A	31.53	37.96
South Sudan	160	39.32	35.65	N/A	37.49
Chad	161	32.18	N/A	37.83	35.00

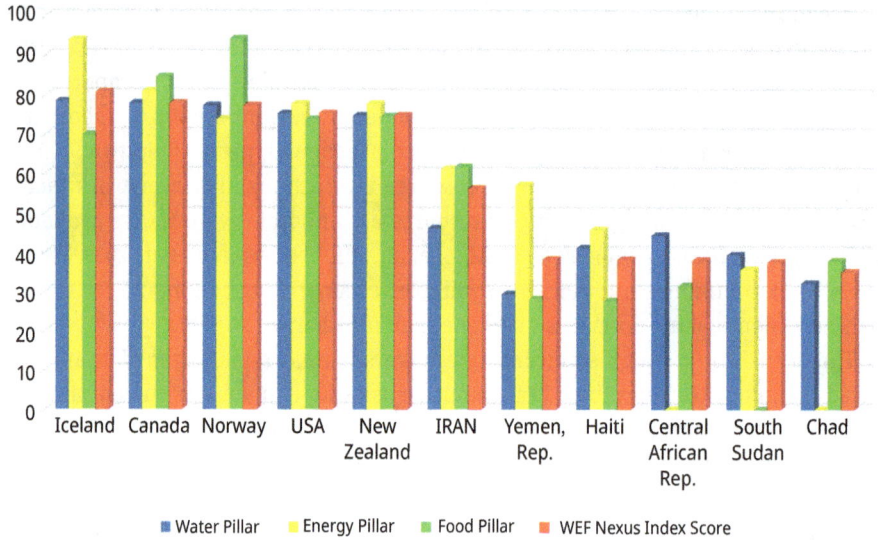

Figure 1: The status of world countries in terms of the Global Nexus Index in 2022.

Table 2 presents an overview of previous studies that explore the nexus. These studies examine a variety of aspects, including both external and internal factors, the elements that influence and are affected by the nexus, and performance measurement indices related to the dimensions and resources of the triple nexus. The breadth of this research highlights the complex nature of understanding and managing the interconnectedness among water, energy, and food systems.

Table 2: Previous nexus studies.

Research title	Results	References
Water-energy-food nexus in the Yarlung Tsangpo-Brahmaputra River Basin: Impact of mainstream hydropower development	Integrating hydrological insights improves water reservoir storage, reduces flood damage in Bangladesh, and supports hydropower development.	Lyu et al. (2023)
Evaluating policy options for adopting a water-energy-food nexus pattern in farming: Optimization and agent-based model applications	The proposed model boosts energy efficiency by 67% and, with its adaptability to government policies, has the potential to drive economic development.	Mirzaei et al. (2023)
Water, energy, and food nexus in Pakistan: Parametric and nonparametric analysis	External factors like urban populations and industrial zones significantly impact the effectiveness of the Nexus and play a key role in enhancing its quality.	Ali et al. (2022)

Table 2 (continued)

Research title	Results	References
Development and validation of management assessment tools considering water, food, and energy security nexus at the farm level	Effective agricultural management serves as a crucial local facilitator for ensuring water, food, and energy security. To assess and enhance this multifaceted approach, indicators have been categorized into three main dimensions: accessibility, availability, and sustainability.	Haghjoo et al. (2022)
Ten years of research on the water-energy-food nexus: An analysis of topics evolution	Over the past decade, the following aspects have been addressed: 1) The concept of nexus and natural resource management, 2) The link between nexus, sustainable development goals (SDGs), and the green economy, 3) Governance and policy integration for nexus applications across time and space, and 4) Nexus components in response to urban community changes and challenges.	Lazaro et al. (2022)
Evaluating the synergy between water-energy-food nexus and decoupling pollution-agricultural growth for sustainable production in the agricultural sector	The concurrent integration of sustainable development and nexus concepts can significantly influence agricultural pollution. The solution lies in decoupling pollution from agricultural development and emphasizing resource management for water and energy within the agricultural sector.	Naghavi et al. (2022)
A new framework of green transition of cultivated land-use for the coordination among the water-land-food-carbon nexus in China	Utilizing this approach, regions are categorized into four groups: pioneers, quality improvement, potential, and flexibility. According to the nexus approach, flexible regions typically experience better development and greater environmental conservation.	Niu et al. (2022)
A system dynamic model to quantify the impacts of water resources allocation on the water-energy-food-society (WEFS) nexus	The proposed scenario elucidates a novel approach, ensuring a specialized and logical relationship between water resources and their utilization. It also fosters a deeper understanding of the commitments and threats associated with the utilization of the new WEFS nexus, ultimately improving resource effectiveness.	Zeng et al. (2022)

Table 2 (continued)

Research title	Results	References
Impacts of climatic change and database information design on the water-energy-food nexus in water-scarce regions	Developing a model based on the WEF nexus in Oman has the potential to significantly enhance climate change mitigation, preserve biodiversity, and promote prosperity in regional fisheries. The model aims to achieve sustainable outcomes.	Boluwade (2021)
Multiscale simulation and dynamic coordination evaluation of water-energy-food and economy for the Pearl River Delta city cluster in China	Consequently, the proposed scenarios for temporal and spatial patterns in the region seek to align urban characteristics with the preservation of the nexus, ultimately reducing the impact of urban vulnerabilities.	Ouyang et al. (2021)
Quantifying and managing the water-energy-food nexus in dry regions food insecurity: New methods and evidence	An optimization-based policy model for sustainable groundwater management can optimize food and energy resource use while addressing environmental concerns and economic efficiency. This approach enhances irrigation efficiency and aligns with nexus objectives.	Radmehr et al. (2021)
A system dynamics model to quantify the impacts of restoration measures on the water-energy-food nexus in the Urmia Lake Basin, Iran	Climate change has far-reaching and irreversible effects across all sectors. Lake Urmia's water level is highly sensitive to climate variations. Reconstruction scenarios hold promise for desirable outcomes until 2040.	Bakhshianlamouki et al. (2020)
Quantifying and predicting the water-energy-food-economy-society environment nexus based on Bayesian networks: A case study of China	The direct impact of water extraction is closely tied to population growth and the rising energy demand. Furthermore, energy requirements significantly influence the nation's gross domestic product (GDP) through population expansion.	Chai et al. (2020)
The food-water-energy nexus governance model: A case study for Iran	To address the water and food crisis in Iran while considering the goals of sustainable development; the nexus approach can foster adaptability and balance across various sectors. This approach can ensure sustainable resource security in Iran, even in the face of environmental challenges.	Norouzi and Kalantari (2020)

Table 2 (continued)

Research title	Results	References
Evaluating the sustainability of water-energy-food nexus using an improved matter-element extension model: A case study of China	By 2015, China had reached a level of general sustainability in its production resources that addressed only basic development needs. To improve overall sustainability, it is crucial to prioritize nexus considerations, which can help better understand resource synergies and enhance resource management.	Wang et al. (2018)
Water-Energy-Food Nexus Tool 2.0: Guiding integrative resource planning and decision-making	This tool can generate realistic and adaptable scenarios to address key questions about the nexus dimensions. Additionally, the new framework can enhance food security and reduce risk levels by leveraging insights from these dimensions.	Daher and Mohtar (2015)
Basin perspectives on the water-energy-food security nexus	Results indicate that cooperation, integration, and earth observation assessments can enhance WEFN applications in security dimensions and are crucial for examining the impact of global changes on current conditions.	Lawford et al. (2013)

Numerous researchers have worked to develop a comprehensive understanding of the WEFN. Their efforts have varied in focus:

1. **Modeling interrelationships**: Some studies have focused on modelling how different factors and dimensions within the nexus interact. For example, Abdel-Aal et al. (2020) and Bakhshianlamouki et al. (2020) have explored these interrelationships to understand how water, energy, and food systems affect each other.

2. **Managing and strategizing**: Other research aimed at managing and strategizing the factors within the nexus. Researchers such as Anser et al. (2020), Daher & Mohtar (2015), Nhamo & Ndlela (2021), and Yue & Guo (2021) have developed strategies to improve the management of these nexus-based factors.

3. **Quantifying relationships**: Additional studies have focused on quantifying and identifying the relationships among nexus components to achieve a comprehensive understanding. This research, including work by Cheng et al. (2023) and Huang et al. (2020), has simulated how nexus components behave under varying conditions and evaluated their interactions.

4. **Developing indicators**: Higher-level investigations have aimed at developing and assessing indicators to measure nexus components and their interactions. Studies by El-Gafy (2017), Haghjoo et al. (2022), Kedir et al. (2022), Saladini et al. (2018), Shu et al. (2021), and Simpson et al. (2022) have contributed to a better

understanding of the nexus process in real-world scenarios, including the impacts of climate change.

Despite these efforts, no single study has yet managed to cover all aspects of the nexus in the context of community development comprehensively. Most research has concentrated on specific aspects and their interactions, revealing a significant knowledge gap and a limitation in the current approach to nexus studies.

The chapter's primary objective is to identify and analyze the key components of the WEFN within the context of agricultural community development. By addressing existing knowledge gaps, the chapter aims to develop a comprehensive classification that integrates all critical elements of the WEFN, ultimately leading to the formulation of policy recommendations for decision-makers.

To ensure clarity and coherence, the chapter is structured around three specific goals, each aligned with the primary objective:

1. **Identification of the main components within each nexus dimension (water, energy, and food):**
   - The first goal is to systematically identify the essential components that define each of the WEFN dimensions. This involves an in-depth analysis of the interactions and influences between water, energy, and food systems within agricultural communities. The chapter explores the unique challenges and opportunities associated with each dimension, laying the foundation for a holistic understanding of the nexus.

2. **Categorization and classification of these components using a qualitative and systematic approach:**
   - After identifying the key components, the chapter focuses on categorizing and classifying them. A qualitative and systematic methodology is employed to ensure that the classification accurately reflects the interdependencies and complexities of the WEFN. This structured classification serves as a critical framework for the subsequent analysis and discussion.

3. **Development of policy recommendations for experts and decision-makers:**
   - The final goal is to translate the research findings into actionable policy recommendations. These recommendations are designed to assist experts and decision-makers in integrating WEFN considerations into their strategies, thereby enhancing the sustainability and resilience of agricultural communities. The policy guidance is rooted in the chapter's comprehensive analysis and classification, ensuring its relevance and applicability.

This structured approach not only clarifies the research objectives but also ensures that the study effectively addresses its primary aim of bridging the knowledge gap in WEFN studies within the context of agricultural community development.

2 Materials and Methods

This study employed a combination of library resource review and literature review methods to extract and identify the constituent concepts of the WEFN for agricultural community development. The population for this study included all articles published in reputable international journals between 2013 and 2023.

Data sources: Relevant articles were searched using platforms such as Science Direct, Scopus, Web of Science, and the official Nexus website (www.water-energy-food. org). Keywords used for article extraction encompassed various aspects of the nexus, including "water-energy-food nexus," "water security," "energy security," "food security," and "water-energy-food indicators," as well as specific nexus components like "water nexus," "energy nexus," and "food nexus." These keywords, or their possible combinations, were used separately in all four databases to extract the most relevant articles. The extraction process continued until no new articles were found, ensuring thorough coverage of available literature while eliminating duplicates.

Article selection and coding procedures: Following the initial extraction, purposeful sampling was employed to select pertinent and accessible articles. Initially, 295 articles were extracted. After removing duplicates, 58 articles were excluded. Subsequently, upon reviewing the subject matter, 81 articles were deemed irrelevant and eliminated, leaving 156 articles for further examination. These remaining articles were scrutinized for clarity of authorship and content, resulting in the removal of 46 articles due to reasons such as unclear authorship or solely literature reviews. Ultimately, 110 relevant articles were selected for detailed review.

A qualitative content analysis was then conducted in three stages:

1. **Open coding:** Concepts were initially extracted from the text through open coding, identifying recurring themes and concepts within the data.
2. **Axial coding:** Related concepts were compared and categorized into secondary categories using axial coding, grouping them based on their relationships and interdependencies.
3. **Selective coding:** Subtopics were finally grouped into three main categories through selective coding, aligning them with the research objectives and ensuring a structured analysis.

Analytical techniques: To visually represent the categorized concepts, a cluster diagram was generated using Sankey 5.1 software. This diagram provided a clear and comprehensive overview of the identified components and their interconnections.

Reliability and validity: To ensure the reliability and validity of the coding process, multiple researchers were involved in the analysis, with regular meetings held to discuss and reconcile any discrepancies. Inter-coder reliability was assessed to maintain consistency in the application of codes across the data set.

By providing these detailed methodological steps, the chapter offers a clear and transparent account of the content analysis process, enhancing the study's transparency and reproducibility. This rigorous and systematic approach ensures that the research effectively addresses its primary objective of bridging the knowledge gap in WEFN studies within the context of agricultural community development (Figure 2).

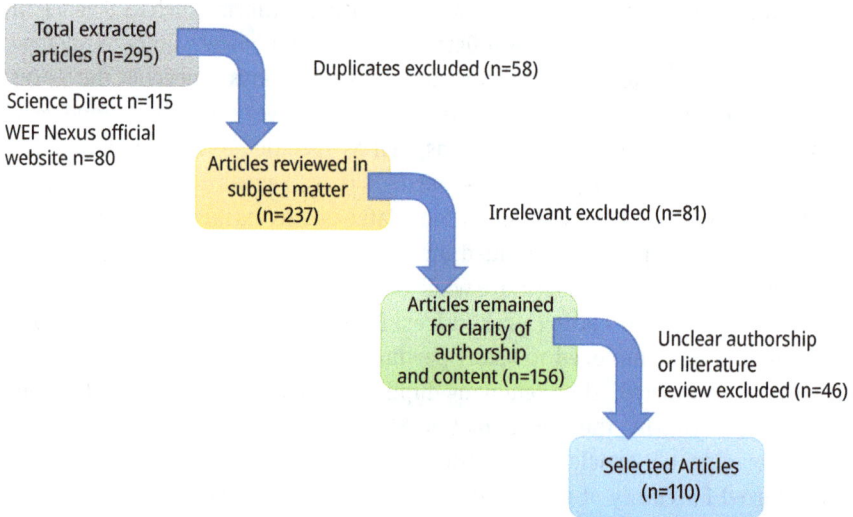

Total extracted articles (n=295)

Science Direct n=115
WEF Nexus official website n=80

Duplicates excluded (n=58)

Articles reviewed in subject matter (n=237)

Irrelevant excluded (n=81)

Articles remained for clarity of authorship and content (n=156)

Unclear authorship or literature review excluded (n=46)

Selected Articles (n=110)

Figure 2: Process of reviewed article selection.

2.1 Rationale for Choosing Content Analysis

Content analysis was chosen as the primary research method for this study because it allows for a systematic, objective, and quantitative examination of communication content, which is essential in exploring complex interrelationships like the WEFN. The WEFN involves multifaceted and interdependent components where the connections between water, energy, and food systems are embedded within various forms of textual data, including policy documents, academic literature, and community reports.

This method is particularly well-suited for this study for several reasons:

1. **Complexity of the WEFN:** The WEFN involves multiple, interconnected sectors that are often discussed in diverse and fragmented forms of communication. Content analysis enables the identification and examination of patterns, themes, and meanings within this complexity, providing insights into how these interdependencies are represented in existing literature and discourse.
2. **Systematic analysis of large textual data:** Given the extensive and varied nature of the data sources involved, content analysis provides a systematic ap-

proach to coding and categorizing large volumes of text, ensuring that the analysis remains rigorous and replicable. This systematic approach helps in revealing underlying trends and relationships that might not be immediately apparent through other qualitative methods.

3. **Objective measurement:** Content analysis allows for the quantification of qualitative data, making it possible to objectively measure the frequency and prominence of specific themes, concepts, and relationships within the text. This quantification is crucial for understanding the relative importance of different components within the WEFN and how they are prioritized in various contexts.

4. **Flexibility and adaptability:** Content analysis offers flexibility in dealing with various types of content, from written documents to media sources. This adaptability is essential for capturing the diverse ways in which the WEF Nexus is discussed across different platforms and by different stakeholders, including policymakers, researchers, and community members.

By using content analysis, this study aims to systematically uncover the underlying narratives and structures that shape the discourse on the WEFN, providing a comprehensive understanding of how these interconnected systems are perceived and addressed within agricultural communities.

In this context, "content" refers to information extracted from raw materials such as data and text, which is then organized into categorized concepts.

Step 1: Determine how many concepts to group and establish a set of categories. In this initial stage, authors must choose between two approaches: allowing flexibility in the coding process or maintaining a focused approach on specific concepts. The qualitative content analysis process begins with "open coding," where all qualitative data is thoroughly reviewed and classified. Researchers aim to create a primary categorization based on insights gained from the literature review.

Step 2: This phase is referred to as "axial coding," during which the extracted concepts are further organized into "subtopics." These subtopics help keep the coding process structured and consistent, ensuring the validity of the analysis as researchers create coherent categories in the next step. Subtopics are grouped based on shared characteristics. For instance, within the water domain, subtopics may include data related to water security, usage, production, and other relevant aspects.

Step 3: This step relies on the researcher's judgment to ensure that the classification of subtopics is carefully aligned with broader categories in what is known as "selective coding." These broader categories represent the main topics under study (Drisko & Maschi, 2016).

3 Results

The constituent concepts of the nexus have been organized into 3 main topics, 29 subtopics, and 384 individual concepts.

3.1 Constituent Concepts of Water Nexus

Table 3 presents the constituent concepts of the water nexus, which is the primary focus of this study. In the initial stage of open coding, 195 relevant concepts were identified. These concepts were then refined and organized into 13 subtopics during the second stage of axial coding. This refinement involved comparing and integrating similar and related concepts. The subtopics cover various aspects such as water resources, production, consumption, recycling, accessibility, and management. The frequency of each concept was determined by counting the number of studies in which one or more related concepts appeared. Notably, "water consumption" emerged as the most frequently studied subtopic, appearing in 72 studies. This subtopic encompasses 44 concepts, including the amount of rainwater consumption, tap water usage, water utilized for irrigation and agriculture, the effective rate of irrigation, and per capita water consumption in both urban and rural areas.

The relationship diagram between the obtained subtopics and the water nexus dimension is depicted in Figure 3 using Sankey 5.1 software. The numbers specified in each subtopic represent the frequency of studies conducted on that subtopic. Figure 4 shows the number of concepts related to each subtopic.

3.2 Constituent Concepts of Energy Nexus

The constituent concepts of the energy nexus (main topic) are presented in Table 4. Initially, 117 relevant concepts were identified in the first stage (open coding), and these concepts were categorized into nine subtopics in the second stage (axial coding) through comparison and integration of similar and common concepts. These subtopics include energy resources, energy production and extraction, energy and fuel consumption, energy indicators, energy demand and needs, energy infrastructure development, access to energy infrastructure, energy resource management and policymaking, energy pollutants and emissions. The frequency of concepts was calculated based on the number of studies in which one or more concepts related to the subtopic were observed. Among the categorized subtopics, the "energy and fuel consumption" subtopic has the highest frequency of studies (60 studies). This subtopic encompasses 36 concepts related to energy consumption across different sectors. It includes the amount of direct and indirect energy used in agriculture, industry, and households, as well as the energy required for processing a specific unit of food. Additionally, it

Table 3: Concepts and subtopics of the water nexus dimension.

Main topic	Subtopics	Concepts	References	Frequency
Water nexus	Water consumption	A reduced and consolidated list: 1. Water consumption: – Total annual water consumption – Total water consumption for food production – Total water consumption for energy production – Total water consumption for water production – Per capita water consumption – Water consumption efficiency 2. Agricultural water use: – Amount of water consumption in irrigated agriculture – Amount of water consumption in rain fed agriculture – Amount of water used for irrigation and agriculture – Amount of water used for livestock and poultry production – Effective rate of irrigation – Intensity of water consumption in the agricultural sector 3. Industrial and urban water use: – Amount of water consumption in urban areas – Amount of water consumption per hectare of crop – Amount of tap water consumption in agriculture and food industry – Amount of tap water consumption for electricity supply and natural gas extraction – Amount of tap water consumption for waste management – Intensity of water consumption in the industrial sector – Proportion of total water consumption in the household – Proportion of total water consumption in the industry	Abdel-Aal et al. (2020), Abulibdeh and Zaidan (2020), Ali et al. (2022), Babel et al. (2023), Lazaro et al. (2021), Boluwade (2021), Caputo et al. (2021), Chai et al. (2020), Cheng et al. (2023), Daher and Mohtar (2015), Daher et al. (2019), Huang et al. (2020), El-Gafy (2017), Fabiani et al., 2020; Ghimire et al., 2022; Ghodsvali et al., 2022; Haghjoo et al., 2022; Ibrahim et al., 2019; Karnib, 2018; Kedir et al., 2022; Khattar et al., 2023; Laspidou et al., 2020; Lazaro et al., 2022; Li et al., 2019; Li et al., 2019a; Li et al., 2019c; Li et al., 2021; Lyu et al., 2023; Mabhaudhi et al., 2019; Mannan et al., 2018; Mannschatz et al., 2016; Martinez et al., 2018; Martinez-Hernandez et al., 2017; McNabola et al., 2022; Mirzaei et al., 2023; Naghavi et al., 2022; Nasrollahi et al., 2021; Ngarava, 2021; Nie et al., 2019; Niu et al., 2022; Niva et al., 2020; Opoku et al., 2022; Ouyang et al., 2021; Purwanto et al., 2019; Purwanto et al., 2021; Qian & Liang, 2021; Radmehr et al., 2021; Rahmani et al., 2023; Ravar et al., 2020; Sadeghi et al., 2020; Saladini et al., 2018; Saray et al., 2022; Sharifinejad et al., 2020; Simpson et al., 2022; Tan et al., 2020; Terrapon-Pfaff et al., 2018;	72

(continued)

Table 3 (continued)

Main topic	Subtopics	Concepts	References	Frequency

		4. Water use in energy production:
		  – Amount of water used to produce a specific unit of energy
		  – Amount of water used to produce hydroelectric energy
		  – Amount of water used to produce thermal energy
		  – Amount of water used to recycle food waste
		  – Amount of water used for desalination of water sources and seawater
		5. Water losses and efficiency:
		  – Amount of water losses in the irrigation network
		  – Net water consumption in irrigation systems
		  – Physical efficiency of water consumption
		6. Water resources and consumption patterns:
		  – Amount of rainwater consumption
		  – Amount of surface water in use
		  – Amount of virtual water consumption in natural gas production and transmission
		  – Average water consumption per cooking time
		  – Frequency of water consumption
		  – Household water consumption per capita
			Wa'el A et al., 2017; Wang et al., 2021; Wang et al., 2023; Wu et al., 2023; Wu et al., 2021; Yi et al., 2020; Yu et al., 2020; Yue & Guo, 2021; Zeng et al., 2022; Zeng et al., 2023; Zhang et al., 2021a; Zhang et al., 2021b; Zhang & Vesselinov, 2017; Zuo et al., 2021	

Water resources		A consolidated summary of the water resources concepts:
		1. Total water resources:
		  – Total available water resources (including surface and underground water)
		  – Total annual renewable water resources
		  – Total portable water
			Abulibdeh & Zaidan, 2020; Babel et al., 2023; Bakhshianlamouki et al., 2020; Lazaro et al., 2021; Chai et al., 2020; Cheng et al., 2023; Correa-Cano et al., 2022; Daher & Mohtar, 2015; Daher et al., 2019; Huang et al., 2020; Enayati et al., 2021; Haghjoo et al., 2022; Hoff et al., 2019; Lawford et al., 2013;	40

	– Total resources of purified water from the sea – Total nonconventional water sources 2. Water storage and management: – Total water storage in tanks – Total effective capacity of irrigation pools – Amount of water storage in drought years – Total storage of dead water in reservoirs – Total net water lost from the storage tank – Total amount of virtual water in tanks – Total resources saved from unused water 3. Water footprint and use: – Amount of virtual water available for each product (water footprint index) – Contribution of local water resources to per capita food production – Per capita renewable freshwater resources 4. Snow and glacial water: – Amount of natural glaciers available for water supply – Amount of snow cover in the effective irrigation area 5. Water efficiency and per capita metrics: – Water resources per capita – Total amount of stored water	Lazaro et al., 2022; Li et al., 2019a; Li et al., 2019b; Lyu et al., 2023; Momblanch et al., 2019; Namany et al., 2019; Niu et al., 2022; Niva et al., 2020; Ouyang et al., 2021; Purwanto et al., 2019; Qian & Liang, 2021; Rahmani et al., 2023; Rahmani et al., 2023; Ravar et al., 2020; Shen et al., 2022; Simpson et al., 2022; Terrapon-Pfaff et al., 2018; Uen et al., 2018; Wang et al., 2018; Wang et al., 2023; Xu et al., 2019; Yi et al., 2020; Yu et al., 2020; Zeng et al., 2023; Zhang et al., 2021c; Zuo et al., 2021
Water resource management and policy-making	A consolidated summary of concepts: 1. Water conservation and efficiency: – Conservation of water resources – Sustainable water use – Water-saving management – Water productivity management 2. Infrastructure and technology: – Management of new technologies in the water industry – Expansion of rainwater storage	Alzaabi & Mezher, 2021; Boluwade, 2021; Correa-Cano et al., 2022; Daher & Mohtar, 2015; Fabiani et al., 2020; Haghjoo et al., 2022; Hoff et al., 2019; Lazaro et al., 2021; Lazaro et al., 2022; Lehmann, 2018; Li et al., 2019; Li et al., 2019a; Lyu et al., 2023; Mabhaudhi et al., 2019; Mannan et al., 2018; Mannschatz et al., 2016; Martinez et al., 2018; Mirzaei et al., 2023; Namany et al., 2019; Niu et al., 2022; Norouzi & Kalantari, 2020; Norouzi, 2022;

33

(continued)

Table 3 (continued)

Main topic	Subtopics	Concepts	References	Frequency
		– Modern irrigation systems (including government support and financial policies)	Ouyang et al., 2021; Purwanto et al., 2019; Qian & Liang, 2021; Rahmani et al., 2023; Ravar et al., 2020; Sadeghi et al., 2020; Saladini et al., 2018; Saray et al., 2022; Shu et al., 2021; Tan et al., 2020; Wu et al., 2023	
		3. Supply and distribution:		
		– Water distribution management		
		– Management of water supply for agriculture and pastures		
		– Balancing water supply and demand		
		4. Groundwater and restoration:		
		– Groundwater restoration and recovery		
		– Regulation of well usage (authorized and unauthorized)		
		5. Governance and security:		
		– Water governance		
		– Water resource security management		
		– Promoting water self-sufficiency		
		6. Training and capacity building:		
		– Applying necessary training in water security		
Water demand and needs		1. Water requirements:	Azzam et al., 2023; Bakhshianlamouki et al., 2020; Ledari et al., 2023; Lazaro et al., 2021; Lazaro et al., 2021; Chen et al., 2020; Daher & Mohtar, 2015; Enayati et al., 2021; Ghimire et al., 2022; Kaddoura & El Khatib, 2017; Khattar et al., 2023; Laspidou et al., 2020; Lehmann, 2018; Li et al., 2019; Li et al., 2019c; Martinez-Hernandez et al., 2017; Nasrollahi et al., 2021; Opoku et al., 2022; Ouyang et al., 2021;	31
		– Total annual water requirement		
		– Per capita water requirement		
		– Monthly water requirements for agriculture, industry, and households		
		– Water needs for livestock and poultry		
		– Total water requirement for domestic use, industrial purposes, and specific products		

			20
	2. Water shortages:	Radmehr et al., 2021; Rahmani et al., 2023; Ravar et al., 2020; Sharifinejad et al., 2020; Spiegelberg et al., 2017; Uen et al., 2018; Wang et al., 2021; Yue & Guo, 2021; Zuo et al., 2021	
	– Total water shortage for agriculture, industry, and domestic use		
	– Average shortage during droughts		
	– Calculated regional water shortage		
	3. Dependency and environmental needs:		
	– Degree of dependence on underground water and water imports		
	– Infiltration of surface water into underground sources		
	– Environmental water needs from surface water		
Physical status of water	– Amount and direction of the river flow	Azzam et al., 2023; Bakhshianlamouki et al., 2020; Huang et al., 2020; Enayati et al., 2021; Haghjoo et al., 2022; Lazaro et al., 2021; Lawford et al., 2013; Lyu et al., 2023; Mannschatz et al., 2016; Martinez-Hernandez et al., 2017; Nasrollahi et al., 2021; Niu et al., 2022; Norouzi, 2022; Radmehr et al., 2021; Rahmani et al., 2023; Rising, 2020; Uen et al., 2018; Wa'el A et al., 2017; Yue & Guo, 2021; Zeng et al., 2023	
	– Amount and direction of underground water flow		
	– Amount of aquifer level drop		
	– Amount of physical reduction of water		
	– Area of reservoirs and ponds		
	– Changing the position of rivers		
	– Consumption water flow rate		
	– Effective water area		
	– Groundwater level		
	– Height difference between the water level of the tank and the level of the flowing water		
	– Intensity of surface water inflow		
	– Intensity of surface water outflow		
	The flow rate of tanks		
	– Outflow rate of tanks		
	– Physical condition of the waterway for navigation		
	– Proportion of the number of wells and the amount of cultivated land		
	– Size, area, and water volume of lakes		
	– Specific weight of water		
	– Water density		
	– Water stress level		

(continued)

Table 3 (continued)

Main topic	Subtopics	Concepts	References	Frequency
	Recycling and reuse of water and wastewater	– Amount of reuse of nonconventional water sources (sewage) – Amount of reuse of water resources – Decarbonization of water resources – Desulfurization of water resources – Purification of water resources – The return rate of water flow after consumption – Treatment and reuse of wastewater – Water recycling rate – Wastewater treatment plant capacity – Water treatment plant capacity	Cheng et al., 2023; Daher et al., 2019; Huang et al., 2020; Haghjoo et al., 2022; Hoff et al., 2019; Ibrahim et al., 2019; Karnib, 2018; Lazaro et al., 2021; Lazaro et al., 2022; Li et al., 2019a; Mannan et al., 2018; Martinez et al., 2018; Ngarava, 2021; Ouyang et al., 2021; Radini et al., 2021; Radmehr et al., 2021; Rahmani et al., 2023; Shu et al., 2021; Tan et al., 2020; Zhang et al., 2021b	20
	Water production and extraction	– Amount of water extraction from the public network – Amount of water extraction from surface water sources – Average efficiency of water production and extraction from the river – Current amount of underground water extraction – Effective supply of water based on existing needs and demands – Raw amount of water extraction – Total volume of water imports – Total water production	Abulibdeh & Zaidan, 2020; Babel et al., 2023; Bazzana et al., 2020; Daher et al., 2019; Lawford et al., 2013; Li et al., 2019c; Mahlknecht et al., 2020; Mannschatz et al., 2016; Momblanch et al., 2019; Ouyang et al., 2021; Purwanto et al., 2021; Rahmani et al., 2023; Tan et al., 2020; Xu et al., 2019; Zuo et al., 2021	15
	Provision and access to water infrastructure	– Access to irrigation networks – Aqueduct dredging in fields – Construction and development of drainage network – Creating mounds and earthen dams along the river – Dam construction – Development of sewage distribution network	Bazzana et al., 2020; Caputo et al., 2021; Haghjoo et al., 2022; Lazaro et al., 2021; Lawford et al., 2013; Mannan et al., 2018; Martinez et al., 2018; Purwanto et al., 2019; Purwanto et al., 2021; Rahmani et al., 2023; Rising, 2020; Tan et al., 2020; Wang et al., 2023; Yue & Guo, 2021; Zhang et al., 2021a	15

Category	Items	References	Count
	– Development of a water distribution network – Development of water supply channels – Reconstruction and repair of the water transmission route – Using pools and artificial ponds to store water		
Chemical status of water	– Amount and type of microorganisms in water (water biology) – Amount of sediments in the water – Drinking water quality in terms of total nutrients – General water quality – Groundwater quality – Hydrography of the river catchment area – Ratio of water quality to the amount of available water – Water quality in terms of salinity and sodium content	Azzam et al., 2023; Daher & Mohtar, 2015; Fabiani et al., 2020; Ghodsvali et al., 2022; Haghjoo et al., 2022; Hoff et al., 2019; Ibrahim et al., 2019; Lazaro et al., 2021; Mannschatz et al., 2016; Martinez et al., 2018; Nhamo & Ndlela, 2021; Purwanto et al., 2019; Shu et al., 2021; Spiegelberg et al., 2017	14
Access and availability of water	– Access to water per capita – Annual water availability in the region – Availability of water in the soil – Freshwater availability for consumption – Net percentage of access to purified and untreated water – Observed water availability – Virtual water availability.Underground water availability – Moving and exchanging water between underground aquifers	Correa-Cano et al., 2022; Lazaro et al., 2021; Martinez-Hernandez et al., 2017; Norouzi, 2022; Purwanto et al., 2019; Purwanto et al., 2021; Radmehr et al., 2021; Sharifinejad et al., 2020; Uen et al., 2018; Wicaksono & Kang, 2019; Zarei, 2020; Zeng et al., 2022; Zhang et al., 2021a	13
Water pollutants	– Amount of minerals in freshwater – Amount of pollutants in freshwater – Amount of pollutants in surface water – Amount of toxic substances in water sources – Amount of water pollution with agricultural toxins – Carbon footprint in water – Environmental pollution of freshwater – Fossil fuels footprint in water	Anser et al., 2020; Chen et al., 2020; Ghodsvali et al., 2022; Haghjoo et al., 2022; Lazaro et al., 2021; Lazaro et al., 2022; Radini et al., 2021; Sharifinejad et al., 2020	8

(continued)

Table 3 (continued)

Main topic	Subtopics	Concepts	References	Frequency
		– GHG footprint in water – Return rate of industrial wastewater flow to surface and underground water resources – Return rate of urban sewage flow to surface and underground water resources – Sewage footprint in water resources		
	Sewage and wastewater	– Amount of accumulated agricultural, industrial, and household waste – Amount of wastewater production – Amount of wastewater production per unit of GDP – Annual output volume of municipal wastewater from treatment plants – Total capacity of urban wastewater treatment plants – Volume of produced urban wastewater – Volume of produced industrial wastewater – Volume of wastewater and treated wastewater – Wastewater disposal rate	Ghodsvali et al., 2022; Laspidou et al., 2020; Lehmann, 2018; Namany et al., 2019; Qian & Liang, 2021; Radini et al., 2021; Wang et al., 2023	7
	Methods of using water	– Drip irrigation – Low irrigation methods – New methods of surface irrigation – Using pressure irrigation methods – Smart irrigation systems – Underground irrigation methods – Traditional methods of surface irrigation	Haghjoo et al., 2022; Hoff et al., 2019; Li et al., 2021; Namany et al., 2019; Nasrollahi et al., 2021	5

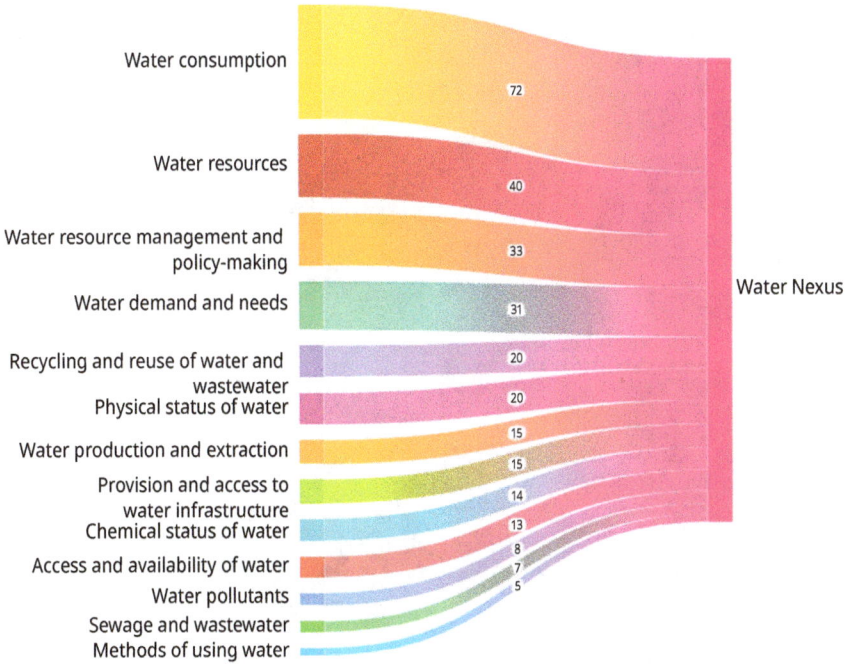

Water consumption 72
Water resources 40
Water resource management and policy-making 33
Water demand and needs 31
Recycling and reuse of water and wastewater 20
Physical status of water 20
Water production and extraction 15
Provision and access to water infrastructure 15
Chemical status of water 14
Access and availability of water 13
Water pollutants 8
Sewage and wastewater 7
Methods of using water 5

Water Nexus

Figure 3: Relationship diagram of subtopics in the water nexus dimension.

addresses the consumption of natural gas for agricultural water pumping and domestic water heating, along with total coal consumption. Collectively, these concepts provide a comprehensive overview of energy usage in various contexts.

Figure 5 illustrates the relationship diagram of subtopics derived and the energy Nexus Dimension, created using Sankey 5.1 software. The numbers indicated in each subtopic reflect the frequency of studies conducted on that topic. Meanwhile, Figure 6 presents the number of concepts associated with each subtopic.

3.3 Constituent Concepts Within the Food Nexus

Table 5 presents the constituent concepts associated with the main topic of the food nexus. In the first stage of analysis (open coding), 56 relevant concepts were identified. These concepts were then categorized into seven subtopics during the second stage (axial coding) by comparing and combining similar ideas. The identified subtopics are food resources, food production, food procurement and access, food consumption, food resource management and policy-making, food waste and contaminants, and dietary patterns. The frequency of each concept was determined based on the number of studies in which it was referenced within the respective subtopic. No-

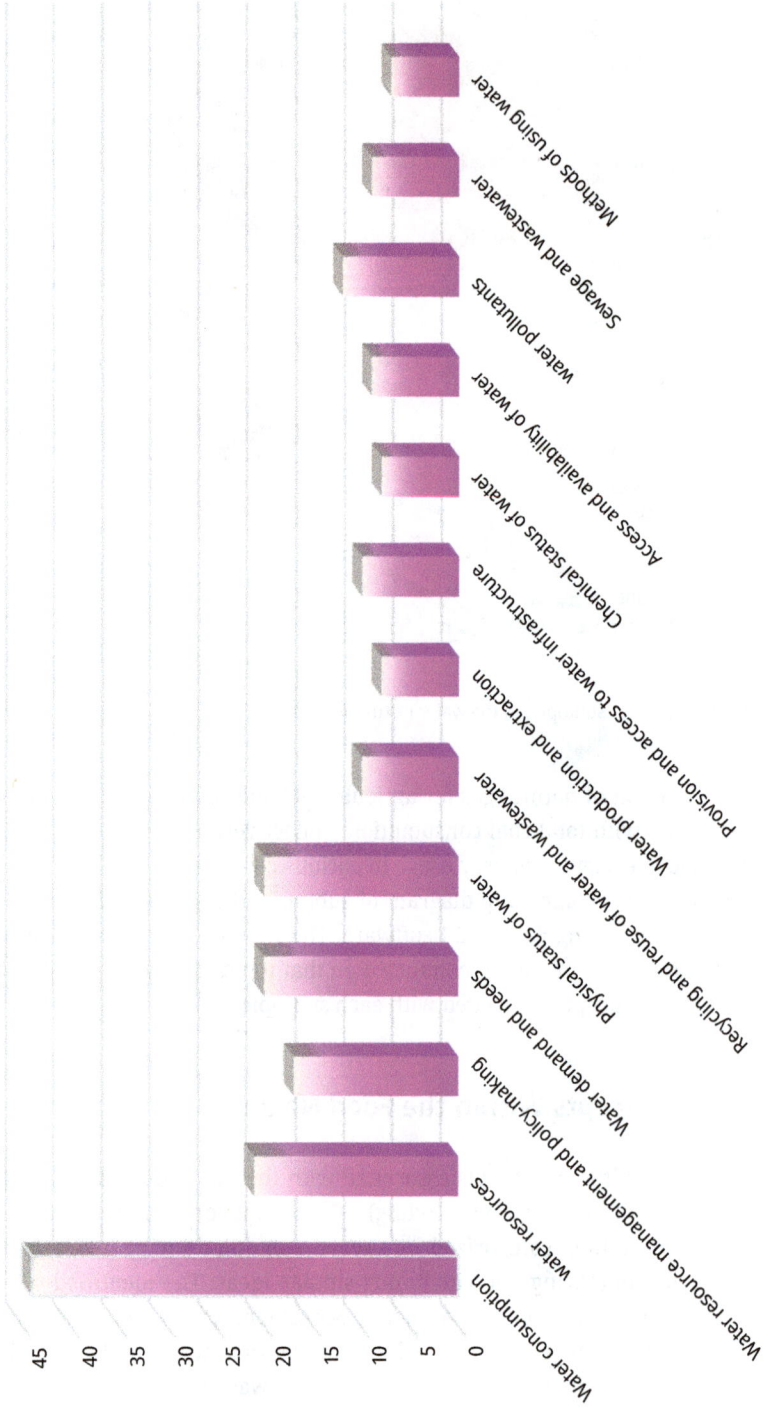

Figure 4: The number of concepts related to each subtopic of the water nexus.

Table 4: Constituent concepts and subtopics of the energy nexus dimension.

Main topic	Subtopics	Concepts	References	Frequency
Energy nexus	Energy and fuel consumption	A more concise list of energy consumption concepts: 1. Total energy consumption: – Total energy consumption (national, local, per capita) – Total fuel consumption – Total imported and exported energy 2. Sector-specific energy use: – Agriculture: Direct and indirect energy consumption, including for water pumping, food production, and crop processing – Industry and households: Energy consumption in industry, households, and for domestic heating 3. Energy efficiency and losses: – Energy consumption per unit of GNP – Energy consumption per unit of food or crops – Energy losses in transmission and distribution 4. Water management: – Energy for water pumping, distribution, desalination, and wastewater treatment 5. Renewable and traditional energy: – Consumption of traditional biofuels – Renewable energy production and per capita consumption 6. Energy production metrics: – Total production of coal, natural gas, crude oil, diesel, and biofuels – Energy production efficiency	Abulibdeh & Zaidan, 2020; Ali et al., 2022; Azzam et al., 2023; Babel et al., 2023; Bakhshianlamouki et al., 2020; Caputo et al., 2021; Chai et al., 2020; Cheng et al., 2023; Daher & Mohtar, 2015; Daher et al., 2019; El-Gafy, 2017; Fabiani et al., 2020; Ghodsvali et al., 2022; Haghjoo et al., 2022; Ibrahim et al., 2019; Kaddoura & El Khatib, 2017; Karnib, 2018; Kedir et al., 2022; Khattar et al., 2023; Lehmann, 2018; Li et al., 2019a; Li et al., 2019b; Li et al., 2019c; Li et al., 2021; Mabhaudhi et al., 2019; Martinez et al., 2018; Mirzaei et al., 2023; Naghavi et al., 2022; Namany et al., 2019; Nasrollahi et al., 2021; Ngarava, 2021; Nie et al., 2019;Norouzi, 2022; Ouyang et al., 2021; Purwanto et al., 2019; Purwanto et al., 2021; Qian & Liang, 2021; Radini et al., 2021; Radmehr et al., 2021; Rahmani et al., 2023; Ravar et al., 2020; Sadeghi et al., 2020; Saray et al., 2022; Sharifinejad et al., 2020; Shu et al., 2021; Simpson et al., 2022; Terrapon-Pfaff et al., 2018; Vaidya et al., 2021; Wang et al., 2018; Wang et al., 2021; Wang et al., 2023; Wu et al., 2021; Wu et al., 2023; Xu et al., 2019; Yi et al., 2020; Yu et al., 2020; Zeng et al., 2022; Zhang et al., 2021a;	60

(continued)

Table 4 (continued)

Main topic	Subtopics	Concepts	References	Frequency
		7. Residential and urban use:	Zhang et al., 2021b; Zhang et al., 2021c	
		– Average annual natural gas consumption of households		
		– Per capita electricity consumption		
		– Fuel consumption for public infrastructure		
Energy production and extraction		Reduced and consolidated list of energy production concepts:	Abulibdeh & Zaidan, 2020; Abdel-Aal et al., 2020; Babel et al., 2023; Caputo et al., 2021; Chai et al., 2020; Cheng et al., 2023; Daher & Mohtar, 2015; Daher et al., 2019; Huang et al., 2020; Enayati et al., 2021; Ghodsvali et al., 2022; Haghjoo et al., 2022; Ibrahim et al., 2019; Kedir et al., 2022; Khattar et al., 2023; Laspidou et al., 2020; Lazaro et al., 2021; Lazaro et al., 2022; Ledari et al., 2023; Lehmann, 2018; Li et al., 2019a; Li et al., 2019b; Li et al., 2021; Mabhaudhi et al., 2019; Martinez et al., 2018; Namany et al., 2019; Ngarava, 2021; Opoku et al., 2022; Ouyang et al., 2021; Purwanto et al., 2019; Purwanto et al., 2021; Ravar et al., 2020; Saray et al., 2022; Sharifinejad et al., 2020; Shu et al., 2021; Simpson et al., 2022; Tan et al., 2020; Uen et al., 2018; Wang et al., 2018; Wang et al., 2021;Wang et al., 2023; Wu et al., 2021; Wu et al., 2023; Xu et al., 2019; Yi et al., 2020; Yu et al., 2020; Zhang & Vesselinov, 2017; Zhang et al., 2021c	48
		1. Total energy production:		
		– Total energy production		
		– Total production of coal, coal gas, crude oil, diesel, kerosene, natural gas, and biofuels		
		2. Energy production by source:		
		– Electrical energy production		
		– Thermal energy production		
		– Energy production from water sources		
		– Renewable energy production (wind, solar, and geothermal)		
		3. Energy production efficiency:		
		– General energy production efficiency		
		– Energy production per unit of GNP		
		– Energy production per unit of water consumed		
		4. Energy import:		
		– Amount of energy import		
		5. Per capita metrics:		
		– Per capita energy production (national and local levels)		
		– Per capita renewable energy production		

	6. Source-specific production: – Amount of electrical energy from biofuels (animal waste, agricultural, and forest waste) – Percentage of energy production from each source		
Energy demand and needs	Reduced and streamlined list of concepts: 1. Energy demand: – Annual energy demand – Per capita energy demand (electric and thermal) 2. Energy dependency: – Dependency on energy exports and imports 3. Energy requirements by sector: – Energy for desalination, agricultural, commercial, household, and industrial activities – Energy for water extraction and wastewater disposal 4. Fuel demand: – Fuel for construction, electricity production, industrial use, and transportation 5. General energy metrics: – Total energy required and total fuel required	25	Almulla et al., 2022; Azzam et al., 2023; Babel et al., 2023; Daher & Mohtar, 2015; Kaddoura & El Khatib, 2017; Khattar et al., 2023; Laspidou et al., 2020; Lehmann, 2018; Li et al., 2019a; Li et al., 2019b; Mabhaudhi et al., 2019; Martinez et al., 2018; Martinez-Hernandez et al., 2017; Opoku et al., 2022; Radmehr et al., 2021; Rahmani et al., 2023; Ravar et al., 2020; Saray et al., 2022; Shu et al., 2021; Terrapon-Pfaff et al., 2018; Wang et al., 2021; Wicaksono & Kang, 2019; Wu et al., 2021; Zeng et al., 2022; Zhang & Vesselinov, 2017
Energy resource management and policy-making	– Applying new technology in energy production – Energy efficiency management – Energy governance – Energy production planning management – Energy security – Energy supply and demand risk management	20	Abulibdeh & Zaidan, 2020; Haghjoo et al., 2022; Karnib, 2018; Kedir et al., 2022; Lazaro et al., 2021; Lehmann, 2018; Li et al., 2019a; Martinez et al., 2018; Mirzaei et al., 2023; Namany et al., 2019; Ouyang et al., 2021; Purwanto et al., 2019; Radmehr et al., 2021; Sadeghi et al., 2020; Saray et al., 2022; Shu et al., 2021;

(continued)

Table 4 (continued)

Main topic	Subtopics	Concepts	References	Frequency
		– Management of obtaining government support for the development of renewable energies – Management of self-sufficiency in energy supply – Necessary training for the development of energy security – Policymaking of energy production, distribution, and consumption – Sustainable energy supply management	Terrapon-Pfaff et al., 2018; Wang et al., 2021; Yi et al., 2020; Zeng et al., 2022	
	Access to energy infrastructure	– Accessibility to the power grid – Accessibility to all types of energy (renewable and conventional) – Availability of energy for extraction from water resources – Availability of energy for food production and processing	Bazzana et al., 2020; Haghjoo et al., 2022; Li et al., 2019c; Mabhaudhi et al., 2019; Purwanto et al., 2019; Purwanto et al., 2021; Radmehr et al., 2021; Simpson et al., 2022; Wicaksono & Kang, 2019; Zarei, 2020	10
	Energy indicators	– Coefficient of elasticity of energy consumption – Efficiency coefficient for comparing water energy to electricity – Energy balance index – Energy intensity coefficient – Energy intensity measured in terms of primary energy and GDP – Energy portfolio index (agricultural, industrial, and household) – Energy security index	Cheng et al., 2023; Daher & Mohtar, 2015; Li et al., 2021; Mabhaudhi et al., 2019; Mannan et al., 2018; Norouzi & Kalantari, 2020; Purwanto et al., 2021; Shu et al., 2021; Yue & Guo, 2021	9
	Energy pollutants and emissions	– Carbon footprint in energy resources – Emission of carbon dioxide for consumption of one unit of energy – Emission of carbon dioxide for consumption of one unit of fossil fuels – Emission of GHGs for consumption of one unit of energy	Almulla et al., 2022; Anser et al., 2020; Daher & Mohtar, 2015; Ghodsvali et al., 2022; Laspidou et al., 2020; Li et al., 2019c; Namany et al., 2019; Spiegelberg et al., 2017; Yue & Guo, 2021	9

	– Fossil fuels footprint in energy resources – GHG footprint in energy resources – Production of inorganic waste for the consumption of a unit of fossil fuel	7
Energy infrastructure development	– Construction and development of a combined cycle power plant – Construction and development of hydroelectric power plant – Construction and development of solar power plant – Construction of energy production units – Development of electricity distribution network – Development of renewable energy infrastructures – The amount of land used for energy production – The number of anaerobic energy production sites	Abdel-Aal et al., 2020; Bazzana et al., 2020; Karnib, 2018; Lazaro et al., 2021; Namany et al., 2019; Purwanto et al., 2019; Rahmani et al., 2023
Energy resources	– Percentage of available energy resources – Total available energy sources – Total capacity of energy resources – Total stock of energy resources	Daher & Mohtar, 2015; Lazaro et al., 2022; Li et al., 2019a; Mabhaudhi et al., 2019; Namany et al., 2019; Wicaksono & Kang, 2019
		6

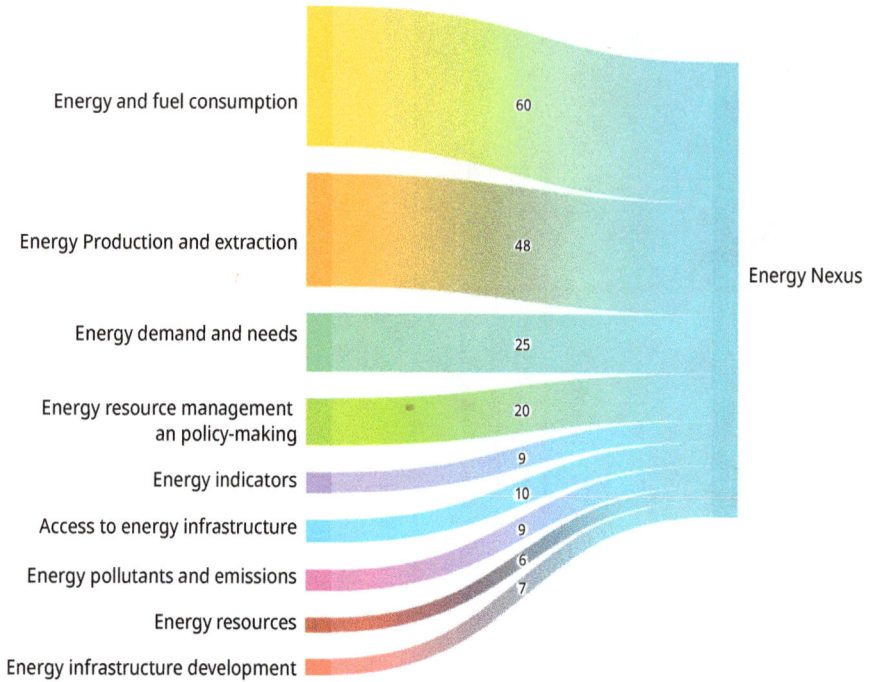

Figure 5: Relationship diagram of subtopics in the energy nexus dimension.Sankey diagram illustrating the flow of various energy-related factors into the energy nexus.

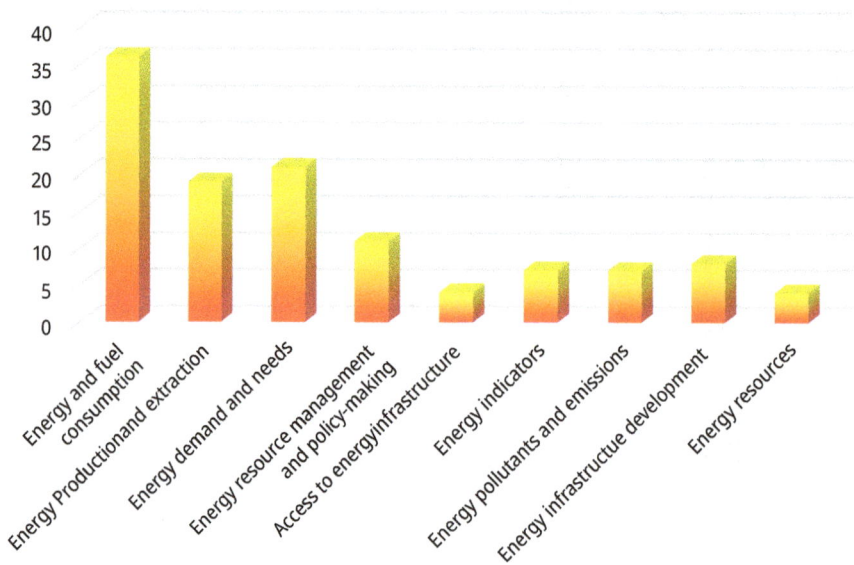

Figure 6: The number of concepts related to each subtopic of the energy nexus.

Table 5: Constituent concepts and subtopics within the food nexus dimension.

Main topic	Subtopics	Concepts	References	Frequency
Food nexus	Food production	– Annual and total food production – Food production efficiency per unit area – Food production in rural areas and urban areas – Food production per capita – Local food production – Targeted food production – Total food imports – Total supply of domestically produced or imported food	Abulibdeh & Zaidan, 2020; Ali et al., 2022; Azzam et al., 2023; Cheng et al., 2023; Huang et al., 2020; Ibrahim et al., 2019; Khattar et al., 2023; Lazaro et al., 2022; Lehmann, 2018; Li et al., 2019b; Li et al., 2021; Mabhaudhi et al., 2019; Mannschatz et al., 2016; Martinez et al., 2018; Namany et al., 2019; Nie et al., 2019; Norouzi, 2022; Opoku et al., 2022; Purwanto et al., 2021; Rahmani et al., 2023; Ravar et al., 2020; Shu et al., 2021; Terrapon-Pfaff et al., 2018; Wang et al., 2021; Xu et al., 2019; Zeng et al., 2022; Zhang & Vesselinov, 2017	27
	Food resource management and policy-making	– Applying necessary training for food security – Food governance – Food health and safety management – Food productivity management – Food security created by individual production – Food security index – Food supply and demand risk management – Food waste reduction management – Self-sufficiency in food production	Alzaabi & Mezher, 2021; Bazzana et al., 2020; Daher & Mohtar, 2015; Haghjoo et al., 2022; Hoff et al., 2019; Karnib, 2018; Lazaro et al., 2022; Lehmann, 2018; Li et al., 2019a; Mabhaudhi et al., 2019; Mannan et al., 2018; Niu et al., 2022; Purwanto et al., 2019; Purwanto et al., 2021; Radmehr et al., 2021; Ravar et al., 2020; Shu et al., 2021; Terrapon-Pfaff et al., 2018; Wang et al., 2021; Zhang et al., 2021a	20

(continued)

Table 5 (continued)

Main topic	Subtopics	Concepts	References	Frequency
	Food consumption	– Amount of change in per capita consumption of food – Amount of food consumed in a household meal – Consumption of one unit of food to produce one unit of energy – Consumption of one unit of food to produce one unit of food – Consumption of one unit of food to produce one unit of GNP – Consumption of non-homemade food – Food consumption of urban and rural households – Food consumption per capita – Total food consumption – Total food exports	Abulibdeh & Zaidan, 2020; Chai et al., 2020; Cheng et al., 2023; Ghodsvali et al., 2022; Ibrahim et al., 2019; Karnib, 2018; Li et al., 2019a; Purwanto et al., 2019; Purwanto et al., 2021; Qian & Liang, 2021; Rahmani et al., 2023; Spiegelberg et al., 2017; Wa'el A et al., 2017; Wang et al., 2018; Yi et al., 2020	15
	Food procurement and access	– Annual demand and purchase of food in urban and rural households – Preparing food as part of providing food security – The level of access to food in urban and rural households – The potential of real access to food in urban and rural households	Bazzana et al., 2020; Khattar et al., 2023; Li et al., 2021; Martinez et al., 2018; Ngarava, 2021; Opoku et al., 2022; Purwanto et al., 2019; Purwanto et al., 2021; Rahmani et al., 2023; Ravar et al., 2020; Wicaksono & Kang, 2019; Wu et al., 2021; Zarei, 2020	13
	Dietary patterns	– Amount and variety of diets – Amount of food quality based on the nutrient index – Amount of shortage and deficit of food at the individual and household level – Average adequacy of energy supply in the diet – Average consumption of protein in the household – Intensity of food consumption	Abulibdeh & Zaidan, 2020; Caputo et al., 2021; Haghjoo et al., 2022; Kaddoura & El Khatib, 2017; Ngarava, 2021; Norouzi & Kalantari, 2020; Purwanto et al., 2019; Ravar et al., 2020; Simpson et al., 2022; Spiegelberg et al., 2017; Wang et al., 2021; Wu et al., 2021; Zeng et al., 2022	13

Category	Indicators	References	
	– Percentage of children under 5 years of age suffering from malnutrition, stunted growth, and death due to lack of food security – Prevalence of malnutrition among age groups – Rate of obesity in the adult population		
Food resources	– Food available resources for consumption – Production of food resources to consume one unit of energy – Production of food resources to consume one unit of water – Total food available resources – Total storage of food resources	Namany et al., 2019; Nie et al., 2019; Ravar et al., 2020; Spiegelberg et al., 2017; Wicaksono & Kang, 2019; Zhang et al., 2021c	6
Food waste and contaminants	– Amount of discarded food waste – Amount of food waste produced in food stores – Amount of food waste produced in restaurants – Annual amount of household waste – Carbon footprint in food – Fossil fuels footprint in food – GHG footprint in food – Inorganic waste produced from food packaging – Total food waste	Abdel-Aal et al., 2020; Anser et al., 2020; Ghodsvali et al., 2022; Haltas et al., 2017; Lehmann, 2018; Shu et al., 2021; Wa'el A et al., 2017	6

tably, among the classified subtopics, "food production" had the highest frequency, with 27 studies focusing on this area. This subtopic includes nine concepts, such as food production in rural areas, local food production, and the total supply of domestically produced or imported food.

Using Sankey 5.1 software, Figure 7 illustrates the relationship between the derived subtopics and the food nexus dimension. The numbers within each subtopic indicate the frequency of studies conducted on that particular aspect. Figure 8 shows the number of concepts associated with each subtopic.

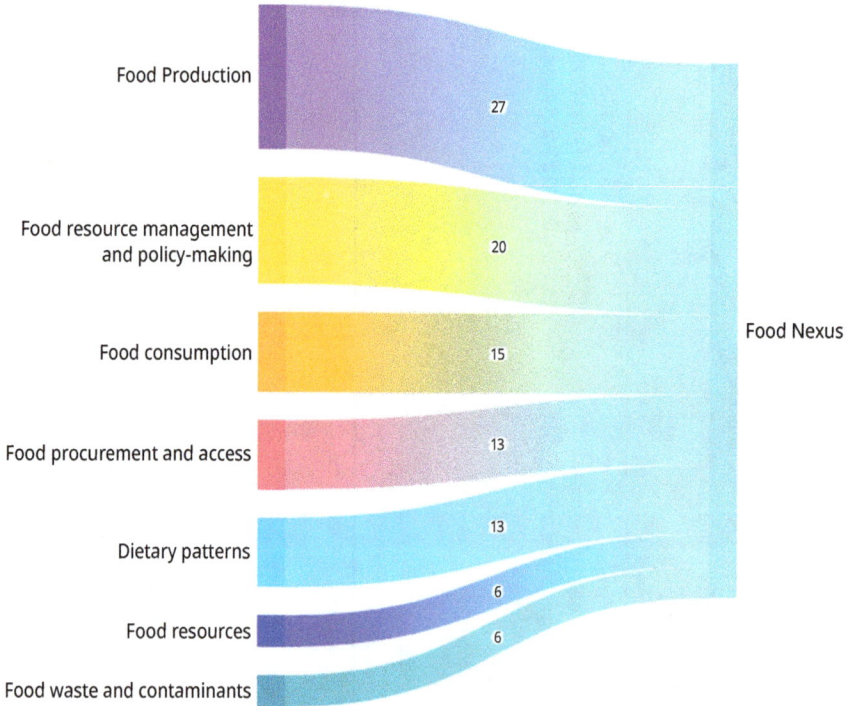

Figure 7: Diagram of relationships between subtopics of the food nexus dimension.

The study aimed to conduct a comprehensive review of the literature concerning the WEFN within a specified timeframe. It explored various studies examining the characteristics and impacts of nexus components, focusing on factors such as their intensity, duration, and modes of influence. Robust nexus tools, including tripartite production resources, emerging technology considerations for nexus alignment, and solid theoretical frameworks, have significantly broadened the scope for addressing both internal and external influential factors. These factors notably impact agricultural communities and respond to societal, economic, and environmental demands for effective crisis management. Over time, the nexus approach has evolved from a basic

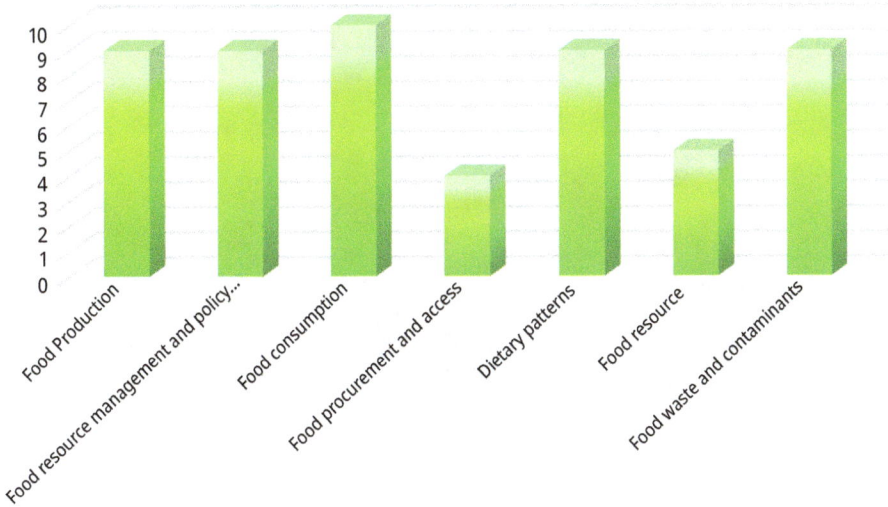

Figure 8: The number of concepts related to each subtopic of the food nexus.

scientific theory to a dynamic, self-contained, and practical solution. By leveraging the evolving capabilities and ongoing feedback from the nexus, it becomes feasible to address internal and external challenges, such as climate change and human impacts on the environment, allowing for the development of appropriate adaptation or mitigation strategies.

This study specifically focuses on identifying the core concepts of the nexus. From a total of 295 articles, a purposive sampling technique was employed to select and review 110 articles. In the qualitative content analysis phase, relevant concepts were initially extracted using keyword-based coding. During the axial coding phase, these concepts were grouped based on common characteristics within each subtopic. Efforts were made to minimize common coding errors and redundancies through pairwise comparison and reevaluation. In the selective coding phase, three main topics emerged: the water nexus, energy nexus, and food nexus (see Figure 9).

4 Concluding Remarks

The purpose of this research was to identify the main components of the WEFN dimensions in the context of agricultural community development and to establish a comprehensive classification of all nexus elements found in relevant articles. This research offers several strengths compared to other studies. First, it is not restricted to a specific geographical area, unlike the work of Adeola et al. (2022). Second, it does not focus on particular population groups; for instance, Zhang et al. (2019) emphasize

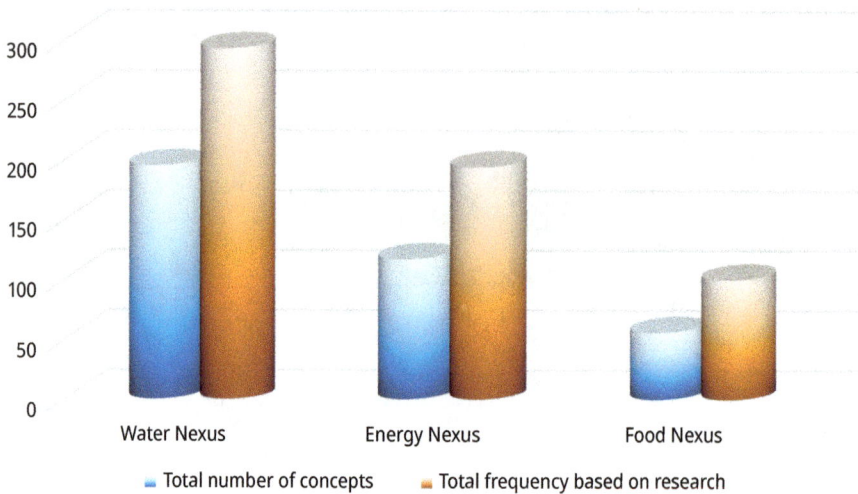

Figure 9: Frequency and ranking of key primary nexus topics shaping agricultural community development.

identifying nexus dimensions in urban settings. Finally, while most researchers, such as Albrecht et al. (2018), analyze the methods, assessment features, and other components of the nexus in isolation, this study aims to consider all variables related to the nexus, conducting a comprehensive analysis to identify its fundamental components.

Our findings emphasize the predominant role of the water nexus in nexus studies, particularly regarding its significance for agricultural community development. The focus on subtopics within the water nexus, especially water consumption-related concepts, underscores its critical importance. Concepts such as total water usage, water consumption for energy production, and water efficiency are essential for understanding the intricate dynamics of the nexus.

Following closely is the energy nexus, with significant emphasis on energy consumption. This highlights the interconnectedness of water and energy resources, particularly in sectors vital to agricultural communities, such as agriculture and industry. While the food nexus ranks third, it remains crucial due to its direct association with water and energy resources. Food production levels are key indicators for evaluating nexus interactions and for developing future community strategies tailored to agricultural needs.

Based on the findings, the following policy implications are suggested:

5 Policy Implications

This research underscores the importance of a holistic approach to managing the WEFN in the context of agricultural community development. By identifying the main components of the WEFN dimensions and providing a comprehensive classification of nexus elements, several critical policy implications emerge:

1. **Prioritization of water management in policy frameworks:** Given the predominant role of the water nexus in agricultural community development, policies should prioritize water management strategies. This includes the development of comprehensive water usage policies that address both direct agricultural needs and the water requirements of associated energy production processes. Water efficiency should be a key consideration in all agricultural policies, with a focus on optimizing water use to sustain agricultural productivity while minimizing environmental impact.

2. **Integration of energy efficiency in agricultural policies:** The significant emphasis on the energy nexus highlights the need for energy-efficient practices in agriculture. Policies should encourage the adoption of energy-saving technologies and the transition to renewable energy sources. This will not only reduce the environmental footprint of agriculture but also ensure the sustainability of energy resources critical to the sector.

3. **Development of cross-sectoral policies:** The interconnectedness of water, energy, and food resources calls for policies that are cross-sectoral in nature. Agricultural policies should not be developed in isolation but should instead consider the implications for water and energy resources. Cross-sectoral collaboration among government agencies, private sectors, and research institutions is essential for the development of integrated policies that address the multifaceted challenges of the WEFN.

4. **Focused investment in sustainable food production:** The food nexus, while ranking third, is crucial for the overall sustainability of agricultural communities. Policies should support sustainable food production practices that are closely aligned with water and energy efficiency goals. Investment in technologies and practices such as precision farming, drip irrigation, and sustainable land management can ensure that food production remains viable without compromising water and energy resources.

5. **Comprehensive nexus-based planning:** Policies should incorporate nexus-based planning that considers the intricate interactions between water, energy, and food resources. This involves the use of nexus indicators and comprehensive assessments to guide policy decisions. Such an approach will enable the development of more resilient agricultural communities that can adapt to changing environmental and resource conditions.

6. **Localized policy adaptation:** Although this study is not limited to a specific geographic area, it emphasizes the importance of localized policy adaptation. Policy-

makers should customize nexus management strategies to address the unique needs and conditions of local agricultural communities, acknowledging that different regions may require distinct approaches to managing their water, energy, and food resources.

In conclusion, the findings of this research suggest that a comprehensive and integrated approach to WEFN management is essential for the sustainable development of agricultural communities. Policymakers are encouraged to adopt strategies that address the interdependencies of water, energy, and food resources, thereby ensuring long-term sustainability and resilience in the face of growing environmental and resource challenges.

6 Future Research Directions

Building on the findings of this chapter, several avenues for future research can be identified that will further explore and underscore the ongoing relevance of the WEFN.

Identification of key components: Future studies should focus on identifying the critical components influencing the WEFN at various spatial scales. This could involve a detailed analysis of regional disparities and localized factors that impact the nexus.

Understanding trade-offs and synergies: There is a pressing need to deepen our understanding of the trade-offs and synergies among the three main resources of the nexus – water, energy, and food. Research could investigate how these elements interact with each other, potentially revealing strategies for optimizing resource use and enhancing sustainability.

Comparative studies in an interdisciplinary context: Conducting comparative studies that examine the components affecting the nexus across different geographic and socio-economic contexts would be beneficial. An interdisciplinary approach could reveal insights from various fields, such as economics, environmental science, and sociology, leading to a more holistic understanding of the WEFN.

Addressing research limitations: While this study provides significant insights, it is important to acknowledge its limitations. The research was constrained by limited access to certain databases due to sanctions, time constraints, and the focus on literature published between 2013 and 2023. Addressing these limitations in future research could yield even more comprehensive and robust findings, ensuring that the study of the WEFN remains a dynamic and evolving field.

Figure 10 illustrates the main and subtopics constituting the WEFN, visually representing its multifaceted nature and highlighting areas for further research and policy intervention.

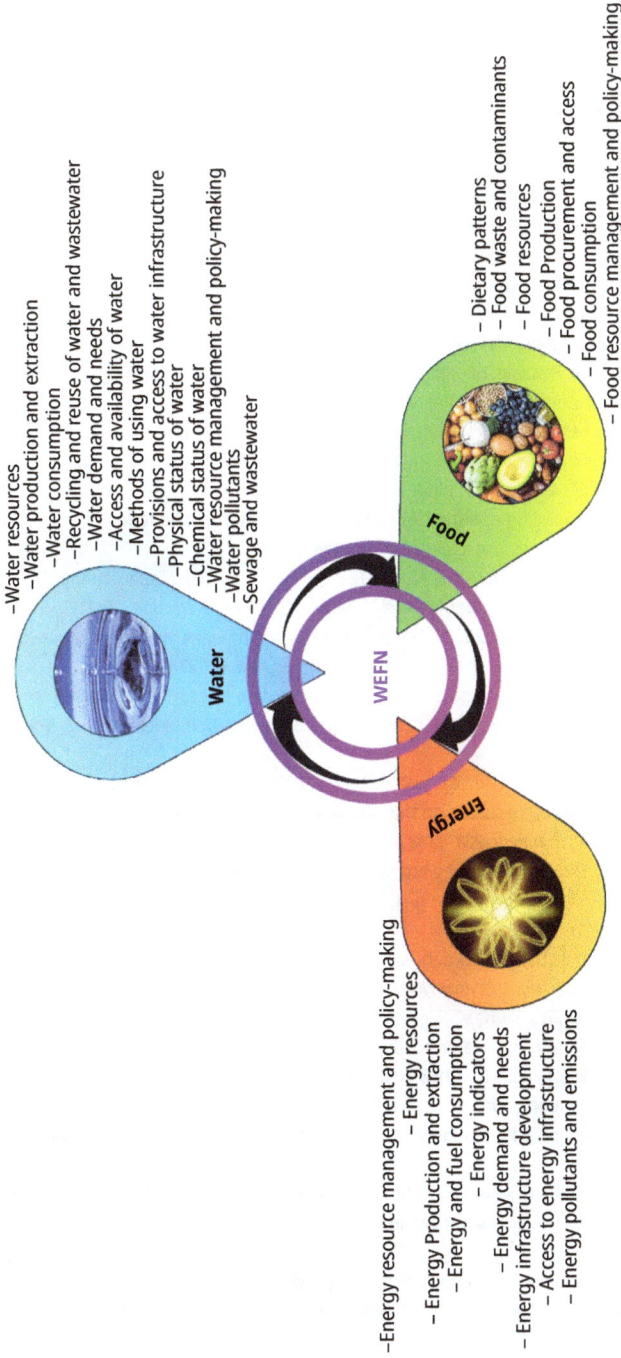

Figure 10: Illustration of the main and subtopics constituting the water-energy-food nexus for agricultural communities' development.

References

Abdel-Aal, M., Haltas, I., & Varga, L. (2020). Modeling the diffusion and operation of anaerobic digestions in Great Britain under future scenarios within the scope of the water-energy-food nexus. *Journal of Cleaner Production, 253*, 119897.

Abulibdeh, A. & Zaidan, E. (2020). Managing the water-energy-food nexus on an integrated geographical scale. *Environmental Development, 33*, 100498.

Albrecht, T. R., Crootof, A., & Scott, C. A. (2018). The Water-Energy-Food Nexus: A systematic review of methods for nexus assessment. *Environmental Research Letters, 13*(4), 043002.

Adeola, O. M., Ramoelo, A., Mantlana, B., Mokotedi, O., Silwana, W., & Tsele, P. (2022). Review of publications on the water-energy-food nexus and climate change adaptation using bibliometric analysis: A case study of Africa. *Sustainability, 14*(20), 13672.

Ali, M., Anjum, M. N., Shangguan, D., & Hussain, S. (2022). Water, Energy, and Food Nexus in Pakistan: Parametric and Non-Parametric Analysis. *Sustainability, 14*(21), 13784.

Almulla, Y., Ramirez, C., Joyce, B., Huber-Lee, A., & Fuso-Nerini, F. (2022). From participatory process to robust decision-making: An Agriculture-water-energy nexus analysis for the Souss-Massa basin in Morocco. *Energy for Sustainable Development, 70*, 314–338.

Alzaabi, M. S. A. & Mezher, T. (2021). Analyzing existing UAE national water, energy, and food nexus-related strategies. *Renewable and Sustainable Energy Reviews, 144*, 111031.

Anser, M. K., Yousaf, Z., Usman, B., Nassani, A. A., Abro, M. M. Q., & Zaman, K. (2020). Management of water, energy, and food resources: Go for green policies. *Journal of Cleaner Production, 251*, 119662.

Azzam, A., Samy, G., Hagras, M. A., & ElKholy, R. (2023). Geographic information systems-based framework for water–energy–food nexus assessments. *Ain Shams Engineering Journal*, 102224.

Babel, M. S., Rahman, M., Budhathoki, A., & Chapagain, K. (2023). Optimization of economic return from water using water-energy-food nexus approach: A case of Karnafuli Basin, Bangladesh. *Energy Nexus, 10*, 100186.

Bakhshianlamouki, E., Masia, S., Karimi, P., Van der Zaag, P., & Sušnik, J. (2020). A system dynamics model to quantify the impacts of restoration measures on the water-energy-food nexus in the Urmia Lake Basin, Iran. *Science of the Total Environment, 708*, 134874.

Bazzana, D., Zaitchik, B., & Gilioli, G. (2020). Impact of water and energy infrastructure on local well-being: An agent-based analysis of the water-energy-food nexus. *Structural Change and Economic Dynamics, 55*, 165–176.

Boluwade, A. (2021). Impacts of climatic change and database information design on the water-energy-food nexus in water-scarce regions. *Water-Energy Nexus, 4*, 54–68.

Caputo, S., Schoen, V., Specht, K., Grard, B., Blythe, C., Cohen, N., Fox-Kämper, R., Hawes, J., Newell, J., & Poniży, L. (2021). Applying the food-energy-water nexus approach to urban agriculture: From FEW to FEWP (Food-Energy-Water-People). *Urban Forestry & Urban Greening, 58*, 126934.

Chai, J., Shi, H., Lu, Q., & Hu, Y. (2020). Quantifying and predicting the Water-Energy-Food-Economy-Society-Environment Nexus based on Bayesian networks-a case study of China. *Journal of Cleaner Production, 256*, 120266.

Chen, C. F., Feng, K. L., & Ma, H. W. (2020). Uncover the interdependent environmental impacts associated with the water-energy-food nexus under resource management strategies. *Resources, Conservation and Recycling, 160*, 104909.

Cheng, Y., Wang, J., & Shu, K. (2023). The coupling and coordination assessment of food-water-energy systems in China based on sustainable development goals. *Sustainable Production and Consumption, 35*, 338–348.

Correa-Cano, M., Salmoral, G., Rey, D., Knox, J. W., Graves, A., Melo, O., Foster, W., Naranjo, L., Zegarra, E., & Johnson, C. (2022). A novel modelling toolkit for unpacking the Water-Energy-Food-Environment (WEFE) nexus of agricultural development. *Renewable and Sustainable Energy Reviews, 159*, 112182.

Daher, B., Lee, S.-H., Kaushik, V., Blake, J., Askariyeh, M. H., Shafiezadeh, H., Zamaripa, S., & Mohtar, R. H. (2019). Towards bridging the water gap in Texas: A water-energy-food nexus approach. *Science of the Total Environment*, *647*, 449–463.

Daher, B. T. & Mohtar, R. H. (2015). Water–energy–food (WEF) Nexus Tool 2.0: Guiding integrative resource planning and decision-making. In Bhaduri, A., Ringler, C., Dombrowsky, I., ohtMar, R. H., & Scheumann, W. ((eds)). *Water International* (Vol. 40, pp. 748–771). London: Routledge.

Drisko, J. W. & Maschi, T. (2016). *Content Analysis: Pocket Guide to Social Work Research*. Oxford University Press.

El Gafy, I., Grigg, N., & Reagan, W. (2017). Water-food-energy nexus index to maximize the economic water and energy productivity in an optimal cropping pattern. *Water International*, *42*(4), 495–503.

Enayati, M., Bozorg-Haddad, O., Fallah-Mehdipour, E., Zolghadr-Asli, B., & Chu, X. (2021). A robust multiple-objective decision-making paradigm based on the water-energy-food security nexus under changing climate uncertainties. *Scientific Reports*, *11*(1), 20927.

Fabiani, S., Vanino, S., Napoli, R., Zajíček, A., Duffková, R., Evangelou, E., & Nino, P. (2020). Assessment of the economic and environmental sustainability of Variable Rate Technology (VRT) application in different wheat-intensive European agricultural areas. A water energy food nexus approach. *Environmental Science & Policy*, *114*, 366–376.

FAO. (2022). *World Food and Agriculture-Statistical Yearbook 2022*. https://doi.org/10.4060/cc2211en. Accessed on October 6, 2022.

Ghimire, U., Piman, T., Shrestha, M., Aryal, A., & Krittasudthacheewa, C. (2022). assessment of climate change impacts on the water, food, and energy sectors in Sittaung River Basin, Myanmar. *Water*, *14*(21), 3434.

Ghodsvali, M., Dane, G., & De Vries, B. (2022). The nexus social-ecological system framework (NexSESF): A conceptual and empirical examination of transdisciplinary food-water-energy nexus. *Environmental Science & Policy*, *130*, 16–24.

Haghjoo, R., Choobchian, S., Morid, S., & Abbasi, E. (2022). Development and validation of management assessment tools considering water, food, and energy security nexus at the farm level. *Environmental and Sustainability Indicators*, *16*, 100206.

Haltas, I., Suckling, J., Soutar, I., Druckman, A., & Varga, L. (2017). Anaerobic digestion: A prime solution for water, energy, and food nexus challenges. *Energy Procedia*, *123*, 22–29.

Hoff, H., Alrahaife, S. A., El Hajj, R., Lohr, K., Mengoub, F. E., Farajalla, N., Fritzche, K., Jobbins, G., Özerol, G., & Schultz, R. (2019). A nexus approach for the MENA region – From concept to knowledge to action. *Frontiers in Environmental Science*, *7*, 48.

Huang, D., Li, G., Sun, C., & Liu, Q. (2020). Exploring interactions in the local water-energy-food nexus (WEF-Nexus) using a simultaneous equations model. *Science of the Total Environment*, *703*, 135034.

Ibrahim, M. D., Ferreira, D. C., Daneshvar, S., & Marques, R. C. (2019). Transnational resource generativity: Efficiency analysis and target setting of water, energy, land, and food nexus for OECD countries. *Science of the Total Environment*, *697*, 134017.

Kaddoura, S. & El Khatib, S. (2017). Review of water-energy-food Nexus tools to improve the Nexus modeling approach for integrated policy-making. *Environmental Science & Policy*, *77*, 114–121.

Karnib, A. (2018). Bridging science and policy in water-energy-food nexus: Using the Q-Nexus model for informing policymaking. *Water Resources Management*, *32*, 4895–4909.

Kedir, Y., Berhanu, B., & Alamirew, T. (2022). Analysis of water–energy–crop nexus indicators in irrigated sugarcane of Awash Basin, Ethiopia. *Environmental Systems Research*, *11*(1), 1–19.

Khattar, R., Mansour, F., Abou Najm, M., Al-Hindi, M., Yassine, A., Chamas, Z., & Geisseler, D. (2023). Incorporating Nitrogen in the water-energy-food nexus: An optimization approach. *Cleaner and Circular Bioeconomy*, 100036.

Laspidou, C. S., Mellios, N. K., Spyropoulou, A. E., Kofinas, D. T., & Papadopoulou, M. P. (2020). Systems thinking on the resource nexus: Modeling and visualization tools to identify critical interlinkages for resilient and sustainable societies and institutions. *Science of the Total Environment*, *717*, 137264.

Lawford, R., Bogardi, J., Marx, S., Jain, S., Wostl, C. P., Knüppe, K., Ringler, C., Lansigan, F., & Meza, F. (2013). Basin perspectives on the water–energy–food security nexus. *Current Opinion in Environmental Sustainability*, *5*(6), 607–616.

Lazaro, L. L. B., Bellezoni, R. A., Puppim de Oliveira, J. A., Jacobi, P. R., & Giatti, L. L. (2022). Ten years of research on the water-energy-food nexus: An analysis of topics evolution. *Frontiers in Water*, *4*, 859891.

Lazaro, L. L. B., Giatti, L. L., Bermann, C., Giarolla, A., & Ometto, J. (2021). Policy and governance dynamics in the water-energy-food-land nexus of biofuels: Proposing a qualitative analysis model. *Renewable and Sustainable Energy Reviews*, *149*, 111384.

Ledari, M. B., Saboohi, Y., & Azamian, S. (2023). Water-food-energy-ecosystem nexus model development: Resource scarcity and regional development. *Energy Nexus*, *10*, 100207.

Lehmann, S. (2018). Implementing the Urban Nexus approach for improved resource-efficiency of developing cities in Southeast Asia, City. *Culture, and Society*, *13*, 2018.

Li, J., Cui, J., Sui, P., Yue, S., Yang, J., Lv, Z., Wang, D., Chen, X., Sun, B., & Ran, M. (2021). Valuing the synergy in the water-energy-food nexus for cropping systems: A case in the North China Plain. *Ecological Indicators*, *127*, 107741.

Li, G., Huang, D., Sun, C., & Li, Y. (2019a). Developing interpretive structural modeling based on factor analysis for the water-energy-food nexus conundrum. *Science of the Total Environment*, *651*, 309–322.

Li, G., Wang, Y., & Li, Y. (2019b). Synergies within the water-energy-food nexus to support the integrated urban resources governance. *Water*, *11*(11), 2365.

Li, M., Fu, Q., Singh, V. P., Ji, Y., Liu, D., Zhang, C., & Li, T. (2019c). An optimal modeling approach for managing agricultural water-energy-food nexus under uncertainty. *Science of the Total Environment*, *651*, 14–16, 1434.

Lyu, H., Tian, F., Zhang, K., & Nan, Y. (2023). Water-energy-food nexus in the Yarlung Tsangpo-Brahmaputra River Basin: Impact of mainstream hydropower development. *Journal of Hydrology: Regional Studies*, *45*, 101293.

Mabhaudhi, T., Nhamo, L., Mpandeli, S., Nhemachena, C., Senzanje, A., Sobratee, N., Chivenge, P. P., Slotow, R., Naidoo, D., & Liphadzi, S. (2019). The water–energy–food nexus as a tool to transform rural livelihoods and well-being in Southern Africa. *International Journal of Environmental Research and Public Health*, *16*(16), 2970.

Mahlknecht, J., González-Bravo, R., & Loge, F. J. (2020). Water-energy-food security: A Nexus perspective of the current situation in Latin America and the Caribbean. *Energy*, *194*, 116824.

Mannan, M., Al-Ansari, T., Mackey, H. R., & Al-Ghamdi, S. G. (2018). Quantifying the energy, water, and food nexus: A review of the latest developments based on life-cycle assessment. *Journal of Cleaner Production*, *193*, 300–314.

Mannschatz, T., Wolf, T., & Hülsmann, S. (2016). Nexus Tools Platform: Web-based comparison of modeling tools for analysis of water-soil-waste nexus. *Environmental Modeling & Software*, *76*, 137–153.

Martinez-Hernandez, E., Leach, M., & Yang, A. (2017). Understanding water-energy-food and ecosystem interactions using the nexus simulation tool NexSym. *Applied Energy*, *206*, 1009–1021.

Martinez, P., Blanco, M., & Castro-Campos, B. (2018). The water–energy–food nexus: A fuzzy-cognitive mapping approach to support nexus-compliant policies in Andalusia (Spain). *Water*, *10*(5), 664.

McNabola, A., Mérida García, A., & Rodríguez Díaz, J. A. (2022). The role of micro-hydropower energy recovery in the water-energy-food nexus. *Environmental Sciences Proceedings*, *21*(1), 27.

Mirzaei, A., Ashktorab, N., & Noshad, M. (2023). Evaluation of the policy options to adopt a water-energy-food nexus pattern by farmers: Application of optimization and agent-based models. *Frontiers in Environmental Science*, *11*, 1139565.

Momblanch, A., Papadimitriou, L., Jain, S. K., Kulkarni, A., Ojha, C. S., Adeloye, A. J., & Holman, I. P. (2019). Untangling the water-food-energy-environment nexus for global change adaptation in a complex Himalayan water resource system. *Science of the Total Environment, 655*, 35–47.

Naghavi, S., Mirzaei, A., Sardoei, M. A., & Azarm, H. (2022). Evaluating the synergy between water-energy-food nexus and decoupling pollution-agricultural growth for sustainable production in the agricultural sector. *Research Square, 4*, 1–21.

Namany, S., Al-Ansari, T., & Govindan, R. (2019). Optimization of the energy, water, and food nexus for food security scenarios. *Computers & Chemical Engineering, 129*, 106513.

Nasrollahi, H., Shirazizadeh, R., Shirmohammadi, R., Pourali, O., & Amidpour, M. (2021). Unraveling the water-energy-food-environment nexus for climate change adaptation in Iran: Urmia Lake Basin case-study. *Water, 13*(9), 1282.

Ngarava, S. (2021). Long-term relationship between food, energy, and water inflation in South Africa. *Water-Energy Nexus, 4*, 123–133.

Nhamo, L. & Ndlela, B. (2021). Nexus planning as a pathway towards sustainable environmental and human health post-Covid-19. *Environmental Research, 192*, 110376.

Nie, Y., Avraamidou, S., Xiao, X., Pistikopoulos, E. N., Li, J., Zeng, Y., Song, F., Yu, J., & Zhu, M. (2019). A Food-Energy-Water Nexus approach for land use optimization. *Science of the Total Environment, 659*, 7–19.

Niu, S., Lyu, X., & Gu, G. (2022). A New Framework of Green Transition of Cultivated Land-Use for the Coordination among the Water-Land-Food-Carbon Nexus in China. *Land, 11*(6), 933.

Niva, V., Cai, J., Taka, M., Kummu, M., & Varis, O. (2020). China's sustainable water-energy-food nexus by 2030: Impacts of urbanization on sectoral water demand. *Journal of Cleaner Production, 251*, 119755.

Norouzi, N. (2022). Presenting a conceptual model of the water-energy-food nexus in Iran. *Current Research in Environmental Sustainability, 4*, 100119.

Norouzi, N. & Kalantari, G. (2020). The food-water-energy nexus governance model: A case study for Iran. *Water-Energy Nexus, 3*, 72–80.

Opoku, E. K., Adjei, K. A., Gyamfi, C., Vuu, C., Appiah-Adjei, E. K., Odai, S. N., & Siabi, E. K. (2022). Quantifying and analyzing water trade-offs in the water-energy-food nexus: The case of Ghana. *Water-Energy Nexus, 5*, 8–20.

Ouyang, Y., Cai, Y., Xie, Y., Yue, W., & Guo, H. (2021). Multi-scale simulation and dynamic coordination evaluation of water-energy-food and economy for the Pearl River Delta city cluster in China. *Ecological Indicators, 130*, 108155.

Purwanto, A., Sušnik, J., Suryadi, F., & De Fraiture, C. (2019). Using group model building to develop a causal loop mapping of the water-energy-food security nexus in Karawang Regency, Indonesia. *Journal of Cleaner Production, 240*, 118170.

Purwanto, A., Sušnik, J., Suryadi, F., & de Fraiture, C. (2021). Quantitative simulation of the water-energy-food (WEF) security nexus in a local planning context in Indonesia. *Sustainable Production and Consumption, 25*, 198–216.

Qian, X. Y. & Liang, Q. M. (2021). Sustainability evaluation of the provincial water-energy-food nexus in China: Evolutions, obstacles, and response strategies. *Sustainable Cities and Society, 75*, 103332.

Radini, S., Marinelli, E., Akyol, Ç., Eusebi, A. L., Vasilaki, V., Mancini, A., Frontoni, E., Bischetti, G. B., Gandolfi, C., & Katsou, E. (2021). Urban water-energy-food-climate nexus in integrated wastewater and reuse systems: Cyber-physical framework and innovations. *Applied Energy, 298*, 117268.

Radmehr, R., Ghorbani, M., & Ziaei, A. N. (2021). Quantifying and managing the water-energy-food nexus in dry regions food insecurity: New methods and evidence. *Agricultural Water Management, 245*, 106588.

Rahmani, M., Jahromi, S. H. M., & Darvishi, H. H. (2023). SD-DSS model of sustainable groundwater resources management using the water-food-energy security Nexus in Alborz Province. *Ain Shams Engineering Journal, 14*(1), 101812.

Ravar, Z., Zahraie, B., Sharifinejad, A., Gozini, H., & Jafari, S. (2020). System dynamics modeling for assessment of water–food–energy resources security and nexus in Gavkhuni basin in Iran. *Ecological Indicators, 108*, 105682.

Rising, J. (2020). Decision-making and integrated assessment models of the water-energy-food nexus. *Water Security, 9*, 100056.

Sadeghi, S. H., Moghadam, E. S., Delavar, M., & Zarghami, M. (2020). Application of water-energy-food nexus approach for designating optimal agricultural management pattern at a watershed scale. *Agricultural Water Management, 233*, 106071.

Saladini, F., Betti, G., Ferragina, E., Bouraoui, F., Cupertino, S., Canitano, G., Gigliotti, M., Autino, A., Pulselli, F. M., & Riccaboni, A. (2018). Linking the water-energy-food nexus and sustainable development indicators for the Mediterranean region. *Ecological Indicators, 91*, 689–697.

Saray, M. H., Baubekova, A., Gohari, A., Eslamian, S. S., Klove, B., & Haghighi, A. T. (2022). Optimization of Water-Energy-Food Nexus considering CO2 emissions from cropland: A case study in northwest Iran. *Applied Energy, 307*, 118236.

Sharifinejad, A., Zahraie, B., Majed, V., Ravar, Z., & Hassani, Y. (2020). Economic analysis of Water-Food-Energy Nexus in Gavkhuni basin in Iran. *Journal of Hydro-environment Research, 31*, 14–25.

Shu, Q., Scott, M., Todman, L., & McGrane, S. J. (2021). Development of a prototype composite index for resilience and security of water-energy-food (WEF) systems in industrialized nations. *Environmental and Sustainability Indicators, 11*, 100124.

Simpson, G. B., Jewitt, G. P., Becker, W., Badenhorst, J., Masia, S., Neves, A. R., Rovira, P., & Pascual, V. (2022). The water-energy-food nexus index: A tool to support integrated resource planning, management and security. *Frontiers in Water, 4*, 825854.

Spiegelberg, M., Baltazar, D. E., Sarigumba, M. P. E., Orencio, P. M., Hoshino, S., Hashimoto, S., Taniguchi, M., & Endo, A. (2017). Unfolding livelihood aspects of the water–energy–food nexus in the dampalit watershed, Philippines. *Journal of Hydrology: Regional Studies, 11*, 53–68.

Tan, A. H. P., Yap, E. H., & Abakr, Y. A. (2020). A complex systems analysis of the water-energy nexus in Malaysia. *Systems, 8*(2), 19.

Terrapon-Pfaff, J., Ortiz, W., Dienst, C., & Gröne, M. C. (2018). Energizing the WEF nexus to enhance sustainable development at the local level. *Journal of Environmental Management, 223*, 409–416.

Uen, T. S., Chang, F. J., Zhou, Y., & Tsai, W. P. (2018). Exploring synergistic benefits of Water-Food-Energy Nexus through multi-objective reservoir optimization schemes. *Science of the Total Environment, 633*, 341–351.

Vaidya, B., Shrestha, S., & Ghimire, A. (2021). Water footprint assessment of food-water-energy systems at Kathmandu University, Nepal. *Current Research in Environmental Sustainability, 3*, 100044.

Wa'el, A., Memon, H., A., F., & Savic, D. A. (2017). An integrated model to evaluate water-energy-food nexus at a household scale. *Environmental Modeling & Software, 93*, 366–380.

Wang, Y., Zhao, Y., Wang, Y., Ma, X., Bo, H., & Luo, J. (2021). Supply-demand risk assessment and multi-scenario simulation of regional water-energy-food nexus: A case study of the Beijing-Tianjin-Hebei region. *Resources, Conservation and Recycling, 174*, 105799.

Wang, Q., Li, S., He, G., Li, R., & Wang, X. (2018). Evaluating the sustainability of water-energy-food (WEF) nexus using an improved matter-element extension model: A case study of China. *Journal of Cleaner Production, 202*, 1097–1106.

WEFNexus. (2022). *WEF results in summary and raw indicators*. https://www.water-energy-food.org/. Accessed on July 12, 2022.

Wicaksono, A., Jeong, G., & Kang, D. (2017). Water, energy, and food nexus: A review of global implementation and simulation model development. *Water Policy, 19*(3), 440–462.

World Economic Forum. (2023). *The global risks report 2023*. Switzerland. https://www.weforum.org/reports/globalrisks-report-2023/. Accessed on January 11,2023.

Wu, L., Elshorbagy, A., & Helgason, W. (2023). Assessment of agricultural adaptations to climate change from a water-energy-food nexus perspective. *Agricultural Water Management, 284*, 108343.

Wu, L., Elshorbagy, A., Pande, S., & Zhuo, L. (2021). Trade-offs and synergies in the water-energy-food nexus: The case of Saskatchewan, Canada. *Resources, Conservation and Recycling, 164*, 105192.

Xu, S., He, W., Shen, J., Degefu, D. M., Yuan, L., & Kong, Y. (2019). Coupling and coordination degrees of the core water–energy–food nexus in China. *International Journal of Environmental Research and Public Health, 16*(9), 1648.

Yi, J., Guo, J., Ou, M., Pueppke, S. G., Ou, W., Tao, Y., & Qi, J. (2020). Sustainability assessment of the water-energy-food nexus in Jiangsu Province, China. *Habitat International, 95*, 102094.

Yu, L., Xiao, Y., Jiang, S., Li, Y., Fan, Y., Huang, G., Lv, J., Zuo, Q. T., & Wang, F. (2020). A copula-based fuzzy interval-random programming approach for planning water-energy nexus system under uncertainty. *Energy, 196*, 117063.

Yue, Q. & Guo, P. (2021). Managing agricultural water-energy-food-environment nexus considering water footprint and carbon footprint under uncertainty. *Agricultural Water Management, 252*, 106899.

Zarei, M. (2020). The water-energy-food nexus: A holistic approach for resource security in Iran, Iraq, and Turkey. *Water-Energy Nexus, 3*, 81–94.

Zeng, Y., Liu, D., Guo, S., Xiong, L., Liu, P., Chen, J., Yin, J., Wu, Z., & Zhou, W. (2023). Assessing the effects of water resources allocation on the uncertainty propagation in the water–energy–food–society (WEFS) nexus. *Agricultural Water Management, 282*, 108279.

Zeng, Y., Liu, D., Guo, S., Xiong, L., Liu, P., Yin, J., & Wu, Z. (2022). A system dynamic model to quantify the impacts of water resources allocation on the water-energy–food–society (WEFS) nexus. *Hydrology and Earth System Sciences, 26*(15), 3965–3988.

Zhang, P., Zhang, L., Chang, Y., Xu, M., Hao, Y., Liang, S., Liu, G., Yang, Z., & Wang, C. (2019). Food-energy-water (FEW) nexus for urban sustainability: A comprehensive review. *Resources, Conservation and Recycling, 142*, 215–224.

Zhang, F., Cai, Y., Tan, Q., Engel, B. A., & Wang, X. (2021a). An optimal modeling approach for reducing carbon footprint in agricultural water-energy-food nexus system. *Journal of Cleaner Production, 316*, 128325.

Zhang, P., Xie, Y., Wang, Y., Li, B., Li, B., Jia, Q., Yang, Z., & Cai, Y. (2021b). Water-Energy-Food system in typical cities of the world and China under zero-waste: Commonalities and asynchronous experiences support sustainable development. *Ecological Indicators, 132*, 108221.

Zhang, P., Zhou, Y., Xie, Y., Wang, Y., Li, B., Li, B., Jia, Q., Yang, Z., & Cai, Y. (2021c). Assessment of the water-energy-food nexus under spatial and social complexities: A case study of Guangdong-Hong Kong-Macao. *Journal of Environmental Management, 299*, 113664.

Zhang, X. & Vesselinov, V. V. (2017). An integrated modeling approach for optimal management of water, energy, and food security nexus. *Advances in Water Resources, 101*, 1–10.

Zuo, Q., Wu, Q., Yu, L., Li, Y., & Fan, Y. (2021). Optimization of uncertain agricultural management considering the framework of water, energy, and food. *Agricultural Water Management, 253*, 106907.

Mitra Bahadori*, Hassan Sadighi

Prospects and Challenges in the Conservation of Medicinal and Aromatic Plants for Biodiversity and Cultural Heritage

Abstract: Medicinal and aromatic plants (MAPs) play a crucial role in traditional medicine and cultural practices, making a significant contribution to biodiversity and the well-being of communities worldwide. This study explores the challenges and prospects of MAP conservation, emphasizing their importance as natural resources that preserve cultural heritage and promote ecological stability. However, despite their substantial value, MAPs face threats from habitat degradation, climate change, and excessive harvesting. To gain a comprehensive understanding of these issues, this study employs inductive qualitative content analysis, a method aimed at providing in-depth insights into both the obstacles and opportunities related to MAP conservation. The findings indicated that environmental and legislation challenges are the most prevalent barriers to conservation. Moreover, integrating environmental and cultural conservation efforts emerges as a key strategy for ensuring sustainable biodiversity preservation and cultural conservation, which underscores the necessity of collaboration among local communities, scientists, and policymakers to develop conservation initiatives that effectively safeguard the ecological integrity of MAPs while supporting the cultural traditions that depend on them. Ultimately, the study highlights the enduring significance of MAPs in both traditional and contemporary contexts.

Keywords: Community-based conservation, cultural heritage, sustainable practice, biodiversity conservation, ecological stability, habitat degradation

1 Introduction

Medicinal and aromatic plants (MAPs) constitute one of the largest groups of plants utilized by humans. Globally, an estimated 50,000–80,000 species of flowering plants are used for medicinal purposes, with the majority being harvested from wild habitats (Labokas & Karpavi, 2023). Increasing evidence highlights the substantial value and potential of MAPs worldwide (Póvoa et al., 2023). These plants serve multiple

*Corresponding author: Mitra Bahadori, Department of Rural Development, College of Agriculture, Isfahan University of Technology, Isfahan, Iran, e-mail: m.bahadori@alumni.iut.ac.ir, ORCID: 0009-0009-3940-8103

Hassan Sadighi, Department of Agricultural Extension and Education, College of Agriculture, Tarbiat Modares University, Tehran, Iran, e-mail: sadigh_h@modares.ac.ir, https://www.scopus.com/authid/detail.uri?authorId=22941821100

https://doi.org/10.1515/9783111563046-003

functions, including disease treatment, health-conscious nutrition, and insect and pest repellency (Mbelebele et al., 2024; Derso et al., 2024). Historically and across cultures, MAPs have been extensively utilized for healthcare and healing practices, continuing to form the foundation of traditional and indigenous health systems, particularly in developing countries (Barata et al., 2016). Beyond their role in healthcare, among various non-wood forest products, MAPs are regarded as a critical component of economic development (Póvoa et al., 2023). MAPs contribute significantly to local economies, cultural preservation, human livelihoods, and overall well-being (Padulosi, 2014; Mittal et al., 2023), especially in remote and underserved rural communities (Rahman & Fakri, 2015). These benefits are particularly vital for marginalized social groups, including the elderly, children, and women (Padulosi, 2014). Since many MAPs are sourced from the wild, their collection and trade provide an additional source of income for impoverished rural households (Rahman & Fakri, 2015; Barata et al., 2016; Karki, 2017; Mbelebele et al., 2024).

These species constitute a significant component of the natural biodiversity assets of numerous countries worldwide (Padulosi, 2014; Mittal et al., 2023). Biodiversity serves as the foundation of life on the Earth, encompassing the diversity of all living organisms, their genetic variation, and the ecosystems they create (Gautam, 2024). Despite extensive efforts at both national and international levels, global biodiversity is experiencing a rapid decline (Brondízio et al., 2019; Petelka et al., 2022). Furthermore, there is an ongoing trend of human migration from traditional cultural landscapes to urban areas (Zerbe, 2022; Petelka et al., 2022). Consequently, these historically multifunctional cultural landscapes are increasingly subjected to intensified land use (Price et al., 2015; Brondízio et al., 2019; Petelka et al., 2022). Therefore, in an era of rapid urbanization and industrialization, which are transforming ecosystems at an unprecedented pace, the urgency of biodiversity conservation has never been greater. While contemporary conservation strategies primarily emphasize scientific methodologies (Rathoure, 2024), there is growing recognition of the critical role that traditional cultural practices play in biodiversity preservation (Aswani et al., 2018, Gautam, 2024). Cultural heritage represents both a legacy from the past and an essential resource for present and future generations. It comprises physical artifacts, cultural assets, and intangible traditions that have been inherited, preserved, and transmitted across generations (Mekenone et al., 2022). Cultural heritage, including indigenous knowledge and traditional ecological practices, often aligns with ecological principles, providing a strong foundation for sustainable conservation efforts (Rathoure, 2024). Indigenous knowledge refers to the cumulative and evolving body of practices, beliefs, and understandings developed by indigenous communities through sustained interaction with their natural environments (Gautam, 2024). This form of ecological wisdom, passed down over generations (Derso et al., 2024; Said et al., 2024), is deeply embedded in indigenous cultural practices and livelihoods. It encompasses not only practical resource management skills but also spiritual and ethical values that promote environmental stewardship (Vikram et al., 2017; Gautam, 2024). It has long been acknowledged as a crucial element in biodiversity conservation, with traditional ecological practices shaped by the close relationship between indigenous communities and their environments offering sustainable methods that complement modern conservation initiatives (Gautam, 2024). The integration of traditional knowledge with environmental conservation represents a key intersection for advancing sustainable development goals (Mbelebele et al., 2024).

Current trends favoring environmentally sustainable agriculture and increased local participation in conservation efforts have led to innovative approaches for enhancing MAP utilization. These approaches prioritize participatory methods that address local needs (Padulosi et al., 2014). The Global Sustainable Development Report has emphasized the importance of indigenous and traditional knowledge in achieving sustainability objectives (Messerli et al., 2019). Furthermore, indigenous knowledge systems and rights have been partially incorporated into international frameworks, such as the Global Biodiversity Framework and the Intergovernmental Science-Policy Platform on Biodiversity and Ecosystem Services (Díaz et al., 2015). The contributions of Indigenous Peoples and Local Communities (IPLCs) to biodiversity conservation have gained global recognition, particularly following the 2003 World Parks Congress in Durban, South Africa (Díaz et al., 2019; Dawson et al., 2021). Involving IPLCs in conservation efforts is considered essential not only for ensuring equity but also for achieving more effective biodiversity outcomes (Dawson et al., 2021). However, indigenous worldviews remain insufficiently acknowledged within mainstream conservation science and policy frameworks (Zheng et al., 2021; Brondízio et al., 2021).

The preservation of biodiversity and cultural heritage extends beyond the protection of individual species or traditions; it encompasses the safeguarding of entire ecological and cultural ecosystems. Recognizing the value of traditional knowledge and cultural practices enables the development of more comprehensive conservation strategies. These strategies should emphasize the empowerment of indigenous and local communities, granting them a central role in managing the lands and resources they have historically stewarded (Rathoure, 2024). Indeed, integrating cultural heritage conservation with natural resource preservation presents a unique opportunity to protect traditional landscapes, an approach that is increasingly gaining recognition among stakeholders and society (Valko et al., 2018). Most research indicating favorable results for both well-being and conservation originate from cases in which local communities are integral participants. This is particularly evident when these groups possess significant authority in decision-making processes or when local institutions governing land tenure are acknowledged as components of the governance framework. Equitable conservation, that empowers the environmental management efforts of local communities, is the key to achieving effective long-term biodiversity conservation, especially when integrated into broader legal and policy structure whether addressing protected areas in biodiversity rich regions or restoring heavily altered ecosystems, conservation efforts can be more successful by emphasizing the type and quality of governance, and promoting solutions that enhance the role, capacity, and rights of local communities (Dawson et al. 2021).

The interdependence of cultural heritage and biodiversity conservation is crucial for fostering sustainable environmental practices. Indigenous knowledge systems not only support biodiversity but also preserve cultural identities. By acknowledging and incorporating these cultural perspectives into contemporary conservation efforts, ecological resilience can be strengthened, ensuring the coexistence of both natural and cultural systems. Ultimately, biodiversity conservation is not solely about protecting species and ecosystems; it also involves preserving the cultural narratives and traditions that connect communities to their natural surroundings. Moving forward, cultivating a deeper appreciation for these cultural practices will be essential in the collective effort to safeguard the planet's rich biodiversity for future generations (Rathoure, 2024; Derso et al., 2024). In this context, this study aims to propose indigenous knowledge a fundamental dimension of cultural heritage as a viable solution to the challenges associated with the conservation of MAPs, complementing modern conservation approaches. The primary objective is to explore how, on one hand, centuries-old indigenous wisdom concerning MAPs can be preserved and respected, while on the other hand, ensuring the conservation of these plants and their habitats for the benefit of future generations. Although previous studies have examined the significance of cultural heritage and biodiversity conservation, none have employed content analysis to investigate the challenges and prospects of MAP conservation while proposing solutions through cultural heritage preservation. Therefore, this study seeks to address the following research questions:

1. What are the key challenges and opportunities in the conservation of MAPs within the context of biodiversity and cultural heritage?
2. How can indigenous knowledge contribute to the sustainable conservation of MAPs and their associated ecosystems?

By addressing these questions, this research aims to provide insights that bridge the gap between traditional ecological knowledge and contemporary conservation strategies, ensuring a more sustainable and inclusive approach to MAP conservation.

2 Study Background

The genetic resources of MAPs serve as crucial reservoirs of biodiversity, providing significant potential for human well-being and rural economic development. Barata et al. (2016) highlighted the necessity of systematically studying and conserving these resources to ensure their benefits for both present and future generations. However, the sustainability of MAPs is increasingly under threat due to overexploitation, inadequate management, and anthropogenic pressures, as emphasized by Igwillo et al.

(2019). To mitigate these threats, conservation strategies must not only focus on biodiversity preservation but also consider the well-being and livelihoods of indigenous and local communities reliant on these resources.

Traditional ecological knowledge plays a crucial role in MAP conservation. Palabaş Uzun and Koca (2020) underscored the importance of documenting traditional knowledge systems, which embody cumulative knowledge, practices, and beliefs shaped through long-standing human-environment interactions. Ethnobotanical research and national and international initiatives should be strengthened to safeguard these knowledge systems. Additionally, as the demand for medicinal plants rises due to environmental concerns and the prevalence of modern diseases, a lack of awareness has led to their unsustainable use. Raising awareness, particularly among younger generations, is essential to ensuring the long-term sustainability of MAPs and the preservation of traditional knowledge.

Dawson et al. (2021) proposed an equitable conservation framework that integrates IPLCs into environmental stewardship efforts. Such an approach, when embedded within legal and policy frameworks, can significantly enhance biodiversity conservation. Whether through the management of protected areas or the restoration of degraded ecosystems, conservation effectiveness is heightened by ensuring inclusive governance, strengthening local capacities, and protecting the rights of IPLCs.

The overexploitation and poor documentation of MAPs have led to the depletion and extinction of numerous species (Kankara et al., 2015; Sher et al., 2022). The study by Sher et al. (2022) further identified that rural communities, particularly women, depend on MAP collection and processing for healthcare and income generation. However, population growth and associated biotic pressures have exacerbated species decline. Addressing these challenges requires active local community engagement in rehabilitation efforts, as well as capacity-building programs for plant collectors and market access support for valuable species.

The erosion of biocultural heritage due to ecological transformations is another pressing concern. The biocultural heritage of a region has evolved and been refined over centuries through complex interactions between human societies and the natural environment (Bridgewater & Rotherham, 2019; Lindholm & Ekblom, 2019). According to Bastos et al. (2022), intensified migration patterns and land-use changes have led to the decline of species, such as brazilwood. To counter this, community-based conservation efforts, policy engagement, and awareness campaigns must be strengthened to protect both natural habitats and cultural landscapes.

Petelka et al. (2022) further argued that participatory research can bridge the gap between biodiversity conservation and cultural identity reinforcement by integrating local knowledge into conservation strategies. Also, Póvoa et al. (2023) identified genetic erosion as a primary driver of biodiversity loss, exacerbated by habitat destruction, invasive species proliferation, and human-induced environmental degradation. In some regions, farm abandonment has emerged as a key factor contributing to genetic erosion, particularly in low-population-density areas. Similarly, the loss of traditional agricultural practices due to urbanization and modern farming systems has altered biodiversity dynamics, necessitating targeted conservation interventions.

Governmental policies play a crucial role in shaping conservation outcomes. Mbelebele et al. (2024) advocated for enhanced governmental and policy support for traditional knowledge and sustainable agricultural practices. Providing smallholder farmers with improved access to land, financial resources, and market development opportunities can contribute to both biodiversity conservation and economic sustainability. Additionally, educational programs should be designed to integrate indigenous knowledge into formal curricula, fostering a deeper understanding of MAPs' ecological and economic significance.

Balancing conservation with commercial utilization is another critical aspect of MAPs' sustainability. Said et al. (2024) emphasized the need for an inclusive approach that aligns the interests of local communities, traditional practitioners, and commercial stakeholders. Their study underscores the vital role of traditional medicine in primary healthcare and advocates for integrating conservation policies with the safeguarding of indigenous knowledge systems.

Finally, educational interventions can play a transformative role in the conservation of MAPs. Vallejo et al. (2024) proposed a biocultural approach that incorporates medical history and ethnobotanical perspectives into healthcare training. By fostering respect for cultural traditions and aligning health policies with traditional medicinal practices, educational reforms can enhance conservation awareness and strengthen community participation. Derso et al. (2024) highlighted the urgent need for conservation strategies that focus on key medicinal plant species, address habitat destruction, and preserve traditional ecological knowledge for future generations.

In summary, the reviewed literature underscores that MAPs are valuable genetic resources with profound socioeconomic, ecological, and cultural significance. Despite their critical importance, challenges, such as overexploitation, habitat degradation, and the erosion of traditional ecological knowledge persist.

As shown in the literature review, several researches have studied the conservation of MAPs, but each of these researches has only focused on a number of challenges and prospects related to the conservation of MAPs. For example, some studies have examined environmental pressures while others have investigated management challenges, the erosion of biocultural heritage, economic challenges, etc. Moreover, most of these studies have ignored the role of indigenous knowledge and local people in the conservation of MAPs and overlooked the urgent need for integrative and community-based conservation strategies that harmonize modern scientific methods with indigenous knowledge systems. This integration not only emphasizes the necessity of safeguarding MAPs for sustainable livelihoods but also calls for a systematic analysis of how traditional ecological knowledge can be effectively integrated into contemporary conservation efforts. To fill this gap, this chapter has attempted to study all these challenges and prospects as much as possible and tried to categorize them. Based on these insights, the next section outlines the research methodology employed in this study, detailing the approaches and analytical tools used in it.

3 Materials and Methods

3.1 Content Analysis

Qualitative content analysis is a research method used for the systematic interpretation of textual data through classification and coding, facilitating the identification of recurring patterns. This methodological approach allows researchers to assess the authenticity and validity of data while adhering to rigorous scientific principles. The objectivity of the results is maintained through a structured coding process. Qualitative content analysis involves the systematic examination of textual content, encompassing both explicit and implicit themes/ patterns (Kleinheksel et al., 2020). There are two primary approaches to qualitative content analysis: inductive and deductive. Inductive content analysis is particularly valuable when there is limited prior knowledge about a phenomenon, as it enables researchers to develop foundational insights. The goal of this approach is to allow research findings to emerge naturally by identifying dominant and recurrent themes within the data. This process involves progressively condensing the collected data to reveal key concepts relevant to the research topic. In contrast to the deductive approach, which begins with a predefined theoretical framework and tests specific hypotheses, inductive content analysis employs a bottom-up, exploratory methodology. In this approach, coding is conducted without predetermined categories, allowing researchers to derive codes (main themes) directly from the dataset (Krippendorff, 2018; Kleinheksel et al., 2020). Given the objectives of the study and the emphasis on analyzing the semantic patterns present in relevant texts and documents, content analysis was chosen as the method of this study. This approach allows for the systematic extraction of hidden themes, interpretation of discourses, and examination of conceptual structures within textual data.

To investigate the prospects and challenges associated with the conservation of MAPs for biodiversity and cultural heritage, a systematic review of academic literature was conducted. This process involved the identification, screening, assessment of eligibility, and inclusion of relevant studies. A comprehensive peer-reviewed literature search was carried out using major academic databases, including ISI Web of Science, Scopus, Nature, Elsevier, Taylor & Francis, and Springer. These databases were selected for their extensive coverage of international journals spanning multiple disciplines, including natural sciences, social sciences, and interdisciplinary research. The title search using keywords, such as "medicinal and aromatic plants," "community involvement in conservation," "cultural heritage," "conservation challenges," "indigenous knowledge," "local ecological knowledge," "biodiversity conservation," and "sustainable use of medicinal plants," the year limit was from 2014 to 2024, but no limitations were placed on the subject area or country of publication, ensuring a broad and inclusive approach to the literature review. Choosing this time frame had several reasons. The selected period encompasses a decade of research and developments in the fields of biodiversity conservation and cultural heritage. This allows for an analysis of the most current trends, and findings that are relevant to the conservation of MAPs. This time frame includes important international and national policy shifts regarding environmental conservation, sustainable practices, and cultural heritage preservation. Analyzing literature from this period can reveal how these policies have influenced conservation strategies and community involvement. The years leading up to 2024 have seen a growing global awareness of biodiversity loss and the importance of cultural heritage. Overall, the 2014 to 2024 time frame is chosen to ensure that the content analysis is relevant, up-to-date, and reflective of the current state of knowledge and practice in the conservation of MAPs for biodiversity and cultural heritage. Priority was given to articles that focused on the challenges and prospects of MAP conservation in relation to biodiversity and cultural heritage. The selection process followed a systematic approach. Initially, 182 articles were identified based on relevant titles. To ensure the credibility and reliability of the sources, articles lacking a valid publisher were excluded from consideration. In the next stages, articles with pertinent abstracts and content were further examined and the numbers of refined papers based on abstracts were 127 papers. In next stage numbers of refined papers based on content were 35 papers. Ultimately, 16 articles met the inclusion criteria and were selected in the process of the content analysis. The process is shown in Figure 1.

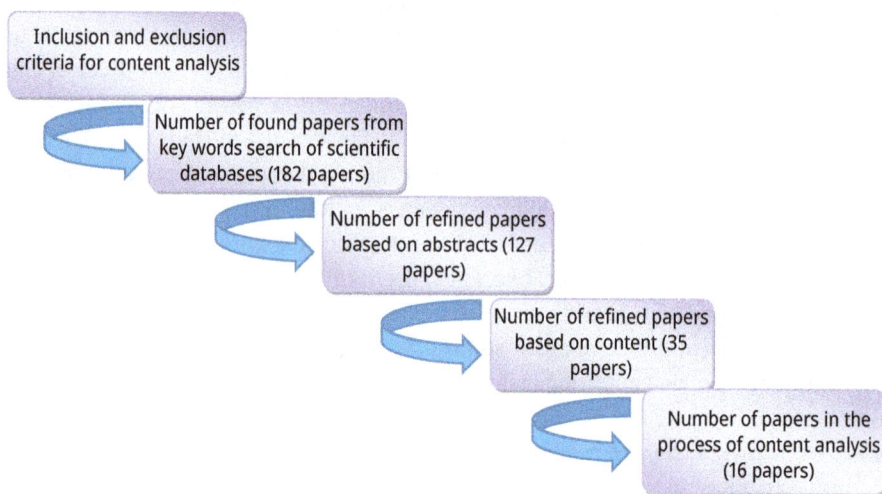

Figure 1: Steps of selecting papers (inclusion and exclusion criteria) in the process of content analysis.

3.1.1 Coding Process

Initially, key concepts related to the challenges, opportunities, and prospects of conserving the cultural heritage and biodiversity of medicinal plants were identified through open coding. This stage involved systematically extracting significant themes from the selected literature. Following open coding, axial coding was conducted to establish relationships between these concepts. In this phase, similar themes were grouped under broader categories based on their relevance and interconnections. Each open code was assigned to a relevant category, facilitating a structured organization of the data and enhancing the coherence of identified patterns. The objective of axial coding was to create meaningful links between the concepts derived in the open coding stage. Finally, selective coding was performed, wherein the extracted challenges and prospects were further refined and integrated into overarching categories. This step involved grouping related elements together based on their meaning, conceptual alignment, and significance. This step of content analysis is the main coding stage, and the main categorization is done. This stage includes the process of integrating and improving the categories.

4 Results and Discussion

The results of this content analysis underscore the intricate relationship between the challenges and opportunities associated with the conservation of MAPs for biodiversity and cultural heritage. While substantial threats persist, there are also promising avenues for sustainable conservation practices that can simultaneously support biodiversity and cultural heritage. The effective integration of community involvement, policy support, and traditional knowledge is essential for developing comprehensive conservation strategies. Figure 2 presents top five challenges according to the study results and Table 1 presents the frequencies, and references of challenges generated during the open, axial, and selective coding phases.

4.1 Challenges in the Conservation of MAPs: From Environmental Threats to Cultural and Intellectual Barriers

As mentioned in Table 1, first, environmental challenges were the most frequently mentioned as the main problems in conservation efforts. Among environmental challenges, habitat loss was the most frequent. This is consistent with the results of Ahmad et al. (2014) and Póvoa et al.'s (2023) conclusion based on their studies. Ahmad et al. (2014) identified unsustainable harvesting practices and habitat loss as main challenges endangering the conservation status of medicinal plants. Their study emphasized the need for government-led initiatives to develop effective conservation measures and strategies. Similarly, Póvoa et al. (2023) highlighted habitat loss, invasive species, and overexploitation primarily driven by human activities, such as watercourse clearing, vegetation management, grazing, and desertification as key challenges. The findings presented in Table 1 further corroborate these concerns, indicating that habitat degradation and other environmental stressors have significantly threatened the availability of MAPs. This decline not only jeopardizes the sustainability of herbal medicine production but also leads to irreversible biodiversity loss and substantial socioeconomic consequences.

Second, challenges based on the study result were legislation issues like illegal activities (land invasion, construction, harvesting, grazing, pasturing), weak management and absence of restoration actions. Illegal grazing was what Bano et al. (2014) reported as a challenge observed in various regions of their study area, which was the increasing pressure on wild plant resources, leading to a concerning decline in the availability of certain essential medicinal plant species. Chandra and Sharma (2018) argued that despite the significant economic potential of the MAP industry, inadequate resource planning, unsustainable harvesting, and trade practices were depleting vital forest resources and posing sustainability challenges for MAP-related businesses. While the expanding MAP trade was fostering new business opportunities in many countries, ensuring that these enterprises effectively conserve forest MAP resources and equitably distribute benefits will necessitate improved regulation and monitoring. Kala (2015) believed that remote rural areas, despite being rich in natural resources, such as forests, face a range of interconnected challenges, including governance deficiencies, market inefficiencies, and resource endowment limitations.

Another challenge in conservation of MAPs were cultural and intellectual challenges like decline of local ecological knowledge, lack of awareness, and cultural degradation. This result is in line with Palabaş Uzun & koca (2020) who believed in recent years, environmental challenges and the emergence of modern diseases have broadened the motivations for utilizing medicinal plants, leading to a significant increase in their consumption. However, this rising demand has also resulted in both overuse and misuse, often due to a lack of adequate knowledge. Raising awareness, particularly among younger generations, was crucial for ensuring the sustainable use of medicinal plants and preserving traditional knowledge. These findings align with the study by Póvoa et al. (2023), which highlighted the ongoing loss of traditional knowledge regarding plant usage. Their research underscored the vital role of ethnobotany in safeguarding the remaining information, emphasizing farmers' expertise in cultivating and utilizing landraces and local spontaneous MAPs essential for the conservation of regional plant genetic resources.

Figure 2 illustrates the percentage of various challenges identified in the study. As shown, environmental challenges have the highest percentage (64%), while economic challenges have the lowest (4%). Other challenges include legislation challenges

Table 1: Selective, axial, and open coding analysis of challenges (main categories, subcategories, and concepts).

Main category (selective coding)	Subcategory (axial coding)	Concepts (open coding)	References	Frequency
Environmental challenges	Ecosystem and habitat degradation	Habitat loss/destruction/degradation	Padulosi et al. (2014), Rotherham, (2015), Sher et al. (2015), Barata et al. (2016), Vikram et al. (2017), Valkó et al. (2018), Dawson et al. (2021), Bastos et al. (2022), Petelka et al. (2022), Mittal et al. (2023), Sinthumule (2023), Derso et al. (2024), Levis et al. (2024), and Mbelebele et al. (2024)	14
		Land use, landscape change, and alteration	Rotherham, (2015), Barata et al. (2016), Cicinelli et al. (2018), Dawson et al. (2021), Bastos et al. (2022), Petelka et al. (2022), Levis et al. (2024), and Mbelebele et al. (2024)	8
		Deforestation, removal of forests, forest loss, and destruction of the forest areas	Rotherham, (2015), Sher et al. (2015), Dawson et al. (2021), Bastos et al. (2022), Levis et al. (2024), and Mbelebele et al. (2024)	6
		Fragmentation	Rotherham, (2015), Dawson et al. (2021), Bastos et al. (2022), and Mittal et al. (2023)	4
				Total = 32
	Overexploitation and unsustainable resource use	Overexploitation and intensification of land (pastures/grazing)	Padulosi et al. (2014), Rotherham, (2015), Sher et al. (2015), Barata et al. (2016), Valkó et al. (2018), Dawson et al. (2021), Bastos et al. (2022), Petelka et al. (2022), Mittal et al. (2023), Levis et al. (2024), and Mbelebele et al. (2024)	11

(continued)

Table 1 (continued)

Main category (selective coding)	Subcategory (axial coding)	Concepts (open coding)	References	Frequency
		Unsustainable practices and overharvesting	Padulosi et al. (2014), Rotherham, (2015), Sher et al. (2015), Barata et al. (2016), Vikram et al. (2017), Dawson et al. (2021), Petelka et al. (2022), Mittal et al. (2023), Sinthumule (2023), Derso et al. (2024), and Mbelebele et al. (2024)	11
			Total = 22	
	Biodiversity loss and threats from invasive species	Biodiversity loss, decline in biodiversity, biodiversity decline, depletion of biodiversity, and species extinction	Padulosi et al. (2014), Rotherham; (2015), Vikram et al. (2017), Valkó et al. (2018), Dawson et al. (2021), Bastos et al. (2022), Petelka et al. (2022), Sinthumule (2023), Derso et al. (2024), Levis et al. (2024), and Mbelebele et al. (2024)	11
		Alien species and invasive species	Rotherham (2015), Cicinelli et al. (2018), Petelka et al. (2022), and Mittal et al. (2023)	4
			Total = 15	
	Impact of extreme weather events on ecosystems, livelihoods, and food security	Flooding: Heavy rainfall increase, river flow pattern changes, water infiltration capacity, destruction of urban and rural infrastructure, loss of agricultural lands, and reduced crop yield	Rotherham (2015), Barata et al. (2016), Vikram et al. (2017), Petelka et al. (2022), Mittal et al. (2023), Levis et al. (2024), and Mbelebele et al. (2024)	7
		Drought: Reduced precipitation, changing rainfall patterns, groundwater depletion, reduced water storage, water shortages, agricultural productivity decline, and rural migration due to resource scarcity		

Category	Sub-item	Description	References	Frequency
		Other extreme weather events: Severe storms, ecosystem damage, heat waves, impact on human and livestock health, frost damage, and reduced agricultural yield		Total = 7

Total frequency of environmental challenges = 76

Category	Sub-item	Description	References	Frequency
Legislation challenges	Unregulated land exploitation	Illegal activities (land invasion, construction, harvesting, grazing, and pasturing) and encroachment	Rotherham, (2015), Sher et al. (2015), Vikram et al. (2017), Valkó et al. (2018), Dawson et al. (2021), Bastos et al. (2022), Levis et al. (2024), and Mbelebele et al. (2024)	8
				Total = 8
	Ineffective governance and policy implementation	Lack of effective management, weaknesses of the current management, governance weaknesses, and absence of restoration actions	Rotherham (2015), Dawson et al. (2021), and Bastos et al. (2022)	3
		The gap between policy and practice, and the gap between social and environmental policies	Dawson et al. (2021), Bastos et al. (2022), and Mbelebele et al. (2024)	3
		Excluding indigenous, local people from decision-making, and top-down conservation governance	Dawson et al. (2021) and Levis et al. (2024)	2

Total frequency of legislation challenges = 16

Category	Sub-item	Description	References	Frequency
Cultural and intellectual challenges	Loss of local ecological knowledge and awareness deficiency	Decline of local ecological knowledge, local ecological knowledge loss, and loss of the knowledge of native species	Rotherham, (2015), Sher et al. (2015), Vikram et al. (2017), Dawson et al. (2021), Bastos et al. (2022), Derso et al. (2024), and Mbelebele et al. (2024)	7
		Lack of awareness	Sher et al. (2015), Bastos et al. (2022), and Mbelebele et al. (2024)	3
		Degradation (cultural and knowledge)	Bastos et al. (2022) and Levis et al. (2024)	2

(continued)

Table 1 (continued)

Main category (selective coding)	Subcategory (axial coding)	Concepts (open coding)	References	Frequency
Total frequency of cultural and intellectual challenges = 12				
Economic transformation process challenges	Urbanization	Reduction of agricultural lands and natural resources for medicinal plant cultivation	Rotherham, (2015), Valkó et al. (2018), Bastos et al. (2022), Petelka et al. (2022), Derso et al. (2024) and Mbelebele et al. (2024)	6
		Conversion of farmland into residential and industrial areas		
		Increased urban pollution negatively affects MAP biodiversity		
		Decrease in native plant diversity in urbanized regions		
		Limited access for farmers to cultivable land		
		Decline in skilled labor for MAP production due to rural-to-urban migration	Rotherham, (2015) and Petelka et al. (2022)	2
		Shift in livelihood patterns and reducing interest in MAP cultivation		
		Increased pressure on natural resources in migration-receiving areas		
				Total = 8
	Industrialization	Industrial pollution affecting the quality and growth of MAPs	Rotherham (2015), Bastos et al. (2022), and Mbelebele et al. (2024)	3
		Reduced water availability for medicinal plant cultivation due to industrial water consumption		

		Expansion of pharmaceutical industries leading to over-commercialization of MAPs		3
		Replacement of traditional farming systems with industrialized production and reducing genetic diversity		
				Total = 3

Total frequency of economic transformation process challenges = 11

Economic challenges	Economic factors	Limited access to financial resources for sustainable conservation programs. Economic instability affecting long-term investment in biodiversity conservation	Padulosi et al. (2014), Rotherham, (2015), Valkó et al. (2018), Levis et al. (2024), and Mbelebele et al. (2024)	5
		High costs of conservation projects, making them financially unfeasible		
		Dependence on short-term funding rather than long-term financial planning		
		Lack of economic incentives for sustainable agricultural and conservation practices		
		Inadequate financial support for conservation and research initiatives		
		Low-budget allocations for biodiversity protection and ecosystem restoration		
				Total = 5

Total frequency of economic challenges = 5

(13%), cultural and intellectual challenges (10%), and economic transformation process challenges (9%).

Top 5 challenges in conservation of MAPs

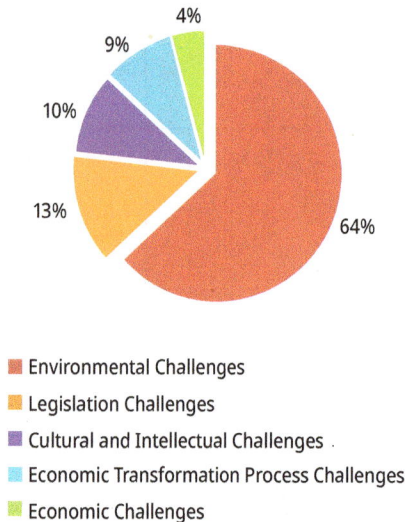

Figure 2: Top five challenges in conservation of MAPs.

4.2 Prospects for Overcoming Challenges in the Conservation of MAPs

The results of the study presented in Table 2 showed that among environmental conservation methods, sacred sites, groves, protected areas and sustainable practices (harvesting, management) had the most frequency. Other methods include foster sustainable utilization; habitat restoration efforts and cultivation of MAPs were placed after them in order. Said et al. (2024) recommended maintaining representative populations of fragile MAP species, which was crucial to establish protected areas or conserve ecosystems where natural ecological processes can continue undisturbed. Based on the results obtained from Said et al.'s (2024) study, achieving a balance between the needs of local communities, traditional medicinal practitioners, and commercial interests was crucial for ensuring the sustainable utilization of medicinal plant species. This approach fosters conservation efforts while supporting livelihoods, preserving traditional knowledge, and maintaining ecological integrity. Sher et al. (2022) had the same point and recommended that providing training for collectors was likely the most effective strategy for ensuring the sustainable management of MAPs. Dajic-Stevanovic and Pljevljakusic (2015) explained that MAPs were natural products environmentally friendly, readily available, cost-effective, and possess therapeutic properties. Their cultivation plays a crucial role in minimizing the risks of plant misidentification and adulteration, ensuring both quality and sustainability in their use. Phondani et al. (2016) also discussed the development of a participatory approach to encourage the cultivation of MAPs as a strategy for both biodiversity conservation and livelihood improvement. They argued that this approach offers farmers' opportunities to enhance their skills, expand their knowledge, and build self-confidence while simultaneously preserving MAP diversity within their natural habitats.

According to the study results among cultural conservation strategies, awareness-raising, documentation of traditional/indigenous knowledge and community-based conservation have the most frequency. This was in line with Rotherhom (2015) findings indicated that raising awareness and expanding education and research in key fields related to biocultural heritage and its conservation, as well as intangible cultural heritage, such as oral traditions and traditional landscape-related knowledge was crucial. It is recommended that more countries investigate the complex processes associated with biocultural heritage, conservation methods, and guiding principles. Bastos et al. (2022) highlighted the potential for biocultural conservation initiatives, emphasizing the importance of social participation, community mobilization, and awareness-raising to enhance the effectiveness of decision-making regarding local environmental conservation.

Mbelebele et al. (2024) suggested that educational campaigns should be implemented to enhance awareness of the ecological and economic significance of indigenous medicinal plants, ensuring their sustainable use and conservation. About the necessity of documentation Adachukwu and Yusuf (2014) and Sher et al. (2022) explained that the majority of MAPs remain undocumented, and in various study areas, numerous medicinal plant species have already become extinct before proper documentation could take place. Urgent action is necessary to systematically record this invaluable knowledge. Lakshman (2016) also emphasized the need for a structured documentation process to preserve farmers' knowledge regarding the diversity and usage of these plants. According to Palabaş Uzun and Koca (2020) recording local traditional ecological knowledge was of utmost importance, as it encompassed the accumulated knowledge, practices, and beliefs developed over time through the interaction between communities and their natural environment. The necessity of community-based conservation aligned with the findings of Dawson et al. (2021), who emphasized that equitable conservation empowering and supporting the environmental stewardship of indigenous peoples and local communities was the most effective approach for ensuring the long-term preservation of biodiversity. When integrated into broader legal and policy frameworks, such conservation strategies become more sustainable. Whether applied to protected areas in biodiversity hotspots or the restoration of highly modified ecosystems, conservation efforts were significantly strengthened by prioritizing governance quality and fostering solutions that enhance the role, capacity, and rights of indigenous peoples and local communities. It was also aligned with Sher et al.'s (2022) study result which they recommended initiating conservation efforts with active involvement from local communities to rehabilitate depleted plant resources effectively.

One of the frequent prospects obtained from our study shown in Table 2 is financial benefits. This was in line with Taghouti et al.'s (2022) study that concluded that the collection and processing of MAPs by rural women have been recognized as a significant source of cash income. Karki (2017) highlighted that the MAP subsector served as a significant source of rural employment. He further emphasized that, in addition to offering various conservation benefits, MAPs possess substantial economic potential and contribute to generating cash incomes. It was also in agreement with El Mekkaoui et al. (2024), who highlighted that, from a socioeconomic perspective, the harvesting of MAPs serves as a means to diversify agricultural production, generate employment opportunities, and create income-generating activities for local populations. Similarly, Sher and Barkworth (2015) noted that farming communities often remain unaware of the high economic value of MAPs, resulting in minimal financial returns from conventional agricultural practices. They emphasized the potential of MAPs to enhance income levels and contribute to broader economic growth.

Additional financial benefit indicated in Table 2 was tourism income benefit. This outcome supported the findings of Vallejo et al. (2024) who suggested that the plant world within cultural landscapes should be valued not only for its contributions to health but also for its economic potential and its role in promoting sustainable tourism, particularly among groups outside rural communities.

Another prospect resulted from Table 2 was increasing well-being and welfare benefits include healthcare. This result supported Sher et al.'s (2022) study findings that concluded MAPs have been recognized as essential for maintaining both human and livestock health. Similarly, the study by Said et al. (2024) emphasized the importance of medicinal plants as a primary healthcare resource for a significant portion of the population, highlighting the necessity of acknowledging their role in traditional and modern healthcare systems.

The notable point from the legislation section of Table 2 was that some researchers advocated for the integration of traditional knowledge into formal education. This was in line with Vallejo et al. (2024), who, based on their findings and an epistemological discussion, proposed an educational intervention aimed at broadening the professional mindset through a biocultural perspective. It was also consistent with Igwillo et al. (2019) who had the same idea in which student should be encouraged to take proactive measures by utilizing education, knowledge, and the recommendations derived from their research to support the sustainability and conservation of MAPs.

Table 2: Selective, axial, and open coding analysis of prospects (main categories, subcategories, and concepts).

Main category (selective coding)	Subcategory (axial coding)	Concepts (open coding)	References	Frequency
Environmental conservation prospects	Conservation efforts	Establishing protected areas, sacred sites, and conservation zones	Rotherham, (2015), Vikram et al. (2017), Valkó et al. (2018), Dawson et al. (2021), Bastos et al. (2022), Sinthumule (2023), and Mbelebele et al. (2024)	8
		Foster sustainable practices (harvesting, management)	Padulosi et al. (2014), Vikram et al. (2017), Mittal et al. (2023), Sinthumule (2023), Levis et al. (2024), and Mbelebele et al. (2024),	7
		Foster sustainable utilization	Padulosi et al. (2014), Bararta et al. (2016), Vikram et al. (2017), Mittal et al. (2023), Sinthumule (2023), and Derso et al. (2024)	6
		Habitat/ecosystem restoration efforts/ reforestation	Rotherham, (2015), Valkó et al. (2018), Dawson et al. (2021), Bastos et al. (2022), Petelka et al. (2022), and Sinthumule (2023)	6
		Cultivation of MAPs	Padulosi et al. (2014), Vikram et al. (2017), Valkó et al. (2018), Mittal et al. (2023), and Sinthumule (2023)	5

Total frequency of environmental conservation prospects = 32

Cultural conservation prospects	Community engagement	Awareness raising, increasing local awareness, environmental and educational awareness/ actions	Padulosi et al. (2014), Rotherham, (2015), Sher et al. (2015), Bararta et al. (2016), Valko et al. (2018), Bastos et al. (2022), Mittal et al. (2023), Mbelebele et al. (2024), and Sinthumule (2023)	9

Table 2 (continued)

Main category (selective coding)	Subcategory (axial coding)	Concepts (open coding)	References	Frequency
		Documentation of traditional, indigenous knowledge	Padulosi et al. (2014), Sher et al. (2015), Bararta et al. (2016), Vikram et al. (2017), Sinthumule (2023), Derso et al. (2024), Levis et al. (2024), and Mbelebele et al. (2024)	8
		Community-based conservation, social participation, place-based conservation, community/social mobilization, involvement of the community in conservation issues, community involvement/recognition, respect and preservation of local knowledge, reinforce the role, capacity, and rights of indigenous peoples and local communities	Padulosi et al. (2014), Sher et al. (2015), Dawson et al. (2021), Bastos et al. (2022), Petelka et al. (2022), Derso et al. (2024), Levis et al. (2024), and Mbelebele et al. (2024)	8
Total frequency of cultural conservation prospects = 25				
Socioeconomic prospects	Economic factors	Financial benefits, viability (income benefits, tourism industry benefits, tourism expansion, maintenance of the rural economy, economic empowerment, ensure financial support	Padulosi et al. (2014), Rotherham, (2015), Vikram et al. (2017), Dawson et al. (2021), Bastos et al. (2022), Petelka et al. (2022), Levis et al. (2024), and Mbelebele et al. (2024)	8
	Social factors	Increasing well-being, welfare benefits	Dawson et al. (2021), Sinthumule (2023), Derso et al. (2024), and Mbelebele et al. (2024)	4
Total frequency of socioeconomics prospects = 12				
Legislation prospects	Policy and governance	Including local people in the decision-making process	Petelka et al. (2022), Sinthumule (2023), and Levis et al. (2024),	3

Table 2 (continued)

Main category (selective coding)	Subcategory (axial coding)	Concepts (open coding)	References	Frequency
		Promote knowledge transfer	Petelka et al. (2022), Derso et al. (2024), and Mbelebele et al. (2024)	3
		Integrating traditional knowledge into formal education	Padulosi et al. (2014), Bastos et al. (2022), and Mbelebele et al. (2024)	3
		Local ecological knowledge should be recognized in law and policy and integrated across scales of governance	Dawson et al. (2021) and Mbelebele et al. (2024)	2

Total frequency of legislation prospects = 11

As shown in Figure 3, the chart represents the percentage distribution of four key prospects analyzed in the study: environmental prospects, cultural conservation, socioeconomic prospects, and legislation prospects. The environmental prospects category has the highest percentage, while legislative prospects have the lowest.

Prospects in conservation of MAPs

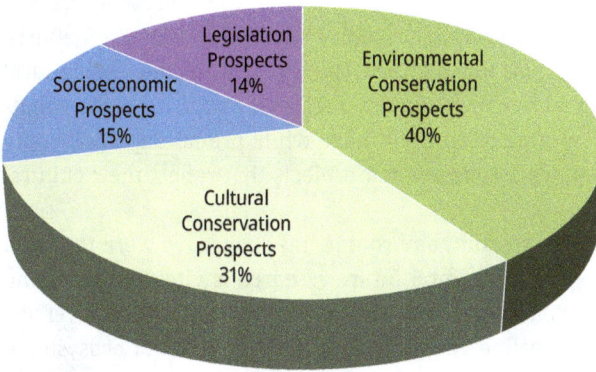

Figure 3: Prospects in conservation of MAPs.

5 Conclusion and Recommendation

It may seem paradoxical in an era of advanced technologies that traditional conservation methods, such as sacred sites, nature reserves, and community-based initiatives remain fundamental to the preservation of MAPs. However, given the vast number of MAP species requiring conservation and the financial constraints associated with ex situ conservation, decentralized, community-driven approaches are likely to play a pivotal role within a complementary conservation framework.

Several key factors reinforce the importance of this perspective. First, sustainable conservation efforts are more effective when they engage and support local communities. Second, preserving the culturally rich and invaluable indigenous knowledge linked to MAPs is essential for both biodiversity conservation and cultural heritage preservation. Finally, as global environmental challenges, such as habitat loss, climate change, and species extinction continue to escalate, it is imperative to adopt diverse, integrative conservation strategies that combine traditional ecological knowledge with modern scientific approaches. By embracing multiple conservation methods, we can enhance the resilience of both natural ecosystems and the cultural traditions that sustain them.

Integrating cultural heritage and biodiversity conservation can be achieved through collaborative approaches that recognize the value of cultural practices in environmental stewardship. This includes incorporating traditional knowledge into conservation strategies and promoting sustainable tourism that respects both cultural and natural resources. For example, utilize indigenous and local knowledge in conservation planning to enhance biodiversity management while engage communities in sharing their cultural practices that promote sustainable land use. Moreover, cultural festivals and events that raise awareness about biodiversity and conservation efforts, and developing educational programs that highlight the relationship between cultural practices and environmental health may be very useful. Finally supporting grassroots movements that focus on preserving cultural heritage while promoting biodiversity encourage local communities to lead conservation projects that reflect their cultural values and practices.

According to the current study result, environmental challenges were the most frequent between other challenges, therefore, future research is vital to understand how habitat degradation, overexploitation, and invasive species threaten biodiversity, as these factors can lead to the decline of species and the disruption of ecosystems, especially as invasive species often outcompete native species for resources, understanding their effects helps maintain ecosystem services and informs effective management strategies. Investigating these impacts also helps develop effective conservation strategies and policies to mitigate risks and promote sustainable resource use in the face of climate change.

On the contrary, the second most frequent challenges category was legislation challenges including unsustainable land exploitation that poses significant threats to biodiversity, ecosystem health, and the livelihoods of communities worldwide. Unsustainable practices, such as deforestation, overgrazing, and urban sprawl, lead to habitat destruction, soil degradation, and loss of biodiversity. Addressing these issues is crucial for ensuring environmental sustainability, food security, and the well-being of future generations and requires a multifaceted approach. According to study results, sustainable land management practices and community engagement, effective policy frameworks, restoration efforts, and research are some of important prospects in conservation of MAPs. By implementing these prospects, we can mitigate the impacts of land exploitation, protect biodiversity, and promote sustainable development.

Ultimately, biodiversity conservation extends beyond the protection of species and habitats; it involves honoring the cultural narratives and traditions that connect communities to their natural surroundings. Moving forward, fostering a greater appreciation for these cultural practices will be essential in our collective effort to safeguard the planet's rich biodiversity for future generations.

5.1 Policy Roadmap for the Sustainable Conservation of MAPs

To effectively implement a sustainable conservation framework, an integrated policy roadmap is required. This roadmap encompasses key phases aimed at enhancing conservation efforts through legal, scientific, economic, and community-driven approaches:

5.1.1 Strategic Planning and Policy Development

- Conducting comprehensive assessments of the current status of MAP conservation, identifying the status of key challenges, and mapping available resources.
- Engaging multiple stakeholders, including local communities, policymakers, researchers, and conservation organizations, to ensure an inclusive decision-making process.
- Establishing a regulatory framework that governs the sustainable harvesting, trade, and conservation of MAPs while balancing economic and environmental interests.

5.1.2 Implementation and Capacity Building

- Enforcing sustainable harvesting guidelines to prevent overexploitation and degradation of natural habitats.
- Integrating indigenous knowledge into formal education systems and organizing community-led workshops to raise awareness of sustainable practices.

- Strengthening research initiatives by promoting innovation in cultivation techniques, ecological restoration, and market-driven conservation strategies.

5.1.3 Monitoring and Evaluation

- Developing systematic monitoring programs using both quantitative and qualitative indicators to assess the effectiveness of conservation efforts.
- Regularly reviewing and refining conservation policies based on environmental, social, and economic changes.
- Enhancing transparency through global cooperation, knowledge-sharing, and participation in international conservation agreements.

5.1.4 Financial Sustainability and Economic Incentives

- Establish dedicated conservation funds and financial incentives to support sustainable MAP harvesting and habitat preservation.
- Developing eco-friendly market structures that empower local communities by promoting value-added MAP products and ethical trade.
- Leveraging ecotourism as a sustainable economic model that simultaneously supports conservation efforts and enhances community livelihoods.

By adopting a multifaceted policy approach that integrates traditional knowledge with modern scientific methods, conservation efforts can be significantly strengthened. A well-structured roadmap not only ensures the long-term viability of MAP species but also fosters social and economic resilience among local communities. Through a collective commitment to sustainability, it is possible to preserve both biodiversity and cultural heritage while securing the benefits of MAPs for future generations.

References

Adachukwu, I. P. & Yusuf, O. N. (2014). A review of the ethnotherapeutics of medicinal plants used in traditional/alternative medicinal practice in Eastern Nigeria. *International Journal of Current Microbiology and Applied Sciences*, ISSN: 2319-7706 3, 675–683.

Ahmad, M., Sultana, S., Fazl-I-Hadi, S., Ben Hadda, T., Rashid, S., Zafar, M., Khan, M. A., Khan, M. P., & Yaseen, G. (2014). An ethnobotanical study of medicinal plants in high mountainous region of Chail valley (District Swat- Pakistan). *Journal of Ethnobiology and Ethnomedicine*, 10, 36. https://doi.org/10.1186/1746-4269-10-36.

Aswani, S., Lemahieu, A., & Sauer, W. H. H. (2018). Global trends of local ecological knowledge and future implications. *PLoS ONE*, 4, e0195440. https://doi.org/10.1371/journal.pone.0195440.

Bano, A., Ahmad, M., Ben Hadda, T., Saboor, A., Sultana, S., Zafar, M., Khan, M. P., Arshad, M., & Ashraf, M. A. (2014). Quantitative ethnomedicinal study of plants used in the skardu valley at high altitude of Karakoram-Himalayan range; Pakistan. *Journal of Ethnobiology and Ethnomedicine, 10*, 43. https://doi. org/10.1186/1746-4269-10-43.

Barata, A. M., Rocha, F., Lopes, V., & Carvalho, A. M. (2016). Conservation and sustainable uses of and Products. *88*, 8–11. https://doi.org/10.1016/j.indcrop.2016.02.035.

Bastos, J. G., Kury, L., Hanazaki, N., Capozzi, R., & Fonseca-Kruel, V. (2022). A Biodiversity Hotspot Losing Its Biocultural Heritage: The Challenge to Biocultural Conservation of Brazilwood (Paubrasilia echinata). *Frontiers in Forests and Global Change, 5*, 696757. https://doi.org/10.3389/ffgc.2022.696757.

Bridgewater, P. & Rotherham, I. D. (2019). A critical perspective on the concept of biocultural diversity and its emerging role in nature and heritage conservation. *People and Nature, 1*(3), 291–304. https://doi. org/10.1002/pan3.10040.

Brondízio, E. S., Aumeeruddy-Thomas, Y., Bates, P., Carino, J., Fernández-Llamazares, Á., Ferrari, M. F., Galvin, K., Reyes-García, V., McElwee, P., Molnár, Z., & Samakov, A. (2021). Locally based; regionally manifested; and globally relevant: Indigenous and local knowledge; values; and practices for nature. *Annual Review of Environment and Resources, 46*(1), 481–509. https://doi.org/10.1146/annurev-environ -012220-012127.

Brondízio, E. S., Settele, J., Diaz, S., & Ngo, H. T. (2019). *Global Assessment Report on Biodiversity and Ecosystem Services of the Intergovernmental Science-Policy Platform on Biodiversity and Ecosystem Services.* https://doi.org/10.5281/zenodo.3831673.

Chandra, P. & Sharma, V. (2018). Strategic Marketing Prospects for Developing Sustainable Medicinal and Aromatic Plants Businesses in the Indian Himalayan Region. *Small-scale Forestry, 17*, 423–441. https://doi.org/10.1007/s11842-018-9396-3.

Chukwuma, E. C., Soladoye, M. O., & Feyisola, R. T. (2015). Traditional medicine and the future of medicinal Plants in Nigeria. *Journal of Medicinal Plants Studies, 3*(4), 23–29.

Cicinelli, E., Salerno, G., & Caneva, G. (2018). An assessment methodology to combine the preservation of biodiversity and cultural heritage: The San Vincenzo al Volturno historical site (Molise; Italy). *Biodiversity and Conservation, 27*, 1073–1093. https://doi.org/10.1007/s10531-017-1480-z.

Dajic-Stevanovic, Z. & Pljevljakusic, D. (2015). Challenges and decision-making in cultivation of medicinal and aromatic plants. *Medicinal and Aromatic Plants of the World: Scientific; Production; Commercial and Utilization Aspects*, 145–164. https://doi.org/10.1007/978-94-017-9810-5_8.

Dawson, N. M., Coolsaet, B., Sterling, E. J., Loveridge, R., Gross-Camp, N. D., Wongbusarakum, S., Sangha, K. K., Scherl, L. M., Phuong Phan, H., Zafra-Calvo, N., Lavey, W. G., Byakagaba, P., Idrobo, C. J., Chenet, A., Bennett, N. J., Mansourian, S., & Rosado, F. J. (2021). The role of Indigenous peoples and local communities in effective and equitable conservation. *Ecology and Society, 26*(3), 19. https://doi. org/10.5751/ES-12625-260319.

Derso, Y. D., Kassaye, M., Fassil, A. et al. (2024). Composition; medicinal values; and threats of plants used in indigenous medicine in Jawi District; Ethiopia: Implications for conservation and sustainable use. *Scientific Reports, 14*, 23638. https://doi.org/10.1038/s41598-024-71411-5.

Díaz, S., Demissew, S., Carabias, J., Joly, C., Lonsdale, M., Ash, N., Larigauderie, A., Adhikari, J. R., Arico, S., Báldi, A., & Bartuska, A. (2015). The IPBES Conceptual Framework – Connecting nature and people. *Current Opinion in Environmental Sustainability, 14*, 1–16. Available at https://www.ipbes.net/document-library-catalogue/ipbes-conceptual-framework-connecting-nature-and-people.

El Mekkaoui, A., Khamar, M., Benlakhdar, S., Ngadi, M., Slimani, C., Louafi, B., Nounah, A., Cherkaoui, E., Balafrej, T., & Rais, C. (2024). Traditional knowledge and biodiversity of medicinal plants in the Taounate region for treating human diseases: An ethnobotanical perspective. *Ethnobotany Research and Applications, 29*, 1–22. https://doi.org/10.32859/era.29.30.1-22.

Gautam, S. K. (2024). The role of indigenous knowledge in biodiversity conservation: Integrating traditional practices with modern environmental approaches. *Environmental Reports*. https://doi.org/10.51470/ER.2019.1.2.01.

Igwillo, U. C., Ola-Adedoyin, A. T., Abdullahi, M. M., & Chukwuemeka, A. E. (2019). A review of opportunities and challenges in conservation and use of medicinal and aromatic plants in Nigeria. *International Journal of Advanced Research*, 7(4), 770–778. https://dx.doi.org/10.21474/IJAR01/8885.

Kala, C. P. (2015). Medicinal and aromatic plants: Boon for enterprise development. *Journal of Applied Research on Medicinal and Aromatic Plants*, 2(4), 134–139. https://doi.org/10.1016/j.jarmap.2015.05.002.

Kankara, S. S., Ibrahim, M. H., Mustafa, M., & Go, R. (2015). Ethnobotanical survey of medicinal plants used for traditional maternal healthcare in Katsina state, Nigeria. *South African Journal of Botany*, 97, 165–175. https://doi.org/10.1016/j.sajb.2015.01.007.

Karki, M. B.; (2017). Challenges; opportunities and trade-offs in commercialization of medicinal and aromatic plants in South Asia Region. In Invited paper presented at the workshop on current challenges and recommendations. Available at: https://www.academia.edu/12863952/Challenges_opportunities_and_tradeoffs_in_commercialization_of_medicinal_and_aromatic_plants_in_South_Asia_Region.

Kleinheksel, A. J., Rockich-Winston, N., Tawfik, H., & Wyatt, T. R. (2020). Demystifying content analysis. *American Journal of Pharmaceutical Education*, 84(1), 7113. https://doi.org/10.5688/ajpe7113.

Krippendorff, K. (2018). *Content Analysis: An Introduction to Its Methodology* 5th ed. California; CA: Sage Publications.

Labokas, J. & Karpavičien E, B. (2023). On the Prospects of In Situ Conservation of Medicinal- and Aromatic-Plant Genetic Resources at Ancient-Hillfort Sites: A Case Study from Lithuania. *Plants*, 12, 861. https://doi.org/10.3390/plants12040861.

Lakshman, C. D. (2016). Bio-diversity and conservation of medicinal and aromatic plants. *Advances in Plants and Agricultural Research*, 5(4), 561–566. doi: 10.15406/apar.2016.05.00186.

Levis, C., Flores, B. M., Campos-Silva, J. V., Peroni, N., Staal, A., Padgurschi, M. C., Dorshow, W., Moraes, B., Schmidt, M., Kuikuro, T. W., & Kuikuro, H. (2024). Contributions of human cultures to biodiversity and ecosystem conservation. *Nature Ecology & Evolution*, 8(5), 866–879. https://doi.org/10.1038/s41559-024-02356-1.

Lindholm, K.-J. & Ekblom, A. (2019). A framework for exploring and managing biocultural heritage. *Anthropocene*, 2019, 100195. https://doi.org/10.1016/j.ancene.2019.100195.

Mbelebele, Z., Mdoda, L., Ntlanga, S. S., Nontu, Y., & Gidi, L. S. (2024). Harmonizing Traditional Knowledge with Environmental Preservation: Sustainable Strategies for the Conservation of Indigenous Medicinal Plants (IMPs) and Their Implications for EconomicWell-Being. *Sustainability*, 16, 5841. https://doi.org/10.3390/su16145841.

Mekonnen, H., Bires, Z., & Berhanu, K. (2022). Practices and challenges of cultural heritage conservation in historical and religious heritage sites: Evidence from North Shoa Zone; Amhara Region; Ethiopia. *Heritage Science*, 10(1), 172. https://doi.org/10.1186/s40494-022-00802-6.

Messerli, P., Murniningtyas, E., Eloundou-Enyegue, P., Foli, E. G., Furman, E., Glassman, A., Hernández Licona, G., Kim, E. M., Lutz, W., Moatti, J. P., & Richardson, K.; (2019). Global sustainable development report 2019: the future is now–science for achieving sustainable development. Available at https://sustainabledevelopment.un.org/content/documents/24797GSDR_report_2019.pdf.

Mittal, M. K., Suthar, M. K., & Das, M. (2023). Conservation of medicinal and aromatic plants in India: Current status and future prospects. *Indian Horticulture*, 68(5), 25–28. Available at https://epubs.icar.org.in/index.php/IndHort/article/view/144771.

Padulosi, S., Leaman, D., & Quek, P. (2014). Challenges and Opportunities in Enhancing the Conservation and Use of Medicinal and Aromatic Plants. *Journal of Herbs, Spices & Medicinal Plants*, 9(4), 243–267. https://doi.org/10.1300/J044v09n04_01.

Palabaş Uzun, S. & Koca, C. (2020). Ethnobotanical survey of medicinal plants traded in herbal markets of Kahramanmaraş. *Plant Diversity*, Dec 29, *42*(6), 443–454. https://doi.org/10.1016/j.pld.2020.12.003.

Petelka, J., Bonari, G., Soumel, I., Plagg, B., & Zerbe, S. (2022). Conservation with local people: Medicinal plants as cultural keystone species in the Southern Alps. *Ecology and Society*, *27*(4), 14. https://doi.org/10.5751/ES-13510-270414.

Phondani, P. C., Bhatt, I. D., Negi, V. S., Kothyari, B. P., Bhatt, A., & Maikhuri, R. K. (2016). Promoting medicinal plants cultivation as a tool for biodiversity conservation and livelihood enhancement in Indian Himalaya. *Journal of Asia-Pacific Biodiversity*, *9*(1), 39–46. https://doi.org/10.1016/j.japb.2015.12.001.

Póvoa, O., Lopes, V., Barata, A. M., & Farinha, N. (2023). Monitoring Genetic Erosion of Aromatic and Medicinal Plant Species in Alentejo (South Portugal). *Plants*, *12*, 2588. https://doi.org/10.3390/plants12142588.

Price, B., Kienast, F., Seidl, I., Ginzler, C., Verburg, P. H., & Bolliger, J. (2015). Future landscapes of Switzerland: Risk areas for urbanisation and land abandonment. *Applied Geography*, *57*, 32–41. https://doi.org/10.1016/j.apgeog.2014.12.009.

Rahman, M. M. & Fakir, M. S. A. (2015). Biodiversity of medicinal plants in Bangladesh: Prospects and problems of conservation and utilization. *International Journal of Minor Fruits; Medicinal and Aromatic Plants*, *1*(1).

Rathoure Ak. (2024). Cultural practices to protecting biodiversity through cultural heritage: preserving nature, preserving culture. Biodiversity Int J. 7(2): 71–75. https://doi.org/10.15406/bij.2024.07.00213.

Rotherham, I. D. (2015). Bio-cultural heritage and biodiversity: Emerging paradigms in conservation and planning. *Biodiversity and Conservation*, *24*, 3405–3429. https://doi.org/10.1007/s10531-015-1006-5.

Said, M. A., Ibrahim, M. M., Beyzi, E., & Ilbas, A. I. (2024). Biodiversity and Utilization Patterns of Medicinal and Aromatic Plants in Africa. *Journal of Erciyes Agriculture and Animal Science*, *7*(2), 39–46. https://doi.org/10.55257/ethabd.1450876.

Sher, H., Aldosari, A., Ali, A., & De Boer, H. J. (2015). Indigenous knowledge of folk medicines among tribal minorities in Khyber Pakhtunkhwa; northwestern Pakistan. *Journal of Ethnopharmacology*, *166*, 157–167. https://doi.org/10.1016/j.jep.2015.03.022.

Sher, H., Ali, A., Ullah, Z., & Sher, H. (2022). Alleviation of Poverty through Sustainable Management and Market Promotion of Medicinal and Aromatic Plants in Swat; Pakistan: Alleviation of Poverty through Sustainable Management . *Ethnobotany Research and Applications*, *23*, 1–19. https://doi.org/10.32859/era.23.16.1-19.

Sher, H. & Barkworth, M. E. (2015). Economic development through medicinal and aromatic plants (MAPs) cultivation in Hindu Kush Himalaya mountains of District Swat; Pakistan. *Journal of Mountain Science*, *12*, 1292–1301. https://doi.org/10.1007/s11629-014-3247-2.

Sinthumule, N. I. (2023). Traditional ecological knowledge and its role in biodiversity conservation: A systematic review. *Frontiers in Environmental Science*, *11*, 1164900. https://doi.org/10.3389/fenvs.2023.1164900.

Taghouti, I., Cristobal, R., Brenko, A., Stara, K., Markos, N., Chapelet, B., Hamrouni, L., Burši´c, D., & Bonet, J.-A. (2022). The Market Evolution of Medicinal and Aromatic Plants: A Global Supply Chain Analysis and an Application of the Delphi Method in the Mediterranean Area. *Forests*, *13*, 808. https://doi.org/10.3390/f13050808.

Valkó, O., Tóth, K., Kelemen, A., Miglécz, T., Radócz, S., Sonkoly, J., Tóthmérész, B., Török, P., & Deák, B. (2018). Cultural heritage and biodiversity conservation – Plant introduction and practical restoration on ancient burial mounds. *Nature Conservation*, *24*, 65–80. https://doi.org/10.3897/natureconservation.24.20019.

Vallejo, J. R., Baptista, G. C. S., Arco, H., Gonzalez, J. A., Santos-Fita, D., & Postigo-Mota, S. (2024). Traditional Knowledge and Biocultural Heritage about Medicinal Plants in a European Transboundary

Area (La Raya: Extremadura; Spain – Alentejo; Portugal): Transdisciplinary Research for Curriculum Design in Health Sciences. *Heritage*, *7*, 225–258. https://doi.org/10.3390/heritage7010012.

Negi, V. S., Pathak, R., Sekar, K. C., Rawal, R. S., Bhatt, I. D., Nandi, S. K., & Dhyani, P. P. (2017). Traditional knowledge and biodiversity conservation: A case study from Byans Valley in Kailash Sacred Landscape; India. *Journal of Environmental Planning and Management*. https://doi.org/10.1080/09640568.2017.1371006.

Zerbe, S. (2022). *Restoration of Multifunctional Cultural Landscapes: Merging Tradition and Innovation for a Sustainable Future* (Vol. 30). Springer Nature. https://doi.org/10.1007/978-3-030-95572-4.

Zheng, X., Wang, R., Hoekstra, A. Y., Krol, M. S., Zhang, Y., Guo, K., Sanwal, M., Sun, Z., Zhu, J., Zhang, J., & Lounsbury, A. (2021). Consideration of culture is vital if we are to achieve the Sustainable Development Goals. *One Earth*, *4*(2), 307–319. https://doi.org/10.1016/j.oneear.2021.01.012.

Nyong Princely Awazi*

Green Architecture and Engineering in the Congo Basin: Challenges and Prospects

Abstract: The need for green architecture and engineering has never been more dire than in the present dispensation of global warming and climate change. The Congo Basin subregion is still lagging behind as far as the green architecture and engineering trend is concerned, attributable to a plethora of challenges. This chapter dissects the situation of green architecture and engineering in the Congo Basin while uncovering some challenges and prospects. Findings indicate that green architecture and engineering in the Congo Basin is still at its infancy with limited use of renewable energy, energy efficiency, water conservation techniques, as well as limited implementation of forest conservation strategies. The main challenges to attaining green architecture and engineering goals in the Congo Basin are governance, weak institutional and policy frameworks, high costs involved, green grabbing, no standard certification systems, poor urban and rural planning, lack of data on energy and water use efficiency in buildings, and lack of awareness and knowledge of green architecture and engineering. The prospects of green architecture and engineering in the Congo Basin, however, look bright as green economy/green growth is being championed in the subregion: The national determined contributions (NDCs) of Congo Basin countries factor in green architecture and engineering; the long-term national development strategies of Congo Basin countries are favorable to the emergence of green architecture and engineering; and Congo Basin countries are working hard to adhere to the UN Sustainable Development Goals that factor in green architecture and engineering, notably goal number 7 on clean and affordable energy; goal number 9 on industry, innovation, and infrastructure; goal number 11 on sustainable cities and communities; and goal number 13 on climate action.

Keywords: Green architecture, green engineering, green buildings, sustainable cities and communities, clean and affordable energy, Congo Basin

*Corresponding author: Nyong Princely Awazi, Department of Forestry and Wildlife Technology, College of Technology (COLTECH), University of Bamenda, Bamenda, Cameroon, e-mail: awazinyong@uniba.cm

https://doi.org/10.1515/9783111563046-004

1 Introduction and General Overview of Green Architecture and Green Engineering

Sustainability has progressively emerged as a crucial focus for countless businesses and corporations since it was initially highlighted in the United Nations' 1987 report (Brundtland, 1987). According to a 2013 survey involving 5,300 corporate executives and managers from 118 countries, nearly two-thirds of respondents identified social and environmental concerns as pivotal factors for maintaining competitiveness in the market (Kiron et al., 2013). Corporations recognize sustainability as a determinant of company success, incorporating environmental, economic, and social dimensions into their corporate strategies to pursue long-term prosperity. In the architecture, engineering, and construction (AEC) industry, sustainability initiatives have significantly reshaped industry norms, influencing project design and operational processes. This transformation has led to a proliferation of green certifications for major projects, such as the Leadership in Energy and Environmental Design (LEED) assessment in America, the Building Research Establishment's Environmental Assessment Method (BREEAM) in Britain, GreenMark in Singapore, Envision™ for sustainable infrastructure assessment, and the Greenroads Rating System for transportation projects. However, the adoption of green practices by AEC organizations often hinges on the demands of external decision-makers rather than internal motivations. This external pressure creates a bias wherein organizations prioritize the delivery of green projects over the sustainability of their own operations.

As AEC organizations continue to undertake numerous green projects at the behest of owners or investors, it is imperative for them to reassess what constitutes "true" sustainable development and to reevaluate their intrinsic motivations and performance. Academic research echoes this sentiment, indicating a disparity between the extensive knowledge available for green construction projects and the limited understanding of sustainable responsibility within AEC organizations. Ragheb et al. (2016), for example, discuss the growing interest in sustainability across various disciplines, particularly in the context of achieving sustainable development. They define green architecture as the practice of designing and constructing buildings in harmony with environmental principles to minimize resource consumption and environmental harm. The chapter emphasizes the importance of implementing "green building systems" to address these challenges comprehensively. Clark II and Cooke (2014) advocate for a shift away from fossil fuels and nuclear power toward renewable energy sources and smart grid technologies to combat environmental degradation. They urge nations, including the United States, to embrace the Green Industrial Revolution and transition to energy-independent, carbon-neutral communities. Abioso (2019) explores the concept of the "invisible" in architecture, referring to aspects often overlooked in green design. The study highlights the importance of considering human behavior, spatial order, and architectural principles to reduce inefficiency and waste in building design, ultimately enhancing the green value of architecture.

Other studies on green architecture and green engineering have trumpeted the need for more sustainability in the construction of buildings. Bai and Qian (2021) focus on the importance of understanding the reliability of human cognitive behavior in green sustainable construction systems to ensure quality and safety. They analyze factors affecting performance reliability and propose a novel hybrid computational intelligence method to assess factor validity and reliability, aiming to improve the overall reliability of the system. Masood et al. (2017) discuss the evolution of sustainable development efforts since the 1992 Earth Summit, emphasizing the importance of incorporating sustainability principles into building and town designs. They highlight the emergence of green architecture as a response to environmental challenges, advocating for energy-efficient designs and the utilization of renewable energy sources to create environmentally friendly buildings. Elshafei et al. (2021) examine the economic and environmental significance of green structures and propose a state-of-the-art review and practices for optimal green solutions. They employ genetic algorithms to identify the most suitable green arrangement models and procedures, emphasizing the importance of modern technologies in achieving sustainable urban environment Mohammed (2021) addresses the challenges in sustainable design strategies and green building rating systems. They aim to enhance the integration of sustainability aspects into rating systems to better guide designers in generating sustainable design concepts. Their research assesses the Green Pyramid Rating System and formulates a sustainable design strategy to promote sustainable green thought in architectural design. Bielek and Bielek (2012) introduce a concept for the development of technology in architecture for a sustainable society. They categorize modern architecture into three technical levels: low energy architecture, green architecture, and sustainable architecture. Their design strategy emphasizes the role of renewable energy sources in achieving environmentally friendly building designs.

Some studies highlight the multifaceted approaches and considerations in green architecture and sustainable building practices. Yuan et al. (2017) emphasize the importance of bionic building energy efficiency and bionic green architecture in achieving harmony with the natural environment. They discuss various applications of bionic technologies in building design, highlighting innovations, such as passive solar construction and the use of natural materials for self-regulation and maintenance. Li et al. (2012) investigate the external actors crucial for the success of green building projects, identifying clients, government bodies, qualified material suppliers, and knowledgeable consultants as significant partners for the AEC firms. Their findings provide insights for AEC firms seeking to gain competitive advantages in the green building market. Saleh and Saied (2017) delve into the emergence of green architecture as a response to environmental concerns and the neglect of ecological aspects in building designs, particularly in developing countries like Egypt. They focus on the environmental treatments and architectural principles of historical Cairo as a model for modern green architecture, highlighting the potential of traditional designs to enhance thermal comfort in hot arid climates. Chen et al. (2012) discuss the urgent need for energy-saving measures in residential buildings to mitigate CO_2 emissions. They analyze the impact of government policies and criteria on green architecture development, proposing an assessment framework to prioritize criteria for implementing green architecture in residential building projects. Their research underscores the importance of considering various factors in promoting sustainable building practices. These studies underline the importance of integrating bionic technologies, external partnerships, historical architectural practices, and government policies to advance green architecture and sustainable building practices, emphasizing the need for holistic approaches to address environmental challenges and achieve long-term sustainability goals.

In the same light, some studies offer insights into various aspects of green architecture, sustainability, and the challenges and opportunities in implementing eco-friendly practices. Merenkov et al. (2019) highlight the increasing global interest in "green" architecture and the need for architects and engineers to effectively incorporate sustainable development principles into their designs. Drawing from overseas experiences, they articulate basic principles for the development of green architectural objects. Bielek (2016) discusses the economic and environmental implications of human activities, emphasizing the importance of transitioning to low-energy, low-emission technologies and renewable energy sources. They underscore the significance of green architecture in conserving material, energy, and water resources, promoting sustainable development. Lu and Zhang (2016) propose a regenerative sustainability framework for AEC organizations, emphasizing the need for a holistic approach to drive sustainability initiatives. Their framework integrates social, environmental, and economic dimensions, highlighting the importance of addressing social stakeholders and governance in addition to environmental concerns. Mahdavinejad et al. (2014) analyze architectural projects in the Middle East for their sustainability and green criteria compatibility, noting discrepancies between sustainable design intentions and actual outcomes. They evaluate the effectiveness of rating systems like LEED and advocate for design-oriented patterns to move toward truly green architecture. Well and Ludwig (2020) examine blue-green infrastructure projects and their impact on urban environments, emphasizing the synergies between water and vegetation elements. They identify the need for better integration of blue and green aspects in planning processes to optimize resource use and enhance urban resilience. Stauskis (2013) provides a comprehensive analysis of green building theory and practices, tracing its historic roots and evaluating complex quality assessment methodologies. They highlight the impact of green architecture on architectural practice through case studies in urban design and landscape architecture. These studies underscore the importance of integrating sustainability principles into architectural practices, addressing environmental, social, and economic dimensions to achieve truly green and resilient built environments.

Chemicals of concern (CoC) in the building and construction sector have long been a significant issue (Molla et al., 2021). Notably, childhood lead poisoning and chronic lung disease resulting from asbestos inhalation are well-documented consequences of these chemicals within the sector. Building and construction represent one of the most chemical-intensive sectors downstream of the chemical industry, and it stands as the largest end market for chemicals, generating the highest chemical revenue across all sectors. With urbanization rapidly on the rise, the global construction sector is projected to grow by 3.5% annually, with its chemicals market expanding by 6.2% annually between 2018 and 2023 (Wei et al., 2024). Unlike other consumer products, such as textiles, electronics, and toys, building and construction products are exclusively used within the built environment context, directly impacting the entire life cycle of buildings – from manufacturing and construction to usage and eventual demolition, recycling, or disposal (Huang et al., 2022). Moreover, buildings typically have longer life cycles, spanning decades or even centuries, compared to products in other sectors. This prolonged life cycle results in a significant time gap between the design and manufacturing phases and the end-of-life stages, during which understanding of chemicals and their associated risks may evolve, alongside their health and environmental impacts. Upon reaching the end of their life cycle, building and construction products contribute to the waste stream as construction and demolition waste, often comprising the largest portion of total waste generated in a country and posing significant risks if not managed properly (Ogungbile et al., 2021; Ahunbaev et al., 2024). It is therefore imperative to promote green architecture and green buildings to limit the use of CoC in the building industry. In the Congo Basin, little has been done to promote green infrastructure and green buildings. This chapter therefore examines what has been done already in the domain of green infrastructure and green buildings, and proposes future pathways for future empirical research in the field of green infrastructure and green buildings.

2 State of the Art of Green Architecture and Engineering in the Congo Basin

The Congo Basin has one of the largest green energy potentials in the world, which makes it a suitable destination for green architecture and green engineering. Unfortunately, there has been limited progress in the development of green architecture and engineering in the region. The current state of green architecture and engineering in the Congo Basin reveals the following: limited use of renewable energy, limited energy efficiency, limited water conservation techniques, limited implementation of forest conservation strategies (due particularly to wanton urbanization), limited durability, and construction of buildings in wetlands and other ecologically fragile zones.

2.1 Limited Use of Renewable Energy

Renewable energy is primordial for the development of green infrastructure and green engineering. The current situation of green infrastructure and green engineering in the Congo Basin indicates the limited use of renewable energy. Different studies expound on this including Kusakana (2016) and Kenfack et al. (2017) who shed light on the energy landscape in the Democratic Republic of Congo (DRC) and Central Africa, emphasizing the urgent need for harnessing renewable energy sources. Kusakana underscores the country's heavy reliance on wood fuel and limited access to electricity, despite possessing vast renewable energy potential. Initiatives, such as decentralized micro hydropower stations and rehabilitation of existing hydropower plants are highlighted as steps toward addressing energy challenges. Similarly, Kenfack et al. stress the underutilization of renewable energy resources in Central Africa due to insufficient government support. They advocate for promoting sustainable energy infrastructure and policies to unlock the region's renewable energy potential, focusing on solar, biomass, and hydropower. Aquilas and Atemnkeng (2022) and Deshmukh et al. (2018) offer insights into the global efforts to mitigate greenhouse gas emissions and transition to renewable energy sources. Aquilas and Atemnkeng's study analyzes the impact of climate-related development financing and renewable energy consumption on greenhouse gas emissions in the Congo Basin, emphasizing the importance of sustained financial support and effective utilization of climate funds for renewable energy projects. Meanwhile, Deshmukh and colleagues assess the feasibility of renewable energy alternatives to large-scale hydropower projects in the DRC, highlighting the cost-effectiveness of wind and solar photovoltaics mixed with natural gas to meet future energy demand. They underscore the abundant renewable energy potential in the DRC and advocate for investment in renewable energy infrastructure as a viable alternative to conventional hydropower projects. These studies emphasize the importance of prioritizing renewable energy development in Africa to address energy access challenges, mitigate climate change, and foster sustainable economic development.

Meanwhile, Kenfack et al. (2011, 2014) highlight the untapped renewable energy potential in Central Africa, particularly in solar, hydro, and biomass resources. They emphasize the region's lack of development in utilizing these resources due to insufficient government commitment and dedication. Both studies advocate for promoting decentralized and renewable energy initiatives to address energy challenges, reduce reliance on thermal plants, and mitigate climate change. They propose actions to optimize the utilization of solar and hydro resources, improve energy infrastructure, and foster renewable energy development, drawing insights from case studies and lessons learned from ongoing initiatives, particularly in Cameroon. Petrovic and Zoucha (2023) provide historical context to the DRC, highlighting its tumultuous past marked by colonization, exploitation, and conflict. Despite its rich energy potential, the DRC continues to face significant challenges in electricity access and governance due to historical and contemporary factors. Dalder et al. (2024) offer a forward-looking perspective on renewable energy policy in the DRC, utilizing advanced modeling techniques to evaluate policy pathways for boosting growth and investment in renewable energy technologies (RETs). Their study recommends market-based instruments, such as RET subsidies and fossil fuel taxes, to incentivize renewable energy deployment and reduce reliance on hydropower. They advocate for reevaluating the energy mix and promoting the DRC as a viable market for solar home systems to address electricity shortages. Akele-Sita and Chowdhury (2015) address the electricity crisis in the DRC by proposing renewable energy solutions, particularly focusing on solar, wind, and biomass resources integrated with energy storage systems. They advocate for decentralized power generation and efficient transmission systems to ensure reliable, cost-effective, and sustainable electricity supply across the country, thereby mitigating blackouts and enhancing energy access for all communities.

2.2 Limited Energy Efficiency

Besides the limited use of renewable energy, there is also limited energy efficiency in the Congo Basin. Different studies have highlighted limited energy efficiency in the Congo Basin, including Kenfack et al. (2011, 2017) both underscore the substantial renewable energy potential in Central Africa, which remains largely untapped due to inadequate government commitment and support. They advocate for optimizing the utilization of solar, biomass, and hydropower resources through improved access to sustainable and affordable energy services, infrastructure enhancements, and promotion of renewable energy initiatives. By identifying obstacles and proposing clean energy policies, these studies aim to elevate the status of renewable energy and combat climate change, with a specific focus on the case study of Cameroon. Mounmemi et al. (2022) investigate strategies to increase charcoal production while reducing carbon emissions, particularly in Cameroon. Through stochastic frontier analysis, they identify opportunities for producers to enhance efficiency and decrease emissions in different socio-ecological zones, suggesting compensation mechanisms to encourage adoption of low-carbon charcoal production methods. Modeste et al. (2015) provide a broader perspective on global energy trends and policies, emphasizing the growing energy needs worldwide and the challenges faced by developing countries, including limited access to electricity and reliance on traditional biomass. Their study highlights the importance of energy efficiency and the under-exploitation of renewable energy resources in Africa, particularly in Cameroon, where residential energy consumption is disproportionately high compared to the global average.

2.3 Limited Water Conservation Techniques

Water conservation is an important element for the development of green infrastructure and green engineering. The current scenario in the Congo Basin shows that there are limited water conservation techniques with different studies attesting to this. The studies by Kayembe et al. (2018) and Chishugi et al. (2021) both address water quality and supply issues in the DRC but from different perspectives. Kayembe et al. focus on the contamination of urban rivers and shallow wells in Kinshasa, highlighting the severe pollution levels and associated health risks due to inadequate sanitation and wastewater management. On the contrary, Chishugi et al. investigate water quality deterioration in the Yangambi Biosphere Reserve, attributing it to land use/land cover changes within the reserve. Both studies underscore the urgent need for effective water management policies to safeguard water resources and public health. Sonwa et al. (2020) provide a broader regional perspective by examining the impact of climate change and human activities on water resources in the Congo Basin. They emphasize the importance of these resources for biodiversity and livelihoods, highlighting the need for sustainable development practices to mitigate adverse effects on water availability and quality. Nsah (2022) takes a literary approach to analyze the ecopolitical dimensions of water pollution and land use in the Congo Basin, drawing insights from plays to critique governance failures and their ecological consequences. By examining the portrayal of water pollution and urban sprawl in literary works, Nsah highlights the intersection of environmental degradation and political negligence, advocating for improved governance and ecological stewardship. Boisson et al. (2010) and Alsdorf et al. (2016) present empirical studies on household water treatment and hydrological dynamics in the Congo Basin, respectively. Boisson et al. evaluate the effectiveness of a novel filtration device in preventing diarrheal diseases, highlighting challenges in achieving consistent use and health impact. Alsdorf et al. provide insights into the hydrology of the Congo Basin, emphasizing its role in regional carbon dynamics and the need for further research to understand the complex interactions shaping water resources in the region. Finally, Nagabhatla et al. (2021) offer a comprehensive analysis of water stress-induced migration and conflict dynamics in the Congo Basin, emphasizing the interconnectedness of water management, migration, and conflict resolution. Their study underlines the importance of hydrodiplomacy and inclusive governance frameworks in addressing the complex sociopolitical challenges exacerbated by water stress and climate change in the region.

2.4 Limited Implementation of Forest Conservation Strategies (Wanton Urbanization)

Although the Congo Basin is one of the most endowed regions in terms of forests, there is currently limited implementation of forest conservation strategies due particularly to urbanization. Different studies attest to this. The studies by Tegegne et al. (2016), Shapiro et al. (2021), and Endamana et al. (2010) collectively shed light on the drivers of deforestation and forest degradation (DD) in the Congo Basin and propose strategies to address these challenges. Tegegne et al. emphasize the importance of institutional and policy factors in combating DD, highlighting the need for land use policies prioritizing degraded lands. In contrast, Shapiro et al. introduce a novel metric, forest condition (FC), to quantify forest degradation and advocate for its integration into conservation planning to protect biodiversity and ecosystem services. Endamana et al., however, reveal the limited progress in achieving conservation and development objectives in the Congo Basin due to weak institutions and external economic pressures. Similarly, Fobissie et al. (2014) and Pyhälä et al. (2016) explore the complexities of REDD + implementation and conservation efforts in the Congo Basin. Fobissie et al. compare the national engagement and progress of REDD+ initiatives in Cameroon and the DRC, highlighting the disparities in political ownership and technical advancement. Pyhälä et al. examine the effectiveness of protected areas and conservation partnerships, revealing the persistence of biodiversity loss and social tensions despite substantial investments in conservation projects. Additionally, Brown et al. (2011) and Clay (2016) investigate the perceptions and outcomes of REDD+ and landscape conservation approaches in Central African countries, emphasizing the opportunities for economic development and biodiversity conservation, as well as the challenges of complex governance dynamics and marginalization of local communities. Finally, Kankeu et al. (2020) and Atyi (2021) provide insights into the transfer of knowledge and monitoring of DD in the Congo Basin. Kankeu and colleagues identify weaknesses in knowledge transfer and utilization of monitoring, reporting, and verification systems, emphasizing the need for improved coordination and integration between global and local stakeholders. Atyi highlights the scale of forest degradation in Central Africa and underscores the role of land-use policies in mitigating deforestation, emphasizing the importance of protected areas and community engagement in conservation efforts.

2.5 Limited Durability

The Congo Basin, a region of immense ecological importance, faces significant challenges related to road infrastructure, conflict, and sustainable development. Green infrastructure and green engineering need more durability for proper functioning. The current scenario in the Congo Basin indicates limited durability, affirmed by different studies carried out in the region. Kleinschroth et al. (2019) highlight the rapid expansion of logging roads in the Congo Basin, emphasizing their detrimental impacts on forest ecosystems and wildlife vulnerable to hunting. They advocate for road decommissioning as a crucial step in mitigating the negative consequences of timber extraction. Similarly, Kleinschroth et al. (2015) examine the effectiveness of road closure strategies in logging concessions, revealing significant differences in vegetation recovery rates depending on environmental conditions, suggesting the need for tailored forest management plans. In contrast, Ali et al. (2015) address the role of transport infrastructure in conflict-prone regions, particularly focusing on the DRC. They emphasize the importance of reducing transport costs to stimulate economic growth and peace-building efforts, although they acknowledge that improvements in infrastructure may not always lead to anticipated benefits during intense conflict situations. Ulimwengu et al. (2009) further underscore the significance of road investments in promoting agricultural and rural development in the DRC, highlighting the need for targeted infrastructure interventions to address existing constraints. Megevand and Mosnier (2013) provide a comprehensive overview of the Congo Basin's ecological and economic significance, emphasizing the critical role of its forests in providing livelihoods for millions of people and mitigating climate change through carbon storage. They highlight the selective and extensive nature of logging activities in the region, underscoring the importance of sustainable forest management practices. Hammons et al. (2000) discuss the potential for regional integration of electrical energy resources in Africa, with a particular focus on the DRC's vast hydroelectric potential. They outline proposals for interconnecting power systems and exporting electricity to neighboring continents, presenting a vision for the future expansion of Africa's power infrastructure. These studies offer insights into the complex interplay between infrastructure development, conflict, and conservation efforts in the Congo Basin, highlighting the need for integrated approaches to promote sustainable development/ durability and environmental conservation in the region.

2.6 Construction of Buildings in Wetlands and Other Ecologically Fragile Zones

For green infrastructure and green engineering to take root, buildings need to be constructed in safe and ecologically stable environments. In the Congo Basin, there is increasing construction of buildings in wetlands and ecologically fragile zones. Different studies attest to this. The vulnerability of the Congo Basin's forests is a multifaceted issue addressed by Zhang et al. (2006) and Kleinschroth et al. (2019) from different perspectives. Zhang et al. (2006) utilize a geographic information system (GIS) platform to assess the future distribution of forests, considering variables, such as population growth, road density, logging concessions, and forest fragmentation. Their study predicts a significant reduction in forest cover over the next 50 years, with current contiguous forests fragmenting into three large blocks and numerous smaller patches. This integrated GIS assessment highlights the complex interplay between social and economic development and tropical deforestation, providing insights into future forest distribution. Kleinschroth et al. (2019) focus on the expansion of roads in the Congo Basin and their impact on forest ecosystems. They observe a substantial increase in road networks, particularly within logging concessions, which has led to heightened deforestation rates, posing significant threats to carbon storage and wildlife habitats. Their study emphasizes the importance of road decommissioning after logging to mitigate the adverse effects of timber extraction on forest ecosystems. Schouten et al. (2022) contribute to this discussion by examining the political and material dimensions of road building projects in eastern DR Congo, conceptualizing the region as an "infrastructural frontier." They highlight the transient nature of authority and control along these roads, which perpetuates a state of flux between spaces of control and resistance. The patchy quality of road infrastructure, exacerbated by the region's clayish soils, contributes to the collapse of traditional dichotomies between presence and absence of transport infrastructure, shaping the region's identity as an infrastructural frontier. These studies provide complementary perspectives on the challenges facing the Congo Basin, illustrating the complex interactions between infrastructure development, forest conservation, and political dynamics in the region. They emphasize the importance of integrated approaches to address these challenges and promote sustainable development in one of the world's most critical ecosystems.

3 Challenges to Green Architecture and Greening in the Congo Basin

In the Congo Basin, the pursuit of green architecture and greening initiatives faces formidable challenges, prominently shaped by governance issues, institutional weaknesses, and socioeconomic complexities. Studies underscore how poor governance hampers sustainable development efforts, with corruption and mismanagement hindering effective forest and resource management. Weak institutional frameworks exacerbate these challenges, undermining efforts to foster socioeconomic well-being and biodiversity conservation. Moreover, transitioning to green infrastructure encounters high costs and ethical dilemmas, including "green grabbing" practices and the absence of standardized certification systems, which undermine sustainability and transparency. Addressing these multifaceted hurdles necessitates robust governance reforms, strengthened policy frameworks, and heightened awareness of green architecture's potential in the region's sustainable future.

3.1 Poor Governance

Governance is vital to the development of green infrastructure and greening in the Congo Basin. Different studies attest to this. The DRC grapples with multifaceted challenges in natural resource governance, as elucidated by Burnley (2011), Trefon (2010), Majambu et al. (2019), Piabuo et al. (2021), and Majambu et al. (2021). Burnley (2011) highlights the persistence of large-scale armed violence in certain provinces despite their resource richness, attributing this phenomenon to additional tensions, such as those between different mining sectors and local socioeconomic factors. This underscores the complexity of conflict dynamics beyond environmental scarcity or resource wealth. Trefon (2010) underscores the importance of innovative forest management strategies in addressing socioeconomic well-being, biodiversity conservation, and climate change mitigation in the DRC. However, corruption and mismanagement hinder the effectiveness of recent forestry initiatives. Majambu et al. (2019) shed light on the role of traditional leaders in forest governance, revealing how vested interests and weak governance structures contribute to illegal hunting and forest exploitation. Piabuo et al. (2021) empirically explore the nexus between governance, illegal logging, and carbon emissions, highlighting the increasing trend of illegal logging and poor governance effectiveness in Congo Basin countries. They emphasize the need for multi-stakeholder engagement and capacity-building to combat corruption and enforce laws effectively. Finally, Majambu et al. (2021) analyze the concept of "good governance" in the DRC's forestry sector, noting the intrusive role of international organizations and the exploitation of governance failures by political and administrative actors for personal gain. These studies stress on the urgent need for comprehensive governance reforms, multilevel stakeholder engagement, and effective enforcement mechanisms to address the intricate challenges facing natural resource management in the DRC.

3.2 Weak Institutional and Policy Frameworks

The Congo Basin faces significant challenges due to weak institutional and policy frameworks, as highlighted by scholars, such as Burnley (2011), Trefon (2010), Majambu et al. (2019), and Piabuo et al. (2021). Burnley (2011) discusses how prolonged instability in the DRC has led to poor natural resource management. Weak institutional structures and corruption among political actors hinder effective governance, exacerbating conflicts over resources. The presence of stakeholders profiting from instability further undermines efforts to improve governance and resource management. Trefon (2010) emphasizes the challenges of forest governance in the DRC, where corruption and mismanagement impede efforts to promote socioeconomic well-being and biodiversity conservation. The political, institutional, and social environment in the country is not conducive to improved forest governance, posing obstacles to sustainable development. Majambu et al. (2019) highlight how weak governance in the DRC contributes to illegal logging and forest resource exploitation. State bureaucracies' inefficiency, corruption, and institutional weaknesses allow vested interests to exploit forest resources unlawfully, exacerbating poverty and conflicts. Traditional leaders' involvement often perpetuates unsustainable resource management practices. Piabuo et al. (2021) explore the link between governance, illegal logging, and carbon emissions in Congo Basin countries. They find increasing trends of illegal logging, poor governance effectiveness, and corruption, particularly in the DRC. Weak governance structures fail to curb illegal logging, exacerbating environmental degradation and carbon emissions. Overall, these studies highlight the urgent need for strengthening institutional and policy frameworks in the Congo Basin. Enhancing transparency, accountability, and the rule of law, as well as promoting community involvement and sustainable practices are essential for effective governance and natural resource management in the region.

3.3 High Transition Costs

The transition to green infrastructure often comes with a high cost, as highlighted by various scholars and experts in the field (Usman and Khamidi, 2012). Some of these costs include financial investment, technological innovation, policy and regulation, transitioning existing infrastructure, and social and economic impacts (Milburn, 2014; Church and Crawford, 2018; Sachs et al., 2022). Transitioning to green infrastructure requires significant financial investment in renewable energy sources, sustainable transportation systems, and eco-friendly buildings. These investments may include the installation of solar panels, wind turbines, electric vehicle charging stations, and the retrofitting of existing infrastructure for energy efficiency. Developing and implementing green infrastructure technologies involves research and development costs. Innovation in renewable energy, energy storage, smart grids, and sustainable construction materials requires substantial investment in technology and infrastructure. Transitioning to green infrastructure often necessitates changes in policy and regulation to incentivize sustainable practices and discourage environmentally harmful activities. Governments may need to implement carbon pricing mechanisms, renewable energy subsidies, and environmental regulations, which can incur administrative and enforcement costs. Retrofitting or replacing existing infrastructure to meet green standards can be costly and challenging. Upgrading outdated power plants, replacing fossil fuel-based transportation systems with electric alternatives, and improving the energy efficiency of buildings require significant financial resources and technical expertise. The transition to green infrastructure may also have social and economic impacts, such as job displacement in traditional industries and potential disparities in access to clean energy and transportation options. Addressing these impacts may require additional investment in workforce training, social safety nets, and equitable distribution of green infrastructure benefits. While the transition to green infrastructure offers long-term environmental and economic benefits, it requires substantial upfront investment and careful planning to overcome the high costs associated with technological, policy, and social transitions.

3.4 Green Grabbing

The discourse surrounding the emergence of a new form of colonialism in Africa in general and the Congo Basin in particular, termed "dirty green colonialism," highlights the complexities and ethical dilemmas inherent in the global transition to green technologies. While the world seeks cleaner energy alternatives, the pursuit of green resources, such as cobalt, lithium, and rare earth minerals has led to exploitative practices reminiscent of historical colonialism (Marijnen et al., 2019; Wong et al., 2022). In the DRC, rich in cobalt deposits, foreign actors, including China and corporations like Tesla, have been accused of perpetuating a cycle of exploitation. Despite promises of infrastructure development, mining operations have resulted in child labor, environmental degradation, and interethnic conflict, mirroring past colonial practices. Moreover, green grabbing extends beyond mineral resources to include land for renewable energy projects. European countries, seeking energy security in the wake of geopolitical tensions, are striking deals in Africa in general and the Congo Basin in particular for oil and gas extraction (Anseeuw, 2013; Nhamo and Chekwoti, 2014). However, these projects risk exacerbating local conflicts and driving migration as communities reliant on fishing are displaced. The emergence of dirty green colonialism underscores the urgent need for ethical and equitable approaches to the global transition to green technologies. Without addressing the systemic injustices inherent in resource extraction and energy production, the pursuit of sustainability risks perpetuating exploitation and inequality in the Congo Basin.

3.5 No Standard Certification Systems

The Congo Basin grapples with a significant challenge: the absence of standardized certification systems for its valuable natural resources. This dearth of certification mechanisms extends across various sectors, including timber, minerals, and agricultural products, exacerbating issues related to sustainability, transparency, and fairtrade practices (POuLSEN and CLARK, 2010; Mazalto, 2010; Franken et al., 2012; Diangha and Wiegleb, 2014; Elad, 2014; Karsenty, 2019; Ndoumbe Berock and Ongolo, 2019). In the timber industry, the lack of standardized certification systems means that consumers and stakeholders have limited assurance regarding the legality and sustainability of timber products sourced from the region. Without robust certification schemes, such as the Forest Stewardship Council or the Programme for the Endorsement of Forest Certification (PEFC), there is a heightened risk of illegal logging, deforestation, and habitat destruction. Similarly, in the mineral sector, the absence of standardized certification systems poses challenges in ensuring responsible sourcing practices. Minerals, such as cobalt, tantalum, and rare earth elements, abundant in the Congo Basin, are essential components in various industries, including electronics and renewable energy. However, without reliable certification mechanisms to verify ethical mining practices, there is a risk of exploitation, human rights abuses, and environmental degradation associated with mineral extraction. Moreover, the agricultural sector in the Congo Basin faces similar issues due to the lack of standardized certification systems for products, such as palm oil, cocoa, and coffee. Without recognized certifications like Fair Trade or Rainforest Alliance, it is challenging for consumers to support ethically produced goods and for producers to access premium markets that prioritize sustainability and social responsibility. Overall, the absence of standard certification systems in the Congo Basin undermines efforts to promote sustainable resource management, protect biodiversity, and support local communities' livelihoods. Addressing this gap requires collaboration among governments, industry stakeholders, and civil society to develop and implement robust certification schemes tailored to the region's unique environmental and social contexts. Such initiatives can help enhance transparency, accountability, and market access while fostering responsible resource exploitation and economic development in the Congo Basin.

3.6 Poor Urban and Rural Planning

The Congo Basin grapples with significant challenges stemming from poor urban and rural planning, which have profound implications for sustainable development, environmental conservation, and community well-being. In urban areas within the Congo Basin, poor planning manifests in several ways, including inadequate infrastructure, informal settlements, and insufficient provision of essential services, such as water, sanitation, and healthcare (Lateef et al., 2010; Binsangou et al., 2018; Tongele, 2021; Muhaya et al., 2022). Rapid urbanization has outpaced the capacity of local authorities to plan and manage urban growth effectively, leading to overcrowding, traffic congestion, and environmental degradation. Moreover, informal settlements often lack secure land tenure, leaving residents vulnerable to eviction and limiting their access to basic services and economic opportunities. Rural areas in the Congo Basin also suffer from inadequate planning, particularly concerning land use and resource management. Deforestation, unsustainable agriculture, and extractive activities threaten the region's valuable ecosystems and biodiversity. Poor land tenure systems and weak regulatory frameworks contribute to land grabbing, conflict over natural resources, and displacement of indigenous communities. Additionally, limited access to education, healthcare, and market opportunities hinders rural development and exacerbates poverty and inequality. The consequences of poor urban and rural planning in the Congo Basin extend beyond social and economic challenges to environmental degradation and climate change. Deforestation and habitat loss contribute to biodiversity loss and carbon emissions, exacerbating global warming and undermining the region's resilience to climate impacts, such as droughts and floods. Moreover, unplanned urbanization and rural development increase vulnerability to natural disasters and public health risks, particularly in the context of climate variability and extreme weather events. Addressing the issues of poor urban and rural planning in the Congo Basin requires a holistic approach that integrates environmental sustainability, social equity, and economic development. This includes strengthening governance mechanisms, enhancing land tenure security, and promoting participatory planning processes that engage local communities and stakeholders. Investing in infrastructure, education, and healthcare can improve living conditions in both urban and rural areas, while sustainable land use practices and conservation efforts are essential for preserving the region's unique ecosystems and biodiversity. Collaboration among governments, civil society, and the private sector is crucial to overcoming the complex challenges of urban and rural planning in the Congo Basin and building resilient, inclusive, and sustainable communities for the future.

3.7 Lack of Data on Energy and Water Use Efficiency in Buildings

The lack of comprehensive data on energy and water use efficiency in buildings presents a significant challenge in the Congo Basin, hindering efforts to address resource consumption, promote sustainability, and mitigate climate change impacts. In the absence of robust data, it is difficult for policymakers, urban planners, and stakeholders to accurately assess the current state of energy and water use in buildings, identify trends, and develop targeted interventions to improve efficiency (Modeste et al., 2015; De Angelis et al., 2021; Nematchoua and Reiter, 2021). This lack of information hampers decision-making processes and limits the effectiveness of policies and initiatives aimed at reducing resource consumption and promoting sustainable development. Several factors contribute to the dearth of data on energy and water use efficiency in buildings in the Congo Basin. These include inadequate monitoring and reporting mechanisms, limited technical capacity and expertise, and insufficient investment in data collection and analysis infrastructure. Additionally, informal construction practices, unregulated building codes, and a lack of awareness about the importance of energy and water efficiency further exacerbate the data gap. The consequences of this data deficiency are far-reaching. Without accurate information on energy and water consumption patterns, it is challenging to design and implement effective strategies to reduce resource use, lower greenhouse gas emissions, and enhance resilience to climate change. Moreover, the lack of data impedes efforts to quantify the potential benefits of energy and water efficiency measures, such as cost savings, improved comfort, and environmental co-benefits. Addressing the lack of data on energy and water use efficiency in buildings requires a multifaceted approach. This includes investing in data collection and monitoring systems, building technical capacity among relevant stakeholders, and raising awareness about the importance of sustainable building practices. Governments, international organizations, and the private sector can play a critical role in supporting data collection efforts through funding, technical assistance, and knowledge sharing initiatives. Furthermore, promoting data transparency and accessibility can facilitate collaboration and knowledge exchange among stakeholders, enabling more informed decision-making and policy development. By prioritizing data collection and analysis on energy and water use efficiency in buildings, the Congo Basin can better understand its resource consumption patterns, identify opportunities for improvement, and work toward building more sustainable and resilient communities for the future.

3.8 Lack of Awareness and Knowledge of Green Architecture and Engineering

The lack of awareness and knowledge of green architecture and engineering in the Congo Basin poses significant challenges to sustainable development and environmental conservation efforts in the region. One of the primary obstacles is the limited access to information and educational resources on green architecture and engineering practices (Nwajiuba et al., 2015; Trefon, 2017; Vitos et al., 2017; Nago and Krott, 2022). Many architects, engineers, and construction professionals in the Congo Basin may not be aware of the principles and techniques associated with sustainable building design and construction. This lack of awareness hampers the adoption of green building practices and contributes to the continued reliance on conventional construction methods that are often resource-intensive and environmentally damaging. Furthermore, there may be a lack of institutional support and incentives for integrating green building practices into the construction industry. Governments, regulatory bodies, and professional associations may not prioritize or promote sustainable building standards, certifications, or training programs. Without clear guidelines and incentives, there is little motivation for architects, engineers, and developers to invest in green building practices or seek out training opportunities. Another factor contributing to the lack of awareness is the limited availability of demonstration projects and case studies showcasing successful examples of green architecture and engineering in the region. Without tangible examples to reference, stakeholders may be hesitant to adopt new approaches or technologies. Additionally, misconceptions or skepticism about the cost-effectiveness and feasibility of green building practices may persist in the absence of evidence to the contrary.

Addressing the lack of awareness and knowledge of green architecture and engineering in the Congo Basin requires a multi-pronged approach notably through education and training, advocacy and outreach, demonstration projects, as well as knowledge sharing and collaboration. Increasing access to educational resources, training programs, and professional development opportunities on green building practices can help raise awareness and build capacity among architects, engineers, and construction professionals. Engaging with government agencies, industry associations, and community organizations to promote the benefits of green architecture and engineering and advocate for supportive policies and incentives can help create a conducive environment for sustainable building practices. Investing in demonstration proj-ects that showcase the feasibility and benefits of green building practices can help de-bunk myths and build confidence among stakeholders. These projects can serve as valuable learning experiences and inspire replication across the region. Facilitating knowledge sharing and collaboration among stakeholders, including architects, engi-neers, policymakers, academia, and the private sector, can foster innovation, best practices, and collective action toward Sustainable Development Goals (SDGs). By ad-dressing the lack of awareness and knowledge of green architecture and engineering, the Congo Basin can unlock opportunities for sustainable development, reduce envi-ronmental impacts, and enhance resilience to climate change.

4 Prospects for Green Architecture and Greening in the Congo Basin

There are enormous prospects for green architecture and greening in the Congo Basin. First, there is a burgeoning movement toward green economy and growth, spurred by the region's vast natural riches and the pressing need for sustainable prog-ress. This shift presents a dual horizon of promise and challenge. The basin's immense biodiversity, forests, and renewable energy reservoirs promise pathways to economic expansion, job generation, and climate mitigation through ventures, such as sustain-able agriculture and renewable energy initiatives. However, formidable barriers loom large: fragile institutional frameworks, inadequate infrastructure, and profound socioeconomic impacts. Aligning green infrastructure with national strategies and commitments like the NDCs and SDGs highlights the region's resolve to cultivate resil-ience, innovation, and inclusive advancement. Negotiating these intricacies is para-mount to unlocking the transformative potential of green architecture and engineer-ing in the Congo Basin.

4.1 Green Economy/Green Growth Is Being Championed in the Subregion

Green economy/green growth is being promoted in the Congo Basin (Peya, 2018; As-soua and Molua, 2018; Chukwu, 2020). The promotion of a green economy and foster-ing green growth in the Congo Basin presents both opportunities and challenges in the region's quest for sustainable development and environmental conservation (Table 1).

Promoting a green economy and green growth in the Congo Basin requires con-certed efforts from governments, businesses, civil society, and international partners.

Table 1: Opportunities and challenges to green economy/green growth in the Congo Basin.

Opportunities	Challenges
Rich natural resources: The Congo Basin is endowed with vast natural resources, including forests, minerals, water resources, and biodiversity. Leveraging these resources sustainably through green economy initiatives can contribute to economic growth, job creation, and poverty reduction.	**Institutional Capacity:** Weak institutional capacity and governance challenges in the Congo Basin countries may hinder the implementation of green economy policies and strategies. Building institutional capacity, strengthening governance frameworks, and enhancing regulatory enforcement are essential for effective green growth initiatives.
Renewable energy potential: The region has significant untapped renewable energy potential, including hydroelectric, solar, and biomass resources. Investing in renewable energy infrastructure can reduce reliance on fossil fuels, promote energy security, and mitigate greenhouse gas emissions.	**Socioeconomic considerations:** Transitioning to a green economy may have socioeconomic implications, including job displacement, income inequality, and social disruption. Ensuring inclusive and equitable development, as well as providing support for affected communities and workers, is crucial for the success of green growth initiatives.
Ecosystem services: The Congo Basin's forests provide valuable ecosystem services, such as carbon sequestration, water regulation, and biodiversity conservation. Incorporating the value of these services into economic decision-making can incentivize conservation efforts and support sustainable land management practices.	**Infrastructure deficits:** Inadequate infrastructure, particularly in rural areas, can limit the adoption of green technologies and the delivery of essential services. Investing in green infrastructure, such as renewable energy systems, sustainable transport networks, and water supply systems, is essential for supporting green growth objectives.
Green technologies: The transition to a green economy can drive innovation and technological advancement in sectors, such as agriculture, forestry, energy, and transportation. Adopting green technologies and practices can improve resource efficiency, enhance productivity, and reduce environmental impacts.	**Financing constraints:** Limited access to financing and investment capital may impede the implementation of green economy projects and initiatives. Mobilizing domestic and international funding, as well as promoting public-private partnerships and innovative financing mechanisms, is essential for scaling up green investments in the region.
International support: The international community is increasingly recognizing the importance of green growth and sustainable development. The Congo Basin countries can benefit from partnerships, funding, and technical assistance from international organizations, donor agencies, and development partners to support green economy initiatives.	**Climate vulnerability:** The Congo Basin countries are vulnerable to the impacts of climate change, including extreme weather events, deforestation, and loss of biodiversity. Building resilience to climate risks and integrating climate adaptation measures into green growth strategies is essential for long-term sustainability.

By addressing challenges and capitalizing on opportunities, the region can achieve SDGs while preserving its natural heritage for future generations.

4.2 The NDCs of Congo Basin Countries Factor in Green Architecture and Engineering

The NDCs of Congo Basin countries outline their commitments to mitigating climate change and transitioning to low-carbon, resilient economies. These NDCs typically include targets for reducing greenhouse gas emissions, enhancing carbon sinks, and adapting to the impacts of climate change. Green infrastructure plays a crucial role in supporting the implementation of NDCs by providing sustainable and climate-resilient solutions for key sectors, such as energy, transportation, agriculture, and urban development (Fobissie et al., 2019; Ozor et al., 2020; Aquilas and Atemnkeng, 2022). Congo Basin countries are integrating green infrastructure into their NDCs as shown in Table 2.

Overall, integrating green infrastructure into NDCs helps Congo Basin countries achieve their climate goals while promoting sustainable development, biodiversity conservation, and resilience to climate change impacts. By investing in green infrastructure projects, these countries can unlock multiple co-benefits, including improved public health, job creation, and enhanced ecosystem services, while contributing to global efforts to combat climate change.

4.3 The Long-Term National Development Strategies of Congo Basin Countries Are Favorable to the Emergence of Green Architecture and Engineering

The national development strategies of Congo Basin countries encompass a range of policies, plans, and initiatives aimed at promoting economic growth, social development, and environmental sustainability (Sonwa et al., 2012; Somorin et al., 2012; Fobissie et al., 2014). While each country's strategy is unique and tailored to its specific circumstances, there are several common themes and priorities across the region. Some key aspects of the national development strategies of Congo Basin countries are presented in Table 3.

Globally, the national development strategies of Congo Basin countries reflect their commitment to achieving SDGs while addressing pressing socioeconomic and environmental challenges. By pursuing inclusive, resilient, and environmentally sustainable development pathways, these countries aim to improve the quality of life for their citizens and contribute to regional peace, stability, and prosperity.

Table 2: Green architecture and engineering in the NDCs of Congo Basin countries.

Ways Congo Basin countries integrate green infrastructure in their NDCs	Description
Renewable energy development	Many Congo Basin countries are prioritizing the expansion of renewable energy infrastructure, including hydropower, solar, wind, and biomass energy. Green infrastructure projects, such as renewable energy power plants, grid expansion, and energy efficiency improvements are being promoted to reduce reliance on fossil fuels and mitigate greenhouse gas emissions.
Sustainable transport	Improving transportation infrastructure is a key component of NDCs in Congo Basin countries. Green transport initiatives focus on developing public transit systems, promoting non-motorized transport options like cycling and walking, and investing in low-emission vehicles and clean fuels. Green infrastructure projects, such as bus rapid transit (BRT) systems, bicycle lanes, and electric vehicle charging stations contribute to reducing emissions from the transportation sector.
Forest conservation and restoration	Protecting and restoring forests is a priority for many Congo Basin countries' NDCs, given the region's significant forest cover and biodiversity. Green infrastructure for forest conservation includes expanding protected areas, implementing sustainable forest management practices, and supporting community-based forestry initiatives. Investments in forest monitoring technologies, fire prevention measures, and eco-tourism infrastructure contribute to maintaining forest carbon stocks and enhancing resilience to climate change.
Climate-resilient agriculture	Agriculture is a vital sector in the Congo Basin, and promoting climate-resilient farming practices is essential for food security and adaptation to climate change. Green infrastructure projects, such as water management systems, irrigation networks, and agroforestry practices help farmers adapt to changing climatic conditions, improve soil health, and enhance crop productivity while reducing emissions from deforestation and land degradation.
Urban greening and sustainable land use	Green infrastructure in urban areas plays a critical role in enhancing resilience to climate change and improving quality of life. NDCs in Congo Basin countries include initiatives to develop green spaces, promote sustainable urban planning, and integrate nature-based solutions into infrastructure development. Investments in green roofs, urban parks, wastewater treatment plants, and green building standards contribute to reducing urban heat island effects, enhancing biodiversity, and mitigating flood risks.

Table 3: Key aspects of the national development strategies of Congo Basin countries.

Key aspects of the national development strategies of Congo Basin countries	Description
Economic diversification	Many Congo Basin countries recognize the need to diversify their economies away from traditional sectors, such as extractive industries (e.g., oil, gas, and mining) toward more sustainable and inclusive economic activities. National development strategies often prioritize sectors, such as agriculture, tourism, manufacturing, and services, which have the potential to generate employment, promote value addition, and reduce dependence on volatile commodity markets.
Infrastructure development	Improving infrastructure is a fundamental component of national development strategies in Congo Basin countries. This includes investments in transportation (roads, railways, and ports), energy (power generation, transmission, and distribution), water and sanitation, and telecommunications. Infrastructure development aims to enhance connectivity, facilitate trade and investment, and improve access to essential services, particularly in rural and remote areas.
Human capital development	Investing in education, healthcare, and social protection is a priority for Congo Basin countries to build human capital and enhance the well-being of their populations. National development strategies often include initiatives to expand access to quality education and healthcare services, improve skills training and vocational education, and strengthen social safety nets to reduce poverty and inequality.
Sustainable natural resource management	Given the region's rich biodiversity and natural resources, sustainable management of forests, water resources, and ecosystems is a key focus of national development strategies. Countries aim to balance economic development with environmental conservation and climate resilience by promoting sustainable forestry practices, biodiversity conservation, and integrated water resource management.
Climate change adaptation and mitigation	Recognizing the threats posed by climate change, Congo Basin countries are integrating climate adaptation and mitigation measures into their national development strategies. This includes initiatives to enhance resilience to climate impacts (e.g., floods, droughts, and extreme weather events), promote renewable energy and energy efficiency, and reduce greenhouse gas emissions from deforestation, agriculture, and energy sectors.

Table 3 (continued)

Key aspects of the national development strategies of Congo Basin countries	Description
Governance and institutional strengthening	Improving governance, enhancing transparency, and strengthening institutions are critical cross-cutting priorities in the national development strategies of Congo Basin countries. Efforts to combat corruption, promote accountability, and ensure effective public administration are essential for creating an enabling environment for sustainable development and attracting investment.
Regional integration and cooperation	Given the interconnectedness of economies and ecosystems in the Congo Basin region, promoting regional integration and cooperation is a strategic objective for many countries. This includes initiatives to harmonize policies and regulations, facilitate cross-border trade and investment, and address common challenges, such as transboundary water management, wildlife conservation, and security.

4.4 Congo Basin Countries Are Working Hard to Adhere to the UN SDGs Which Factor in Green Architecture and Engineering

Congo Basin countries have committed to meeting the SDGs, particularly goal number 7 on clean and affordable energy; goal number 9 on industry, innovation, and infrastructure; goal number 11 on sustainable cities and communities; and goal number 13 on climate action (Megevand and Mosnier, 2013; Ordway et al., 2019; Tegegne et al., 2019). They are actively working toward meeting SDGs related to green infrastructure, recognizing the importance of environmentally sustainable development in the region. Some ways in which these countries are striving to achieve SDGs related to green infrastructure as presented in Table 4.

By prioritizing investments in green infrastructure and adopting sustainable development practices, Congo Basin countries are taking significant steps toward achieving the SDGs related to environmental sustainability, climate action, and inclusive growth. Collaboration and cooperation among governments, international organizations, and civil society will be essential to accelerate progress toward these goals and build a more sustainable and resilient future for the region.

Table 4: Ways Congo Basin countries are striving to achieve SDGs related to green infrastructure and green engineering.

Ways Congo Basin countries are striving to achieve SDGs related to green infrastructure and green buildings	Description
Renewable energy development	Many Congo Basin countries are investing in renewable energy projects to reduce dependence on fossil fuels and mitigate climate change. This includes the development of hydropower, solar, wind, and biomass energy sources to provide clean and affordable electricity to both urban and rural areas.
Sustainable transport	Improving transportation infrastructure with a focus on sustainability is a priority for Congo Basin countries. This includes the construction of roads, railways, and ports that minimize environmental impacts, promote energy efficiency, and support the movement of goods and people in a sustainable manner.
Forest conservation and restoration	Protecting and restoring forest ecosystems is essential for biodiversity conservation, climate change mitigation, and sustainable development in the Congo Basin. Countries are implementing policies and initiatives to combat deforestation, promote sustainable forestry practices, and enhance the resilience of forest ecosystems to climate change.
Water resource management	Ensuring access to clean water and sanitation services is crucial for public health, economic development, and environmental sustainability. Congo Basin countries are working to improve water resource management through integrated approaches that balance competing demands for water while protecting freshwater ecosystems.
Climate resilience and adaptation	Building resilience to climate change impacts is a key priority for Congo Basin countries, given the region's vulnerability to climate-related risks, such as floods, droughts, and extreme weather events. Green infrastructure projects, such as natural flood management measures, green roofs, and sustainable land use planning, are being implemented to enhance climate resilience and adaptation.

Table 4 (continued)

Ways Congo Basin countries are striving to achieve SDGs related to green infrastructure and green buildings	Description
Green building and urban planning	Promoting green building practices and sustainable urban planning is essential for reducing the environmental footprint of cities and human settlements in the Congo Basin. Countries are adopting building codes and standards that prioritize energy efficiency, use of renewable materials, and sustainable design principles to create healthier and more environmentally friendly built environments.
Community engagement and participation	Engaging local communities and stakeholders in the planning, implementation, and monitoring of green infrastructure projects is critical for ensuring their success and sustainability. Congo Basin countries are fostering partnerships with civil society organizations, indigenous groups, and local communities to promote participatory decision-making processes and ensure that green infrastructure projects meet the needs and priorities of the people they serve.

5 Conclusion

Ensuring the sustainable development in the Congo Basin requires green infrastructure and engineering. However, the adoption of green architecture and engineering practices in the Congo Basin presents both challenges and prospects for sustainable development in the region. The main challenges include lack of awareness and knowledge, limited capacity and expertise, poor urban and rural planning, and lack of data and standards. There is a notable lack of awareness and knowledge of green architecture and engineering principles among stakeholders in the Congo Basin. This hinders the widespread adoption of sustainable building practices and technologies. The region faces a shortage of skilled professionals trained in green architecture and engineering. Without adequate capacity and expertise, it is challenging to design and implement green infrastructure projects effectively. Inadequate urban and rural planning exacerbates environmental degradation and undermines efforts to promote green architecture and engineering. Without proper land use planning and zoning regulations, sustainable development initiatives face significant obstacles. There is a lack of reliable data on energy and water use efficiency in buildings in the Congo Basin, making it difficult to assess the environmental performance of existing infrastructure and set benchmarks for improvement. There are, however, prospects for the development of green infrastructure and engineering in the Congo Basin due mainly to growing interest in sustainability, international support and collaboration, natural resource endowment, as well as policy and regulatory frameworks. There is a growing interest in sustainability among governments, businesses, and civil society organizations in the Congo Basin. This presents an opportunity to raise awareness and promote the adoption of green architecture and engineering practices. International organizations and donor agencies are increasingly supporting initiatives aimed at promoting sustainable development in the Congo Basin. Collaboration with global partners can provide technical expertise, funding, and capacity-building opportunities. The Congo Basin is rich in natural resources, including renewable energy sources, timber, and water resources. Harnessing these resources sustainably can support the transition to green architecture and engineering practices. Developing and implementing robust policy and regulatory frameworks is essential for mainstreaming green architecture and engineering in the Congo Basin. Governments can incentivize sustainable building practices through tax incentives, subsidies, and building codes. While there are significant challenges to overcome, there are also promising prospects for advancing green architecture and engineering in the Congo Basin. By addressing key barriers and leveraging available opportunities, the region can move toward a more sustainable and resilient built environment that contributes to long-term environmental, social, and economic well-being.

References

Abioso, W. S. (2019, November). Invisible in Architecture Confront the Green Architecture. In *IOP Conference Series: Materials Science and Engineering*. (Vol 662, No. 4, p. 042019). IOP Publishing.

Ahunbaev, A., Adakhayev, A., & Chuyev, S. (2024). Petrochemical industry in Eurasia: Opportunities for Deeper Processing. *Available at SSRN 4820754*.

Akele-Sita, A. & Chowdhury, S. P. (2015, September). Optimal allocation of energy storage in a future congolese power system. In *2015 50th International Universities Power Engineering Conference (UPEC)* (pp. 1–6). IEEE.

Ali, R., Barra, A. F., Berg, C. N., Damania, R., Nash, J. D., & Russ, J. (2015). Infrastructure in conflict-prone and fragile environments: Evidence from the Democratic Republic of Congo. *World Bank Policy Research Working Paper, 7273*.

Alsdorf, D., Beighley, E., Laraque, A., Lee, H., Tshimanga, R., O'Loughlin, F., . . . Spencer, R. G. (2016). Opportunities for hydrologic research in the Congo Basin. *Reviews of Geophysics, 54*(2), 378–409.

Anseeuw, W. (2013). The rush for land in Africa: Resource grabbing or green revolution?. *South African Journal of International Affairs, 20*(1), 159–177.

Aquilas, N. A. & Atemnkeng, J. T. (2022). Climate-related development finance and renewable energy consumption in greenhouse gas emissions reduction in the Congo basin. *Energy Strategy Reviews, 44*, 100971.

Assoua, J. E. & Molua, E. L. (2018). Opportunities and Challenges of Sustainable Forest Management for a Green Economy Transition in Cameroon. *Journal of Economics and Sustainable Development, 9*(12), 1–8.

Atyi, R. E. A. (2021). *State of the Congo Basin Forests in 2021: Overall Conclusions* (Vol. 2022, p. 367). Bogor, Indonesia: CIFOR.

Bai, X. P. & Qian, C. (2021). Factor validity and reliability performance analysis of human behavior in green architecture construction engineering. *Ain Shams Engineering Journal, 12*(4), 4291–4296.

Bielek, B. (2016). Green building–towards sustainable architecture. *Applied Mechanics and Materials, 824,* 751–760.

Bielek, B. & Bielek, M. (2012). Environmental Strategies for Design of Sustainable Buildings in Technique of Green Eco-Architecture. *Journal of Civil Engineering and Architecture, 6*(7), 892.

Binsangou, S., Ifo, S. A., Ibocko, L., Louvouandou, L., Tchindjang, M., & Koubouana, F. (2018). Urban growth and deforestation by remote sensing in the Humid tropical forest of Congo Bassin: Case of impfondo in republic Of Congo. *American Journal of Environment and Sustainable Development, 3*(3), 46–54.

Boisson, S., Kiyombo, M., Sthreshley, L., Tumba, S., Makambo, J., & Clasen, T. (2010). Field assessment of a novel household-based water filtration device: A randomised, placebo-controlled trial in the Democratic Republic of Congo. *PloS One, 5*(9), e12613.

Brown, H. C. P., Smit, B., Sonwa, D. J., Somorin, O. A., & Nkem, J. (2011). Institutional perceptions of opportunities and challenges of REDD+ in the Congo Basin. *The Journal of Environment & Development, 20*(4), 381–404.

Brundtland, G. H. (1987). What is sustainable development. *Our Common Future, 8*(9).

Burnley, C. (2011). Natural resources conflict in the Democratic Republic of the Congo: A question of governance. *Sustainable Dev. L. & Pol'y, 12*(7).

Chen, R., Yang, H. C., & Chang, H. C. (2012). The application of green architecture to residential building development. *Applied Mechanics and Materials, 193,* 34–39.

Chishugi, D. U., Sonwa, D. J., Kahindo, J. M., Itunda, D., Chishugi, J. B., Félix, F. L., & Sahani, M. (2021). How climate change and land use/land cover change affect domestic water vulnerability in Yangambi watersheds (DR Congo). *Land, 10*(2), 165.

Chukwu, V. E. (2020). Potentials, drivers and barriers to green economy transition: Implications for Africa. *Advanced Journal of Plant Biology, 1*(1), 7–17.

Church, C. & Crawford, A. (2018). *Green Conflict Minerals*. London: International institute for sustainable development.

Clark, W. W. II & Cooke, G. (2014). The Green Industrial Revolution. *Global Sustainable Communities Handbook: Green Design Technologies and Economics, 13.*

Clay, N. (2016). Producing hybrid forests in the Congo Basin: A political ecology of the landscape approach to conservation. *Geoforum, 76,* 130–141.

Dalder, J., Oluleye, G., Cannone, C., Yeganyan, R., Tan, N., & Howells, M. (2024). Modelling Policy Pathways to Maximise Renewable Energy Growth and Investment in the Democratic Republic of the Congo Using OSeMOSYS (Open Source Energy Modelling System). *Energies, 17*(2), 342.

De Angelis, P., Tuninetti, M., Bergamasco, L., Calianno, L., Asinari, P., Laio, F., & Fasano, M. (2021). Data-driven appraisal of renewable energy potentials for sustainable freshwater production in Africa. *Renewable and Sustainable Energy Reviews, 149,* 111414.

Deshmukh, R., Mileva, A., & Wu, G. C. (2018). Renewable energy alternatives to mega hydropower: A case study of Inga 3 for Southern Africa. *Environmental Research Letters, 13*(6), 064020.

Diangha, M. N. & Wiegleb, G. (2014). The adoption and impact of forest stewardship council standards in the Congo Basin forestry sector. In *Voluntary Standard Systems: A Contribution to Sustainable Development* (pp. 229–241). Berlin, Heidelberg: Springer Berlin Heidelberg.

Elad, C. (2014). Forest certification and biodiversity accounting in the Congo basin countries. In *Accounting for Biodiversity*. Routledge, 189–211.

Elshafei, G., Vilčeková, S., Zeleňáková, M., & Negm, A. M. (2021). An extensive study for a wide utilization of green architecture parameters in built environment based on genetic schemes. *Buildings, 11*(11), 507.

Endamana, D., Boedhihartono, A. K., Bokoto, B., Defo, L., Eyebe, A., Ndikumagenge, C., . . . Sayer, J. A. (2010). A framework for assessing conservation and development in a Congo Basin Forest Landscape. *Tropical Conservation Science*, *3*(3), 262–281.

Fobissie, K., Alemagi, D., & Minang, P. A. (2014). REDD+ Policy Approaches in the Congo Basin: A Comparative Analysis of Cameroon and the Democratic Republic Congo (DRC). *Forests*, *5*(10), 2400–2424.

Fobissie, K., Chia, E., Enongene, K., & Oeba, V. O. (2019). Agriculture, forestry and other land uses in Nationally Determined Contributions: The outlook for Africa. *International Forestry Review*, *21*(1), 1–11.

Franken, G., Vasters, J., Dorner, U., Melcher, F., Sitnikova, M., & Goldmann, S. (2012). Certified trading chains in mineral production: A way to improve responsibility in mining. *Non-Renewable Resource Issues: Geoscientific and Societal Challenges*, 213–227.

Hammons, T. J., Blyden, B. K., Calitz, A. C., Gulstone, A. B., Isekemanga, E., Johnstone, R., . . . Taher, F. (2000). African electricity infrastructure interconnections and electricity exchanges. *IEEE Transactions on Energy Conversion*, *15*(4), 470–480.

Huang, L., Fantke, P., Ritscher, A., & Jolliet, O. (2022). Chemicals of concern in building materials: A high-throughput screening. *Journal of Hazardous Materials*, *424*, 127574.

Kankeu, R. S., Demaze, M. T., Krott, M., Sonwa, D. J., & Ongolo, S. (2020). Governing knowledge transfer for deforestation monitoring: Insights from REDD+ projects in the Congo Basin region. *Forest Policy and Economics*, *111*, 102081.

Karsenty, A. (2019). Certification of tropical forests: A private instrument of public interest? A focus on the Congo Basin. *Forest Policy and Economics*, *106*, 101974.

Kayembe, J. M., Thevenon, F., Laffite, A., Sivalingam, P., Ngelinkoto, P., Mulaji, C. K., . . . Poté, J. (2018). High levels of faecal contamination in drinking groundwater and recreational water due to poor sanitation, in the sub-rural neighbourhoods of Kinshasa, Democratic Republic of the Congo. *International Journal of Hygiene and Environmental Health*, *221*(3), 400–408.

Kenfack, J., Bossou, O. V., & Tchaptchet, E. (2017). How can we promote renewable energy and energy efficiency in Central Africa? A Cameroon case study. *Renewable and Sustainable Energy Reviews*, *75*, 1217–1224.

Kenfack, J., Bossou, O. V., Voufo, J., & Djom, S. (2014). Addressing the current remote area electrification problems with solar and microhydro systems in Central Africa. *Renewable Energy*, *67*, 10–19.

Kenfack, J., Fogue, M., Hamandjoda, O., & Tatietse, T. T. (2011). Promoting renewable energy and energy efficiency in Central Africa: Cameroon case study. In *World Renewable Energy Congress: Linköping* (pp. 2071–1050). *Sweden*.

Kleinschroth, F., Gourlet-Fleury, S., Sist, P., Mortier, F., & Healey, J. R. (2015). Legacy of logging roads in the Congo Basin: How persistent are the scars in forest cover?. *Ecosphere*, *6*(4), 1–17.

Kleinschroth, F., Laporte, N., Laurance, W. F., Goetz, S. J., & Ghazoul, J. (2019). Road expansion and persistence in forests of the Congo Basin. *Nature Sustainability*, *2*(7), 628–634.

Kusakana, K. (2016). A review of energy in the Democratic Republic of Congo. In *Conference: ICDRE, Copenhagen, Denmark*.

Lateef, A. S. A., Fernandez-Alonso, M., Tack, L., & Delvaux, D. (2010). Geological constraints on urban sustainability, Kinshasa City, Democratic Republic of Congo. *Environmental Geosciences*, *17*(1), 17–35.

Li, Y. Y., Chen, P. H., Chew, D. A. S., Teo, C. C., & Ding, R. G. (2012). Exploration of critical external partners of architecture/engineering/construction (AEC) firms for delivering green building projects in Singapore. *Journal of Green Building*, *7*(3), 193–209.

Lu, Y. & Zhang, X. (2016). Corporate sustainability for architecture engineering and construction (AEC) organizations: Framework, transition and implication strategies. *Ecological Indicators*, *61*, 911–922.

Mahdavinejad, M., Zia, A., Larki, A. N., Ghanavati, S., & Elmi, N. (2014). Dilemma of green and pseudo green architecture based on LEED norms in case of developing countries. *International Journal of Sustainable Built Environment*, *3*(2), 235–246.

Majambu, E., Demaze, M. T., & Ongolo, S. (2021). The politics of forest governance failure in the Democratic Republic of Congo (DRC): Lessons from 35 years of political rivalries. *International Forestry Review, 23*(3), 321–337.

Majambu, E., Mampeta Wabasa, S., Welepele Elatre, C., Boutinot, L., & Ongolo, S. (2019). Can traditional authority improve the governance of forestland and sustainability? Case study from the Congo (DRC). *Land, 8*(5), 74.

Marijnen, E. & Schouten, P. (2019). Electrifying the green peace? Electrification, conservation and conflict in Eastern Congo. *Conflict, Security & Development, 19*(1), 15–34.

Masood, O. A. I., Abd Al-Hady, M. I., & Ali, A. K. M. (2017). Applying the principles of green architecture for saving energy in buildings. *Energy Procedia, 115*, 369–382.

Mazalto, M. (2010). Environmental liability in the mining sector: Prospects for sustainable development in the Democratic Republic of the Congo. *Mining, Society, and a Sustainable World*, 289–317.

Megevand, C. & Mosnier, A. (2013). *Deforestation Trends in the Congo Basin: Reconciling Economic Growth and Forest Protection*. World Bank Publications.

Merenkov, A. V., Akchurina, N. S., & Matveeva, T. M. (2019). Basic principles of "green" architecture in foreign realization experience. In *IOP conference series: materials science and engineering* (Vol. 687, No. 5, p. 055058). IOP Publishing.

Milburn, R. (2014). The roots to peace in the Democratic Republic of Congo: Conservation as a platform for green development. *International Affairs, 90*(4), 871–887.

Modeste, K. N., Mempouo, B., René, T., Costa, Á. M., Orosa, J. A., Raminosoa, C. R., & Mamiharijaona, R. (2015). Resource potential and energy efficiency in the buildings of Cameroon: A review. *Renewable and Sustainable Energy Reviews, 50*, 835–846.

Mohammed, A. B. (2021). Sustainable design strategy optimizing green architecture path based on sustainability. *HBRC Journal, 17*(1), 461–490.

Molla, A. S., Tang, P., Sher, W., & Bekele, D. N. (2021). Chemicals of concern in construction and demolition waste fine residues: A systematic literature review. *Journal of Environmental Management, 299*, 113654.

Mounmemi, M., Lins, M. E., & Lovell, C. A. K. (2022). Eco-efficiency and carbon emission abatement cost inthe charcoal value chain in Congo basin: Evidence from Cameroon.

Muhaya, V. N., Chuma, G. B., Kavimba, J. K., Cirezi, N. C., Mugumaarhahama, Y., Fadiala, R. M., . . . Karume, K. (2022). Uncontrolled urbanization and expected unclogging of Congolese cities: Case of Bukavu city, Eastern DR Congo. *Environmental Challenges, 8*, 100555.

Nagabhatla, N., Cassidy-Neumiller, M., Francine, N. N., & Maatta, N. (2021). Water, conflicts and migration and the role of regional diplomacy: Lake Chad, Congo Basin, and the Mbororo pastoralist. *Environmental Science & Policy, 122*, 35–48.

Nago, M. & Krott, M. (2022). Systemic failures in north–south climate change knowledge transfer: A case study of the Congo basin. *Climate Policy, 22*(5), 623–636.

Ndoumbe Berock, I. & Ongolo, S. (2019). Why do logging companies adopt or reject forest certification in the Congo basin? Insights from Cameroon. *International Forestry Review, 21*(3), 341–349.

Nematchoua, M. K., & Reiter, S. (2021). Evaluation of bioclimatic potential, energy consumption, CO2-emission, and life cycle cost of a residential building located in Sub-Saharan Africa; a case study of eight countries. *Solar Energy, 218*, 512–524.

Nhamo, G. & Chekwoti, C. (2014). New generation land grabs in a green African economy. In *Land Grabs in a Green African Economy. Implications for Trade, Investment and Development Policies* (pp. 1–9). Pretoria: Africa Institute of South Africa.

Nsah, K. T. (2022). The ecopolitics of water pollution and disorderly urbanization in Congo-Basin plays. *Orbis Litterarum, 77*(5), 314–332.

Nwajiuba, C., Emmanuel, T. N., & Bangali Solomon, F. A. R. A. (2015). State of knowledge on CSA in Africa: Case studies from Nigeria, Cameroun and the Democratic Republic of Congo. In *Forum for Agricultural Research in Africa, Accra, Ghana ISBN* (pp. 978–9988).

Ogungbile, A. J., Shen, G. Q., Wuni, I. Y., Xue, J., & Hong, J. (2021). A hybrid framework for direct CO2 emissions quantification in China's construction sector. *International Journal of Environmental Research and Public Health, 18*(22), 11965.

Ordway, E. M., Sonwa, D. J., Levang, P., Mboringong, F., Naylor, R. L., & Nkongho, R. N. (2019). *Sustainable Development of the Palm Oil Sector in the Congo Basin: The Need for a Regional Strategy Involving Smallholders and Informal Markets.* Vol. 255. CIFOR.

Ozor, N., Nyambane, A., Onuoha, C. M., Makokha, M. O., & M'mboyi, F. (2020). *Nationally Determined Contributions (Ndcs) Implementation Index, Monitoring and Tracking Tools for Selected Countries in Africa.* https://atpsnet.org/wpcontent/uploads/2020/07/NDC-Implementation-Index-Report.pdf.

Petrovic, S. & Zoucha, J. (2023). Congo. In *World Energy Handbook* (pp. 37–52). Cham: Springer International Publishing.

Peya, M. I. (2018). The green vision of Denis Sassou N'guesso facing a blind world in danger: The gospel of environmental management and sustainable development.

Piabuo, S. M., Minang, P. A., Tieguhong, C. J., Foundjem-Tita, D., & Nghobuoche, F. (2021). Illegal logging, governance effectiveness and carbon dioxide emission in the timber-producing countries of Congo Basin and Asia. *Environment, Development and Sustainability, 23*, 14176–14196.

POuLSEN, J. R. & CLARK, C. J. (2010). 3.1 Congo Basin timber certification and biodiversity conservation. *Biodiversity Conservation in Certified Forests, 55*.

Pyhälä, A., Osuna Orozco, A., & Counsell, S. (2016). *Protected Areas in the Congo Basin: Failing Both People and Biodiversity.* London: Rainforest Foundation-UK.

Ragheb, A., El-Shimy, H., & Ragheb, G. (2016). Green architecture: A concept of sustainability. *Procedia-Social and Behavioral Sciences, 216*, 778–787.

Sachs, J. D., Toledano, P., Dietrich Brauch, M., Mebratu-Tsegaye, T., Uwaifo, E., & Sherrill, B. M. (2022). Roadmap to Zero-Carbon Electrification of Africa by 2050: The Green Energy Transition and the Role of the Natural Resource Sector (Minerals, Fossil Fuels, and Land).

Saleh, H. S. & Saied, S. Z. (2017). Green architecture as a concept of Historic Cairo. *Procedia Environmental Sciences, 37*, 342–355.

Schouten, P., Verweijen, J., Murairi, J., & Batundi, S. K. (2022). Paths of authority, roads of resistance: Ambiguous rural infrastructure and slippery stabilization in eastern DR Congo. *Geoforum, 133*, 217–227.

Shapiro, A. C., Grantham, H. S., Aguilar-Amuchastegui, N., Murray, N. J., Gond, V., Bonfils, D., & Rickenbach, O. (2021). Forest condition in the Congo Basin for the assessment of ecosystem conservation status. *Ecological Indicators, 122*, 107268.

Somorin, O. A., Brown, H. C. P., Visseren-Hamakers, I. J., Sonwa, D. J., Arts, B., & Nkem, J. (2012). The Congo Basin forests in a changing climate: Policy discourses on adaptation and mitigation (REDD+). *Global Environmental Change, 22*(1), 288–298.

Sonwa, D. J., Nkem, J. N., Idinoba, M. E., Bele, M. Y., & Jum, C. (2012). Building regional priorities in forests for development and adaptation to climate change in the Congo Basin. *Mitigation and Adaptation Strategies for Global Change, 17*, 441–450.

Sonwa, D. J., Oumarou Farikou, M., Martial, G., & Félix, F. L. (2020). Living under a fluctuating climate and a drying Congo Basin. *Sustainability, 12*(7), 2936.

Stauskis, G. (2013). Green architecture paradigm: From urban utopia to modern methods of quality assessment. *Mokslas–Lietuvos ateitis/Science–Future of Lithuania, 5*(3), 181–188.

Tegegne, Y. T., Cramm, M., Van Brusselen, J., & Linhares-Juvenal, T. (2019). Forest concessions and the United Nations sustainable development goals: Potentials, challenges and ways forward. *Forests, 10*(1), 45.

Tegegne, Y. T., Lindner, M., Fobissie, K., & Kanninen, M. (2016). Evolution of drivers of deforestation and forest degradation in the Congo Basin forests: Exploring possible policy options to address forest loss. *Land Use Policy, 51*, 312–324.

Tongele, T. N. (2021). Human ways of life and environmental sustainability: Congo Basin case study. *Journal of Civil Engineering and Architecture, 15*(1), 547–559.

Trefon, T. (2010). Forest governance in Congo: Corruption rules?. *U4 Brief.*

Trefon, T. (2017). Forest governance and international partnerships in the Congo Basin. *Forest, 10*(13).

Ulimwengu, J. M., Funes, J., Headey, D. D., & You, L. (2009). Paving the way for development: The impact of road infrastructure on agricultural production and household wealth in the Democratic Republic of Congo.

Vitos, M., Altenbuchner, J., Stevens, M., Conquest, G., Lewis, J., & Haklay, M. (2017). Supporting collaboration with non-literate forest communities in the congo-basin. In *Proceedings of the 2017 ACM Conference on Computer Supported Cooperative Work and Social Computing.* (pp 1576–1590).

Wei, W., Yusong, D., Sanyuan, N., Ruimin, W., Ting, S., Yong, Z., & Xin, L. (2024). *China's Urbanization in the New Round of Technological Revolution, 2020–2050: Impact, Prospect and Strategy.* Taylor & Francis.

Well, F. & Ludwig, F. (2020). Blue–green architecture: A case study analysis considering the synergetic effects of water and vegetation. *Frontiers of Architectural Research, 9*(1), 191–202.

Wong, G. Y., Holm, M., Pietarinen, N., Ville, A., & Brockhaus, M. (2022). The making of resource frontier spaces in the Congo Basin and Southeast Asia: A critical analysis of narratives, actors and drivers in the scientific literature. *World Development Perspectives, 27*, 100451.

Yuan, Y., Yu, X., Yang, X., Xiao, Y., Xiang, B., & Wang, Y. (2017). Bionic building energy efficiency and bionic green architecture: A review. *Renewable and Sustainable Energy Reviews, 74*, 771–787.

Zhang, Q., Justice, C. O., Jiang, M., Brunner, J., & Wilkie, D. S. (2006). A GIS-based assessment on the vulnerability and future extent of the tropical forests of the Congo Basin. *Environmental Monitoring and Assessment, 114*, 107–121.

Nyong Princely Awazi

Green Growth Transition in Africa: Current State, Rationale, and National Strategies

Abstract: The transition to green growth in Africa is increasingly viewed as a pathway to sustainable development, economic growth, and environmental resilience. Green growth emphasizes the integration of environmental sustainability into economic policies, focusing on resource efficiency, low-carbon development, and poverty reduction. This chapter therefore examines green growth transition in Africa focusing on the current state, rationale, and national strategies of green growth on the continent. The chapter is based entirely on a review of empirical literature on green growth in Africa. Findings reveal that green growth has gained traction on the African continent since the early 2000s with the coming into force of the Millennium Development Goals and since 2015 with the Sustainable Development Goals. The rationale for green growth transition in Africa is grounded in the continent's vulnerability to climate change, reliance on natural resources, and the need to diversify economies. Africa is experiencing increasing environmental pressures, such as land degradation, deforestation, and water scarcity, which threaten agricultural productivity and food security. Moreover, rapid urbanization and industrialization raise concerns over air pollution, energy consumption, and waste management. National strategies for green growth in Africa vary but often share common goals including promoting renewable energy, enhancing sustainable agriculture, protecting biodiversity, and fostering green innovation. Several African nations, such as Morocco, Kenya, and South Africa, have introduced ambitious policies and investments to scale up clean energy technologies, improve resource efficiency, and create green jobs. For instance, Morocco's solar energy initiatives and Kenya's geothermal energy investments highlight the potential for renewable energy to drive sustainable growth. Despite these advancements, challenges such as inadequate financing, weak institutional frameworks, and limited technological capacity hinder the widespread implementation of green growth strategies.

Keywords: Green growth strategies, climate change, food security, environmental degradation, Africa

Nyong Princely Awazi, Department of Forestry and Wildlife Technology, University of Bamenda, Cameroon; FOKABS INC.,2500 St. Laurent Blvd, Ottawa, ON, Canada K1H 1B1; Department of Forestry, Faculty of Agronomy and Agricultural Sciences, University of Dschang, Cameroon,
e-mail: nyongprincely@gmail.com

https://doi.org/10.1515/9783111563046-005

1 Introduction

Africa stands at a critical juncture in its development trajectory, where the imperative for sustainable growth intersects with the urgent need to address pressing environmental challenges (Adams, 2008; Satgar, 2014). As the continent grapples with the complexities of rapid urbanization, population growth, and economic development, the concept of green growth has emerged as a guiding framework for fostering inclusive, resilient, and environmentally sustainable development pathways (Okereke et al., 2019; Basubas, 2023). Green growth entails decoupling economic growth from environmental degradation, promoting resource efficiency, and enhancing social inclusivity to achieve sustainable development outcomes (Simbanegavi, 2019). In Africa, the transition toward green growth represents a paradigm shift toward more sustainable and equitable patterns of development that prioritize environmental stewardship, social equity, and economic prosperity (Gu et al., 2018). Africa is endowed with abundant natural resources, diverse ecosystems, and rich biodiversity that underpin its economic development and livelihoods (Steiner, 2008; Archer et al., 2018). However, rapid industrialization, urbanization, and extractive activities have exerted immense pressure on natural ecosystems, leading to environmental degradation, loss of biodiversity, and climate change impacts (Abernethy et al., 2016; Tiamgne et al., 2022). Against this backdrop, the transition to green growth offers a transformative approach to addressing the interconnected challenges of poverty alleviation, environmental sustainability, and socio-economic development. By integrating environmental considerations into economic policies, investment decisions, and development planning processes, African countries can unlock opportunities for green investment, job creation, and innovation while safeguarding natural resources and promoting inclusive growth.

Key pillars of the green growth transition in Africa include renewable energy and low-carbon development, sustainable agriculture and food security, ecosystem conservation and biodiversity protection, circular economy and resource efficiency, green finance and investment, and policy coherence and governance (Burkolter and Perch, 2014; Nhamo and Nhamo, 2014; Chavula and Turyasingura, 2022; Tsegay, 2023). Investing in renewable energy infrastructure and promoting low-carbon technologies are central to Africa's green growth agenda. By harnessing its abundant solar, wind, hydro, and geothermal resources, Africa can expand access to clean and affordable energy, reduce reliance on fossil fuels, and mitigate climate change impacts (Onyeji-Nwogu, 2017; Nwokolo et al., 2023; Chukwuemeka et al., 2023). Adopting sustainable agricultural practices, such as agroecology, conservation agriculture, and climate-smart farming, is critical for promoting food security, enhancing resilience to climate change, and conserving natural resources (Makate et al., 2019; Akanmu et al., 2023). Green growth strategies prioritize sustainable land management, water conservation, and biodiversity conservation to ensure food sovereignty and rural livelihoods (Shilomboleni, 2017). Preserving ecosystems, conserving biodiversity, and restoring degraded landscapes are essential for maintaining ecosystem services, supporting biodiversity conservation, and enhancing climate resilience (Durán-Díaz, 2023). Green growth initiatives promote nature-based solutions, protected area management, and sustainable land use planning to safeguard ecosystems and promote ecotourism and sustainable livelihoods (Dupar et al., 2023). Transitioning toward a circular economy model, which emphasizes reducing, reusing, and recycling resources, is crucial for minimizing waste generation, promoting resource efficiency, and fostering sustainable consumption and production patterns (Thorn et al., 2021; Neves et al., 2022). Green growth strategies encourage waste management, recycling initiatives, and sustainable manufacturing practices to minimize environmental footprints and maximize resource utilization (Moore, 2019; Chisholm et al., 2021). Mobilizing green finance and investment is essential for scaling up green growth initiatives and funding sustainable development projects (Schwerhoff and Sy, 2017; Chirambo, 2018; Mungai et al., 2022). Green growth policies facilitate access to climate finance, private sector investment, and innovative financing mechanisms, such as green bonds, carbon markets, and impact investing, to support renewable energy, sustainable infrastructure, and climate-resilient development (Fonta et al., 2018; Muhammad et al., 2023). Ensuring policy coherence and good governance is critical for mainstreaming green growth principles into national development agendas and sectoral policies (Kararach et al., 2018; Mohamed and Montmasson-Clair, 2018; Afionis et al., 2020). Green growth strategies promote multisectoral coordination, stakeholder engagement, and institutional reforms to align economic, social, and environmental objectives, strengthen regulatory frameworks, and enhance transparency and accountability in decision-making processes.

The transition to green growth in Africa offers a transformative pathway toward sustainable and inclusive development, where economic prosperity, environmental sustainability, and social equity converge. By embracing green growth principles and investing in renewable energy, sustainable agriculture, ecosystem conservation, circular economy, green finance, and governance reforms, Africa can unlock its full potential for sustainable development, prosperity, and resilience in the face of global environmental challenges. This chapter therefore examines green growth transition in Africa focusing on the current state, rationale and national green growth strategies in different countries on the continent.

2 The State of Green Growth in Africa

The state of green growth in Africa reflects a complex landscape characterized by both opportunities and challenges (Swilling et al., 2016; Chandrashekeran et al., 2017). While the continent boasts vast natural resources and renewable energy potentials, it also faces pressing environmental, social, and economic challenges that require urgent attention (AfDB, 2023b). An overview of the current state of green growth in Africa reveals the following opportunities and challenges for green growth: abundant natural resources, rapid urbanization and population growth, environmental degradation and climate change, renewable energy potential, sustainable agriculture and food security, circular economy and resource efficiency, green finance and investment, and policy and governance. Africa is endowed with abundant natural resources including renewable energy sources such as solar, wind, hydro, and geothermal energy (Schoneveld and Zoomers, 2015). The continent's diverse ecosystems, fertile soils, and rich biodiversity provide a strong foundation for sustainable development and green growth initiatives (AfDB, 2016). Africa is experiencing rapid urbanization and population growth, with a significant portion of its population projected to live in urban areas by 2050. Urbanization presents both opportunities and challenges for green growth, as cities become hubs for innovation, economic growth, and sustainable development, but also face pressure on infrastructure, resources, and environmental sustainability (Freire, 2013). Africa is disproportionately affected by environmental degradation and climate change impacts including deforestation, desertification, land degradation, water scarcity, and extreme weather events. These environmental challenges threaten livelihoods, food security, and economic stability, underscoring the urgency of adopting green growth strategies to build resilience and mitigate climate risks (Amin et al., 2016; Bari and Dessus, 2022). Africa has immense renewable energy potential, particularly in solar and hydroelectric power, which can provide clean, affordable, and reliable energy for sustainable development. However, renewable energy deployment remains limited due to investment constraints, policy gaps, and technical challenges, hindering the continent's transition to a low-carbon energy system (AfDB, 2023b). Agriculture remains a key driver of Africa's economy and livelihoods, supporting the majority of the population. Sustainable agriculture practices, such as agroecology, conservation agriculture, and climate-smart farming, are essential for promoting food security, enhancing resilience to climate change, and conserving natural resources (Chuku and Ajayi, 2023; AfDB, 2023a).

Transitioning toward a circular economy model is crucial for minimizing waste generation, promoting resource efficiency, and fostering sustainable consumption and production patterns in Africa. While circular economy initiatives are emerging in some countries, there is a need for greater investment, innovation, and policy support to scale up circular practices across sectors (Burkolter and Perch, 2014). Access to green finance and investment is essential for scaling up green growth initiatives and funding sustainable development projects in Africa. While progress has been made in mobilizing climate finance and attracting private sector investment, there are still gaps in financing mechanisms, risk management, and capacity building that need to be addressed to unlock the full potential of green finance in Africa (CPI, 2022a,b,c; AfDB, 2024). Policy coherence and good governance are critical for mainstreaming green growth principles into national development agendas and sectoral policies (African Natural Resources Management and Investment Centre, 2022). African countries are increasingly integrating green growth considerations into their policy frameworks, but there is a need for stronger institutional capacities, stakeholder engagement, and regulatory frameworks to drive green growth at the national and regional levels. While Africa faces significant environmental, social, and economic challenges, the continent also possesses immense potential for green growth and sustainable development. By harnessing its renewable energy resources, promoting sustainable agriculture, adopting circular economy practices, mobilizing green finance, and strengthening policy and governance frameworks, Africa can chart a path toward a more resilient, inclusive, and sustainable future.

3 Rationale for a Green Growth Transition in Africa

The rationale for a green growth transition in Africa is rooted in the urgent need to address interconnected environmental, social, and economic challenges while fostering sustainable and inclusive development. Several key factors underpin the rationale for prioritizing green growth in Africa including environmental sustainability, climate change mitigation and adaptation, energy access and energy security, sustainable economic development, health and well-being, and resilience and adaptation.

3.1 Environmental Sustainability

Africa is endowed with rich biodiversity, diverse ecosystems, and natural resources that are essential for supporting livelihoods, economic activities, and ecosystem services. However, unsustainable exploitation of natural resources, deforestation, land degradation, pollution, and climate change threaten the continent's environmental sustainability. Transitioning to green growth offers a pathway to safeguarding ecosystems, conserving biodiversity, and mitigating climate change impacts, ensuring the long-term resilience and sustainability of Africa's natural capital. Environmental sustainability serves as a compelling rationale for embracing green growth in Africa due to several key reasons including biodiversity conservation, ecosystem services, climate change mitigation, natural resource management, environmental health, and well-being (AfDB, 2021). Africa is renowned for its rich biodiversity and diverse ecosystems, which are critical for supporting ecosystem services, ecological balance, and human well-being. However, unsustainable land use practices, habitat destruction, and wildlife poaching threaten biodiversity conservation efforts. Green growth emphasizes the protection and restoration of ecosystems, wildlife habitats, and endangered species, promoting biodiversity conservation and preserving the continent's natural heritage for future generations. Ecosystem services, such as water provision, soil fertility, carbon sequestration, and pollination, are essential for supporting agricultural productivity, water security, climate regulation, and human health in Africa (AfDB, 2018). Environmental degradation, including deforestation, soil erosion, and pollution, undermines the provision of ecosystem services, leading to negative impacts on livelihoods and economic activities. Green growth strategies prioritize sus-

tainable land management, watershed protection, and ecosystem restoration to safeguard ecosystem services and enhance resilience to environmental shocks and
stresses. Africa is disproportionately vulnerable to the impacts of climate change, including extreme weather events, droughts, floods, and sea-level rise, which threaten
food security, water resources, and infrastructure. Green growth focuses on climate
change mitigation measures, such as reducing greenhouse gas emissions, promoting
renewable energy, enhancing energy efficiency, and adopting low-carbon technologies, to mitigate climate risks and limit global warming (AfDB, 2018). By investing in
clean and sustainable energy sources, Africa can reduce its carbon footprint, contribute to global climate goals, and enhance climate resilience.

Africa's natural resources, including forests, minerals, water, and land, are vital for
supporting economic activities, livelihoods, and socioeconomic development (African
Natural Resources Management and Investment Centre, 2022). However, unsustainable
exploitation of natural resources, including deforestation, overfishing, and mineral extraction, depletes resource stocks, damages ecosystems, and exacerbates environmental
degradation. Green growth advocates for sustainable natural resource management
practices, such as sustainable forestry, fisheries management, water conservation, and
responsible mining, to ensure the long-term viability of natural resource assets and promote equitable resource sharing (Pausata et al., 2020). Environmental degradation and
pollution have significant implications for public health and well-being in Africa, contributing to a range of health issues including respiratory diseases, waterborne illnesses, and malnutrition. Green growth prioritizes environmental health by promoting
clean air, safe drinking water, and sanitation, reducing exposure to hazardous chemicals and pollutants, and enhancing food security through sustainable agriculture practices. By addressing environmental health risks, green growth improves overall well-
being and quality of life for African communities (AfBD, 2022a). Environmental sustainability provides a compelling rationale for embracing green growth in Africa, as it offers a holistic approach to addressing environmental challenges, mitigating climate
change impacts, conserving biodiversity, promoting sustainable resource management,
and enhancing human well-being. By prioritizing environmental sustainability in development strategies and policies, Africa can achieve long-term resilience, prosperity, and
sustainable development while safeguarding the planet's natural resources and ecosystems for future generations.

3.2 Climate Change Mitigation and Adaptation

Africa is disproportionately vulnerable to the impacts of climate change, including extreme weather events, droughts, floods, and sea-level rise, which exacerbate poverty,
food insecurity, and displacement. Green growth strategies prioritize climate change
mitigation and adaptation measures such as promoting renewable energy, enhancing
energy efficiency, implementing climate-smart agriculture, and strengthening resil-

ience to climate risks (Swilling et al., 2016). By investing in low-carbon technologies and climate-resilient infrastructure, Africa can reduce its greenhouse gas emissions, build climate resilience, and contribute to global efforts to limit global warming. Climate change mitigation and adaptation are crucial rationales for promoting green growth in Africa. This is due to increasing vulnerability to climate change, economic implications, energy access, natural resource management, opportunities for innovation and investment, and global cooperation and commitments. Africa is particularly vulnerable to the impacts of climate change due to its reliance on rain-fed agriculture, limited access to clean water, and high dependence on natural resources for livelihoods. Rising temperatures, changing rainfall patterns, and extreme weather events threaten food security, water availability, and infrastructure stability across the continent. Climate change poses significant economic risks to African nations. Losses in agricultural productivity, damage to infrastructure, increased healthcare costs due to climate-related diseases, and displacement of communities all have substantial economic implications. Green growth strategies can help mitigate these risks by promoting sustainable practices that build resilience and reduce emissions (AfDB, 2018).

Many African countries face energy poverty, with millions of people lacking access to electricity. Traditional energy sources such as biomass and coal not only contribute to deforestation and air pollution but also exacerbate climate change (AfDB 2022a,b). Transitioning to renewable energy sources like solar, wind, and hydroelectric power not only reduces emissions but also enhances energy security and promotes economic development. Africa is endowed with abundant natural resources including forests, minerals, and biodiversity. Unsustainable exploitation of these resources contributes to deforestation, soil degradation, loss of biodiversity, and pollution. Green growth initiatives promote sustainable resource management practices that preserve ecosystems, support biodiversity, and ensure the long-term viability of these resources for future generations. Embracing green growth presents opportunities for innovation, technological advancement, and investment in clean technologies (Popp, 2012). By prioritizing renewable energy, sustainable agriculture, eco-tourism, and green infrastructure, African countries can attract investment, create jobs, and stimulate economic growth while simultaneously addressing climate change. African nations are part of the global community committed to addressing climate change through international agreements like the Paris Agreement. By embracing green growth strategies, African countries not only fulfill their commitments to reduce greenhouse gas emissions but also demonstrate leadership in global efforts to combat climate change. Integrating climate change mitigation and adaptation into green growth strategies in Africa is essential for fostering sustainable development, enhancing resilience, and ensuring a prosperous future for the continent.

3.3 Energy Access and Energy Security

Access to reliable and affordable energy is critical for driving economic growth, improving living standards, and advancing human development in Africa (AfDB, 2022a,b). However, a significant portion of the population lacks access to modern energy services, relying on traditional biomass fuels that contribute to deforestation, indoor air pollution, and health risks. Green growth prioritizes expanding access to clean and sustainable energy sources, such as solar, wind, hydro, and biomass, to enhance energy access, improve energy security, and promote inclusive development. Energy access and energy security are compelling rationales for promoting green growth in Africa. This is due to energy poverty, health benefits, environmental sustainability, energy security, job creation and economic development, rural electrification, and global commitments. Access to reliable and affordable energy remains a significant challenge in many parts of Africa. Millions of people lack access to electricity, relying on traditional and often inefficient energy sources like biomass for cooking and heating. Green growth strategies focus on expanding access to clean and sustainable energy sources such as solar, wind, and hydroelectric power, thereby lifting people out of energy poverty and improving their quality of life (Fadly, 2019). Traditional energy sources like biomass and coal are often used indoors for cooking and heating, leading to indoor air pollution that contributes to respiratory diseases and premature deaths, particularly among women and children. Transitioning to clean energy sources reduces indoor air pollution, improving public health outcomes and reducing healthcare costs. Fossil fuel-based energy production contributes to air and water pollution, deforestation, and greenhouse gas emissions, exacerbating climate change and environmental degradation. Green growth initiatives prioritize renewable energy sources that have minimal environmental impact, thereby promoting environmental sustainability and mitigating the adverse effects of climate change.

Many African countries rely heavily on imported fossil fuels for their energy needs, making them vulnerable to price fluctuations and supply disruptions in global energy markets. Investing in domestic renewable energy sources reduces dependence on imported fuels, enhances energy security, and shields economies from external shocks (IEA, 2022). The transition to green energy sources stimulates economic growth and creates employment opportunities across various sectors including manufacturing, construction, and maintenance of renewable energy infrastructure. By investing in renewable energy technologies and fostering a conducive environment for green businesses, African countries can spur economic development and reduce unemployment rates. Green growth strategies prioritize decentralized energy solutions such as off-grid and mini-grid systems, which are particularly beneficial for rural communities with limited access to centralized electricity grids. These decentralized systems can power agricultural machinery, irrigation pumps, and small-scale enterprises, thereby catalyzing rural development and reducing disparities between urban and rural areas. African countries are signatories to international agreements like the

Paris Agreement, which aim to mitigate climate change by reducing greenhouse gas emissions. By embracing green growth and transitioning to renewable energy sources, African nations demonstrate their commitment to combating climate change and fulfilling their obligations under global climate accords (Kedir, 2014). Promoting green growth in Africa through increased energy access and energy security not only addresses pressing socio-economic challenges but also contributes to environmental sustainability and helps mitigate the impacts of climate change.

3.4 Sustainable Economic Development

Green growth offers opportunities for promoting sustainable economic development, job creation, and poverty reduction in Africa. By investing in renewable energy, sustainable agriculture, eco-tourism, green infrastructure, and clean technologies, Africa can stimulate economic growth, diversify its economy, and create green jobs while minimizing environmental impacts and enhancing resource efficiency. Green growth strategies prioritize inclusive and equitable economic growth that benefits all segments of society, including marginalized communities and vulnerable populations (Vazquez-Brust et al., 2014). Sustainable economic development is a compelling rationale for promoting green growth in Africa. This is driven by the need for long-term prosperity, resource efficiency, diversification of economies, job creation, innovation and technology, resilience to climate change, global market access, and community development. Green growth focuses on promoting economic development that meets the needs of the present without compromising the ability of future generations to meet their own needs. By prioritizing sustainable practices in sectors such as energy, agriculture, transportation, and infrastructure, African countries can build resilient economies that thrive over the long term. Green growth emphasizes the efficient use of resources, minimizing waste and maximizing productivity (AfDB, 2021). This approach not only conserves natural resources but also reduces production costs, improves competitiveness, and enhances overall productivity, leading to sustained economic growth. Many African economies are heavily reliant on a few sectors such as agriculture, mining, and oil extraction, which are often vulnerable to external shocks and price fluctuations. Green growth promotes diversification by investing in renewable energy, eco-tourism, sustainable agriculture, and green technology sectors, reducing dependence on volatile commodities and creating new avenues for economic growth.

The transition to a green economy creates employment opportunities across various sectors including renewable energy, conservation, eco-tourism, and sustainable agriculture. Green jobs tend to be labor-intensive and have a lower environmental impact, making them well-suited for inclusive economic growth and poverty reduction, particularly in rural areas. Green growth fosters innovation and technological advancement by incentivizing research and development in clean technologies, renewable energy, and sustainable practices (Barbier, 2015). Investing in green innovation not only drives economic growth but also enhances competitiveness in the global marketplace, positioning African countries as leaders in sustainable development. Climate change poses significant risks to African economies, affecting agriculture, water resources, infrastructure, and human health. Green growth strategies build resilience by promoting adaptive measures, such as climate-smart agriculture, sustainable water management, and resilient infrastructure, reducing vulnerability to climate-related risks and enhancing overall economic stability (AfDB, 2018). Increasingly, consumers and investors around the world are demanding products and services that are produced sustainably and ethically. By adopting green growth strategies, African countries can access global markets for environmentally friendly goods and services, attracting investment, and enhancing export competitiveness. Green growth prioritizes inclusive and participatory approaches to development, engaging local communities in decision-making processes and ensuring that development benefits are equitably distributed. By investing in sustainable infrastructure, social services, and livelihood opportunities, green growth contributes to poverty reduction and improves the well-being of marginalized populations. Sustainable economic development is a powerful rationale for embracing green growth in Africa. By integrating environmental sustainability, social equity, and economic prosperity, green growth strategies can catalyze inclusive and resilient development pathways that benefit both current and future generations.

3.5 Health and Well-Being

Environmental degradation and pollution have significant implications for public health and well-being in Africa. Poor air quality, contaminated water sources, and inadequate sanitation contribute to a range of health issues including respiratory diseases, waterborne illnesses, and malnutrition (Tearfund, 2022). Green growth promotes clean energy, sustainable water management, and waste reduction measures that improve environmental health and enhance overall well-being, particularly for vulnerable populations. Health and well-being provide a compelling rationale for promoting green growth in Africa owing to reduced air pollution, access to clean water and sanitation, nutrition and food security, physical activity and mental health, climate resilience and disaster preparedness, access to healthcare facilities, promotion of active transportation, community empowerment, and participation. Traditional energy sources such as biomass, coal, and kerosene used for cooking and heating contribute to indoor and outdoor air pollution, leading to respiratory diseases, cardiovascular problems, and other health issues. Transitioning to clean and renewable energy sources like solar and biogas reduces air pollution, improving air quality and public health outcomes. Lack of access to clean water and sanitation facilities is a significant health challenge in many African countries, leading to waterborne diseases such as diarrhea, cholera, and typhoid. Green growth initiatives, such as sustainable water management practices, wastewater treatment systems, and ecological sanitation, promote access to safe drinking water and sanitation, reducing the prevalence of water-related illnesses and improving overall health outcomes. Climate change, environmental degradation, and unsustainable agricultural practices threaten food security and nutrition outcomes in Africa. Green growth approaches, such as sustainable agriculture, agroforestry, and biodiversity conservation, promote resilient food systems that ensure access to nutritious and diverse food sources, contributing to improved health and well-being among communities.

Green spaces, parks, and recreational areas play a vital role in promoting physical activity, mental well-being, and social cohesion. Green growth strategies that prioritize urban green infrastructure, sustainable land use planning, and green transportation options create opportunities for outdoor recreation, exercise, and relaxation, enhancing overall quality of life and mental health outcomes (AfDB, 2021). Climate change-related disasters, such as floods, droughts, and extreme weather events, pose significant threats to human health and well-being in Africa. Green growth interventions, such as early warning systems, climate-resilient infrastructure, and community-based disaster preparedness measures, enhance resilience to climate-related risks, reducing the impact of disasters on health and livelihoods. Green growth strategies can improve access to healthcare facilities and services, particularly in rural and underserved areas. Investing in renewable energy solutions, such as solar-powered clinics and medical refrigeration systems, ensures reliable electricity supply for medical equipment, vaccines, and essential healthcare services, thereby enhancing health outcomes and saving lives. Green transportation options, such as walking, cycling, and public transit, not only reduce greenhouse gas emissions and air pollution but also promote physical activity and reduce the risk of chronic diseases such as obesity, diabetes, and heart disease. By investing in sustainable transportation infrastructure and promoting active modes of travel, green growth contributes to healthier and more livable cities in Africa (Wodajo, 2021; McKinsey & Co, 2021). Green growth approaches prioritize community engagement, empowerment, and participation in decision-making processes related to environmental management, public health, and sustainable development. By involving local communities in the planning, implementation, and monitoring of green projects and initiatives, green growth fosters ownership, resilience, and social cohesion, leading to improved health and well-being outcomes at the grassroots level. Health and well-being provide a compelling rationale for advancing green growth in Africa. By integrating environmental sustainability, public health, and social equity considerations into development strategies, African countries can create healthier, more resilient, and sustainable communities for present and future generations.

3.6 Resilience and Adaptation

Africa faces a range of environmental and socioeconomic risks, including natural disasters, food insecurity, conflict, and economic shocks, which threaten stability and undermine development gains. Green growth emphasizes building resilience and adaptive capacity to cope with and recover from shocks and stresses, strengthening social safety nets, enhancing community resilience, and fostering sustainable livelihoods that reduce vulnerability and enhance adaptive capacity (AfDB, 2018). Resilience and adaptation are crucial rationales for promoting green growth in Africa owing to climate change vulnerability, natural resource management, water scarcity and management, agricultural resilience, disaster risk reduction, health and social resilience, infrastructure and urban resilience, community empowerment, and participation. Africa is highly vulnerable to the impacts of climate change including extreme weather events, shifting rainfall patterns, and rising temperatures. These changes pose significant risks to food security, water availability, infrastructure, and livelihoods across the continent. Green growth strategies prioritize resilience and adaptation measures that help communities and ecosystems withstand and recover from climate-related shocks and stresses. Unsustainable exploitation of natural resources, such as deforestation, overgrazing, and land degradation, exacerbates vulnerability to climate change and undermines ecosystem resilience. Green growth approaches promote sustainable natural resource management practices that conserve biodiversity, protect ecosystems, and enhance their capacity to provide essential services such as clean water, food, and climate regulation (Chuku and Ajayi, 2024). Many African countries face water scarcity and variability due to climate change, population growth, and inadequate water management practices. Green growth interventions focus on sustainable water management strategies, such as rainwater harvesting, watershed protection, and water-efficient irrigation techniques, to enhance water availability, quality, and resilience to droughts and floods.

Agriculture is a critical sector in many African economies, providing livelihoods for millions of people. Green growth approaches promote climate-smart agriculture practices, such as conservation agriculture, agroforestry, and crop diversification, which improve soil health, enhance water efficiency, and increase resilience to climate variability, thereby ensuring food security and livelihoods in the face of changing climatic conditions. Climate change-related disasters, such as floods, droughts, and storms, are becoming more frequent and severe in Africa, causing loss of life, displacement, and economic disruptions. Green growth strategies integrate disaster risk reduction measures, such as early warning systems, resilient infrastructure, and community-based preparedness initiatives, to reduce vulnerability and build adaptive capacity to cope with and recover from disasters (Auktor, 2020). Climate change impacts on health, including heat-related illnesses, vector-borne diseases, and malnutrition, disproportionately affect vulnerable populations such as women, children, and the elderly. Green growth initiatives promote health resilience through climate-resilient healthcare facilities, disease surveillance systems, and community-based health inter-ventions, enhancing the capacity of health systems to respond to climate-related health risks. Rapid urbanization and inadequate infrastructure planning exacerbate vulnerability to climate change in African cities, increasing the risk of flooding, water scarcity, and heatwaves. Green growth strategies prioritize resilient urban infrastruc-ture, green building design, and nature-based solutions, such as green roofs and urban parks, to enhance urban resilience, improve quality of life, and reduce the im-pact of climate-related hazards on urban populations (Arimoro, 2021). Green growth promotes community empowerment and participation in resilience-building efforts, recognizing the knowledge, skills, and resources that local communities possess to adapt to climate change. By engaging communities in decision-making processes, fos-tering social cohesion, and supporting local initiatives, green growth strengthens so-cial capital and builds collective resilience to climate change impacts at the grassroots level. Resilience and adaptation provide compelling rationales for advancing green growth in Africa. By integrating climate resilience, sustainable resource management, and community empowerment into development strategies, African countries can build more resilient, adaptive, and sustainable societies that thrive in a changing climate.

4 National Strategies for Green Growth in Africa

Several African countries have developed national strategies for green growth to pro-mote sustainable development, address environmental challenges, and seize eco-nomic opportunities. Here are examples of national strategies for green growth in Africa.

4.1 Ethiopia

Ethiopia has been a pioneer in promoting green growth through its Climate-Resilient Green Economy (CRGE) strategy. The CRGE strategy, launched in 2011, aims to achieve middle-income status by 2025 while keeping greenhouse gas emissions low. It integra-tes climate change mitigation and adaptation into national development planning and identifies key sectors for green growth interventions (Paul and Weinthal, 2019; Raoux, 2024). The CRGE strategy is structured around four key pillars: building resilience, re-ducing emissions, greening the economy, and enabling institutions and financing. The building resilience pillar focuses on enhancing Ethiopia's capacity to adapt to climate change impacts, particularly in vulnerable sectors such as agriculture, water resour-ces, and natural resource management (Bass et al., 2013; Borchgrevink, 2014; Alba-goury, 2016; Addisu, 2019). The reducing emissions pillar aims to limit the growth of greenhouse gas emissions by promoting sustainable practices in energy, forestry, agriculture, and industry. The greening the economy pillar emphasizes investments in renewable energy, sustainable transport, eco-industries, and green infrastructure to foster economic growth while minimizing environmental degradation (Okereke et al., 2019). The enabling institutions and financing pillar focuses on strengthening institutional capacity, enhancing policy coordination, and mobilizing financial resources for green growth initiatives.

The CRGE strategy identifies priority sectors for green growth interventions including energy, agriculture, forestry, and transport. Promoting the expansion of renewable energy sources such as hydropower, wind, solar, and geothermal energy to reduce reliance on fossil fuels and increase energy access. Implementing climate-smart agriculture practices, such as conservation agriculture, agroforestry, and improved land management, to enhance food security, increase resilience to climate change, and reduce emissions from the agricultural sector (Kassa et al., 2015; Mulugetta, 2016). Scaling up reforestation, afforestation, and sustainable forest management initiatives to conserve biodiversity, protect watersheds, and sequester carbon. Investing in sustainable transport infrastructure, such as public transportation systems and nonmotorized transport options, to reduce emissions from the transport sector and promote urban sustainability (Lambert and Deyganto, 2024). Ethiopia has adopted various policy instruments to support the implementation of the CRGE strategy, including the National Climate Change Policy (NCCP) which provides a framework for mainstreaming climate change considerations into national development planning and sectoral policies; Renewable Energy Policy which promotes investments in renewable energy technologies and establishing incentives for renewable energy development; and the Climate-Resilient Green Economy Facility which mobilizes financial resources and providing technical assistance to support green growth projects and initiatives. On the whole, Ethiopia's CRGE strategy represents a holistic approach to green growth, integrating climate change considerations into national development planning and prioritizing investments in sustainable development pathways that enhance resilience, reduce emissions, and promote inclusive economic growth.

4.2 Rwanda

Rwanda has been proactive in promoting green growth through its Green Growth and Climate Resilience Strategy (GGCRS). The GGCRS, launched in 2011, aims to transform Rwanda into a low-carbon, climate-resilient economy by 2050. It integrates climate change mitigation and adaptation into national development planning and identifies key sectors for green growth interventions. The GGCRS is structured around four key pillars which are sustainable agriculture and land use, sustainable energy, resource-efficient green cities, and ecosystem restoration and conservation (Banerjee et al., 2020). The sustainable agriculture and land use pillar focuses on promoting climate-smart agriculture practices, such as agroforestry, soil conservation, and water management, to enhance food security, increase agricultural productivity, and build resilience to climate change impacts. The sustainable energy pillar aims to increase access to clean and renewable energy sources, such as solar, hydro, and biogas, to reduce reliance on fossil fuels, improve energy security, and mitigate greenhouse gas emissions (Niyigaba et al., 2020). The resource-efficient green cities pillar emphasizes sustainable urban development, green infrastructure, and low-carbon transportation options to promote livable, resilient cities and reduce environmental pollution and congestion. The ecosystem restoration and conservation pillar focus on protecting and restoring ecosystems, such as forests, wetlands, and watersheds, to enhance biodiversity, regulate climate, and provide essential ecosystem services.

The GGCRS identifies priority sectors for green growth interventions including renewable energy, sustainable agriculture, green cities, and ecosystem restoration. Promoting investments in renewable energy technologies, such as solar photovoltaics, small-scale hydropower, and biomass, to increase energy access, reduce greenhouse gas emissions, and stimulate economic development in rural areas. Implementing climate-smart agriculture practices, such as conservation agriculture, crop diversification, and integrated pest management, to enhance food security, improve soil fertility, and reduce agricultural emissions. Promoting sustainable urban planning, green building design, and low-carbon transportation options, such as walking, cycling, and public transit, to create inclusive, resilient, and environmentally friendly cities. Scaling up reforestation, afforestation, and sustainable land management initiatives to restore degraded ecosystems, enhance biodiversity, and sequester carbon to mitigate climate change impacts. Rwanda has adopted various policy instruments to support the implementation of the GGCRS, including the National Green Growth and Climate Resilience Policy which provides a framework for mainstreaming green growth and climate resilience considerations into national development planning and sectoral policies; the Renewable Energy Policy which promotes investments in renewable energy technologies and establishing incentives for renewable energy development and deployment; and the National Adaptation Plan (NAP) which identifies priority adaptation measures and interventions to build resilience to climate change impacts across various sectors (Hudani, 2020). Globally, Rwanda's GGCRS represents a comprehensive approach to green growth, integrating climate change considerations into national development planning and prioritizing investments in sustainable development pathways that enhance resilience, reduce emissions, and promote inclusive economic growth.

4.3 Kenya

Kenya has developed several national strategies and plans for promoting green growth and sustainable development with key initiatives, including the National Climate Change Action Plan (NCCAP), Kenya Vision 2030, National Climate Change Framework Policy (NCCFP), Kenya Green Economy Strategy and Implementation Plan (GESIP), National Renewable Energy Policy (NREP), and the NAP. The NCCAP outlines Kenya's strategy for addressing climate change mitigation, adaptation, and resilience-building. It focuses on various sectors, including energy, agriculture, water resources, forestry, and infrastructure to mainstream climate change considerations into national development planning and policy implementation (King-Okumu, 2015; Owino et al., 2016; Khisa, 2016; Nyika, 2021). Kenya Vision 2030 is the country's long-term development blueprint aimed at transforming Kenya into a newly industrializing, middle-income country by 2030. The vision incorporates green growth principles and objectives, emphasizing sustainable development, environmental conservation, and climate resilience as key pillars for achieving inclusive economic growth and poverty reduction (Muthama, 2021). The NCCAP outlines Kenya's strategy for addressing climate change mitigation, adaptation, and resilience-building. It focuses on various sectors, including energy, agriculture, water resources, forestry, and infrastructure, to mainstream climate change considerations into national development planning and policy implementation.

The NCCFP provides a policy framework for guiding Kenya's climate change response efforts including mitigation, adaptation, capacity-building, and international cooperation. It emphasizes the importance of mainstreaming climate change considerations into sectoral policies and promoting low-carbon development pathways to achieve Sustainable Development Goals (SDGs). The GESIP aims to promote green growth and transition Kenya to a low-carbon, resource-efficient economy. It focuses on sectors such as renewable energy, sustainable agriculture, green infrastructure, waste management, and eco-tourism to stimulate economic growth, create green jobs, and enhance environmental sustainability (Andersen et al., 2022). The NREP provides a policy framework for promoting investments in renewable energy technologies such as solar, wind, geothermal, and biomass, to increase energy access, reduce reliance on fossil fuels, and mitigate greenhouse gas emissions. The NAP identifies priority adaptation measures and interventions to build resilience to climate change impacts across various sectors including agriculture, water resources, health, infrastructure, and disaster risk management. It aims to enhance adaptive capacity, reduce vulnerability, and safeguard livelihoods and ecosystems from climate-related risks and hazards. These national strategies and plans demonstrate Kenya's commitment to promoting green growth, sustainable development, and climate resilience, aligning with global efforts to address climate change and achieve the .

4.4 South Africa

South Africa has implemented various national strategies and policies to promote green growth and transition to a low-carbon, sustainable economy with the most prominent including the National Development Plan (NDP), Green Economy Accord, Integrated Resource Plan (IRP), National Climate Change Response Policy (NCCRP), Renewable Energy Independent Power Producer Procurement Program (REIPPPP), Carbon Tax Act, and the National Biodiversity Strategy and Action Plan (NBSAP). South Africa's NDP is a long-term development blueprint aimed at eliminating poverty and reducing inequality by 2030. The plan identifies green growth as a key pillar for achieving inclusive economic growth, job creation, and environmental sustainability. It emphasizes the need to transition to a low-carbon economy, invest in renewable energy, and promote sustainable development practices across all sectors (Resnick et al., 2012; Death, 2014; Musango et al., 2014; Gu et al., 2018). The Green Economy Accord is a partnership between the South African government, business, labor, and civil society to promote green growth and create green jobs (Borel-Saladin and Turok, 2013). It outlines commitments and actions to advance the green economy agenda including investments in renewable energy, energy efficiency, waste management, and sustainable agriculture (Musvoto et al., 2015). The IRP is a long-term energy planning framework that outlines South Africa's energy mix and investment priorities. The latest version of the IRP (IRP 2019) includes ambitious targets for renewable energy deployment, aiming to increase the share of renewable energy in the country's electricity generation mix to reduce greenhouse gas emissions and enhance energy security.

The NCCRP provides a policy framework for addressing climate change mitigation, adaptation, and resilience-building in South Africa. It sets out sectoral mitigation targets, adaptation priorities, and measures to promote climate resilience across various sectors including energy, transport, agriculture, water resources, and forestry (Hickel and Kallis, 2020; Ofori et al., 2022). The REIPPPP is a flagship initiative aimed at accelerating the deployment of renewable energy in South Africa. It enables private sector participation in renewable energy projects through competitive bidding processes, attracting investment, creating green jobs, and reducing reliance on fossil fuels. South Africa introduced a carbon tax in 2019 to incentivize emissions reductions and promote transition to a low-carbon economy. The carbon tax applies to large emitters across various sectors and aims to internalize the cost of carbon emissions, encourage energy efficiency, and spur investment in cleaner technologies. The NBSAP outlines South Africa's commitments and actions to conserve biodiversity, promote sustainable use of natural resources, and enhance ecosystem resilience (Schoneveld and Zoomers, 2015). It identifies key biodiversity priorities and measures to mainstream biodiversity considerations into national development planning and sectoral policies. These national strategies and policies demonstrate South Africa's commitment to promoting green growth, addressing climate change, and advancing sustainable development objectives in line with international agreements such as the Paris Agreement and the SDGs.

4.5 Morocco

Morocco has implemented several national strategies and initiatives to promote green growth and sustainable development namely Morocco's Green Plan (Plan Maroc Vert), Morocco's Energy Strategy, National Sustainable Development Strategy (SNDD), Morocco's NAP, National Waste Management Strategy, NBSAP, and the National Sustainable Tourism Strategy. Plan Maroc Vert is a national strategy launched in 2008 aimed at modernizing the agricultural sector while promoting sustainable resource management and rural development. It focuses on increasing agricultural productivity, promoting sustainable water management, and enhancing the resilience of rural communities to climate change impacts (Günay et al., 2018; Zaatari, 2022; Zahir et al., 2022; Ainou et al., 2023). Morocco has set ambitious targets for renewable energy development as part of its energy strategy. The country aims to increase the share of renewable energy in its energy mix to 52% by 2030, with investments in solar, wind, and hydroelectric power projects (Rihab, 2019). The Noor Solar Complex, one of the largest concentrated solar power facilities in the world, is a flagship project under this strategy. SNDD is Morocco's overarching framework for promoting sustainable development across various sectors. It integrates environmental, social, and economic objectives to ensure inclusive and environmentally sustainable growth (Badraoui and Dahan, 2011; Elmoukhtar et al., 2022; Azzeddine et al., 2024). The strategy emphasizes the need to transition to a low-carbon economy, reduce greenhouse gas emissions, and promote green technologies and innovation.

Morocco has developed an NAP to address climate change impacts and enhance resilience in key sectors such as agriculture, water resources, coastal areas, and infrastructure. The plan includes measures to improve water management, promote climate-smart agriculture, and strengthen disaster risk management and adaptation planning (Perry, 2020). Morocco has implemented a national strategy to improve waste management and promote recycling and resource efficiency. The strategy aims to reduce waste generation, increase recycling rates, and minimize environmental pollution. Morocco's commitment to waste management is demonstrated through initiatives such as the Oum Azza landfill closure project and the development of waste-to-energy facilities. Morocco has developed an NBSAP to conserve biodiversity, protect ecosystems, and promote sustainable use of natural resources. The plan includes measures to protect endangered species, establish protected areas, and promote community-based conservation initiatives. Morocco has recognized the importance of sustainable tourism as a driver of economic growth while minimizing negative environmental and social impacts. The country has developed a national strategy to promote sustainable tourism practices, protect cultural and natural heritage, and enhance the socioeconomic benefits of tourism for local communities (Mathez and Loftus, 2023). These national strategies and initiatives demonstrate Morocco's commitment to promoting green growth, addressing climate change, and advancing sustainable development objectives in line with international agreements such as the Paris Agreement and the SDGs.

4.6 Ghana

Ghana has implemented several national strategies and initiatives to promote green growth and sustainable development such as National Climate Change Policy Framework and Action Plan (NCCPF/AP), Renewable Energy Master Plan (REMP), National Environmental Policy (NEP), National Climate Change Adaptation Strategy (NCCAS), National REDD + Strategy, National Waste Management Strategy, and the National SDGs framework. The NCCPF/AP provides a comprehensive framework for addressing climate change mitigation, adaptation, and resilience-building in Ghana (Dovie, 2017; Ali et al., 2021). It outlines sectoral priorities, policy interventions, and action plans to mainstream climate change considerations into national development planning and implementation. Ghana's REMP aims to increase the share of renewable energy in the country's energy mix to enhance energy security, reduce greenhouse gas emissions, and promote sustainable development (Ackah and Kizys, 2015). The plan includes targets and strategies for promoting investments in renewable energy technologies such as solar, wind, hydro, and biomass (Amankwah-Amoah and Sarpong, 2016). The NEP provides a policy framework for promoting sustainable environmental management, biodiversity conservation, and ecosystem protection in Ghana. It emphasizes the need to integrate environmental considerations into development planning and decision-making processes to ensure the sustainable use of natural resources and the protection of environmental quality. The NCCAS outlines Ghana's strategy for enhancing resilience to climate change impacts across various sectors including agriculture, water resources, coastal areas, and infrastructure. It includes measures to promote climate-smart agriculture, improve water management, enhance disaster risk management, and strengthen community resilience.

Ghana has developed a National REDD+ (Reducing Emissions from Deforestation and Forest Degradation) Strategy to address deforestation and forest degradation while promoting sustainable forest management and conservation. The strategy aims to reduce emissions from deforestation and forest degradation, enhance carbon sequestration, and improve livelihoods for forest-dependent communities. Ghana has implemented a national strategy to improve waste management and promote recycling and resource efficiency. The strategy aims to reduce waste generation, increase recycling rates, and minimize environmental pollution. Ghana's commitment to waste management is demonstrated through initiatives such as the establishment of waste recycling plants and the promotion of community-based waste management initiatives. Ghana has aligned its national development priorities with the SDGs to promote inclusive and sustainable development. The government has developed a framework for implementing the SDGs, integrating environmental sustainability, social equity, and economic development objectives into national development planning and policy implementation (Jabik and Bawakyillenuo, 2016; Seidu, 2020; Martinelli, 2022; Yeboah et al., 2023). These national strategies and initiatives demonstrate Ghana's commitment to promoting green growth, addressing climate change, and advancing sustainable development objectives in line with international agreements such as the Paris Agreement and the SDGs.

4.7 Nigeria

Nigeria has developed several national strategies and initiatives to promote green growth and sustainable development with the most prominent being the National Climate Change Policy and Response Strategy (NCCPRS), REMP, National Policy on Climate Change and National Adaptation Strategy and Plan of Action (), NBSAP, National Policy on Renewable Energy and Energy Efficiency (NPREEE), NEP, and the National Waste Management Strategy (Okoh, 2014; Oyebanji et al., 2017; Tunji-Olayeni et al., 2018; Ike, 2019; Adejumo and Asongu, 2020; Anabaraonye et al., 2021). The NCCPRS provides a comprehensive framework for addressing climate change mitigation, adaptation, and resilience-building in Nigeria. It outlines sectoral priorities, policy interventions, and action plans to mainstream climate change considerations into national development planning and implementation. Nigeria's REMP aims to increase the share of renewable energy in the country's energy mix to enhance energy security, reduce greenhouse gas emissions, and promote sustainable development (Elum and Mjimba, 2020). The plan includes targets and strategies for promoting investments in renewable energy technologies such as solar, wind, hydro, and biomass. The National Policy on Climate Change and NASPA provide a roadmap for enhancing resilience to climate change impacts across various sectors including agriculture, water resources, coastal areas, and infrastructure. They include measures to promote climate-smart agriculture, improve water management, enhance disaster risk management, and strengthen community resilience.

Nigeria has developed an NBSAP to conserve biodiversity, protect ecosystems, and promote sustainable use of natural resources. The plan includes measures to protect endangered species, establish protected areas, and promote community-based conservation initiatives. The NPREEE provides a policy framework for promoting investments in renewable energy technologies and energy efficiency measures to reduce energy consumption, enhance energy security, and mitigate greenhouse gas emissions (Adetiloye et al., 2022). Nigeria's NEP provides a policy framework for promoting sustainable environmental management, biodiversity conservation, and ecosystem protection. It emphasizes the need to integrate environmental considerations into development planning and decision-making processes to ensure the sustainable use of natural resources and the protection of environmental quality. Nigeria has implemented a national strategy to improve waste management and promote recycling and resource efficiency. The strategy aims to reduce waste generation, increase recycling rates, and minimize environmental pollution. Nigeria's commitment to waste management is demonstrated through initiatives such as the establishment of waste recycling plants and the promotion of community-based waste management initiatives. These national strategies and initiatives demonstrate Nigeria's commitment to promoting green growth, addressing climate change, and advancing sustainable development objectives in line with international agreements such as the Paris Agreement and the SDGs.

4.8 Tanzania

Tanzania has developed several national strategies and initiatives to promote green growth and sustainable development including the National Climate Change Strategy (NCCS), NREP, NBSAP, National Forestry Policy, National Water Policy, National Sustainable Transport Policy, and the National Waste Management Strategy (Buseth, 2017; Bergius et al., 2018; Bergius et al., 2020; Bilame, 2023). Tanzania's NCCS provides a comprehensive framework for addressing climate change mitigation, adaptation, and resilience-building. It outlines sectoral priorities, policy interventions, and action plans to mainstream climate change considerations into national development planning and implementation. Tanzania's NREP aims to increase the share of renewable energy in the country's energy mix to enhance energy security, reduce greenhouse gas emissions, and promote sustainable development. The policy includes targets and strategies for promoting investments in renewable energy technologies such as solar, wind, hydro, and biomass (Mihayo and Swai, 2019). Tanzania has developed an NBSAP to conserve biodiversity, protect ecosystems, and promote sustainable use of natural resources. The plan includes measures to protect endangered species, establish protected areas, and promote community-based conservation initiatives (Theodory, 2022). Tanzania's National Forestry Policy aims to promote sustainable forest management, conservation, and restoration to enhance biodiversity, mitigate climate change, and improve livelihoods for forest-dependent communities. The policy includes measures to combat deforestation, promote afforestation and reforestation, and strengthen community-based forest management.

Tanzania's National Water Policy provides a policy framework for promoting sustainable water management, conservation, and allocation to enhance water security, support ecosystem health, and improve access to safe drinking water and sanitation services (Scherr et al., 2013; Mihayo, 2020). The policy emphasizes integrated water resources management, water conservation, and efficient water use practices. Tanzania has developed a National Sustainable Transport Policy to promote sustainable urban mobility, reduce greenhouse gas emissions, and improve air quality. The policy includes measures to promote nonmotorized transport, public transportation systems, and low-carbon technologies to reduce traffic congestion and pollution in urban areas. Tanzania has implemented a national strategy to improve waste management and promote recycling and resource efficiency. The strategy aims to reduce waste generation, increase recycling rates, and minimize environmental pollution. Tanzania's commitment to waste management is demonstrated through initiatives such as the establishment of waste recycling facilities and the promotion of community-based waste management initiatives. These national strategies and initiatives demonstrate Tanzania's commitment to promoting green growth, addressing climate change, and advancing sustainable development objectives in line with international agreements such as the Paris Agreement and the SDGs.

4.9 Ivory Coast

The Ivory Coast, also known as Côte d'Ivoire, has made strides in promoting green growth and sustainable development through various national strategies and initiatives, the most prominent being the NCCP (Politique Nationale sur le Changement Climatique), National Renewable Energy Development Plan (Plan National de Développement des Energies Renouvelables), NBSAP (Stratégie et Plan d'Action Nationaux pour la Biodiversité), National Forestry Policy (Politique Forestière Nationale), National Water Resources Management Policy (Politique Nationale de Gestion des Ressources en Eau), National Sustainable Transport Policy (Politique Nationale des Transports Durables), and the National Waste Management Strategy (Stratégie Nationale de Gestion des Déchets) (Heldt, 1997; Pedercini et al., 2018; Menkeh, 2021; Degila et al., 2022, 2024). The Ivory Coast has developed a national policy framework to address climate change mitigation, adaptation, and resilience-building. The policy outlines sectoral priorities, policy interventions, and action plans to mainstream climate change considerations into national development planning and implementation. The Ivory Coast has set targets and strategies for increasing the share of renewable energy in its energy mix to enhance energy security, reduce greenhouse gas emissions, and promote sustainable development (Kararach et al., 2018; Asongu and Odhiambo, 2020; Mohsin et al., 2022; Kakou, 2023). The plan focuses on promoting investments in renewable energy technologies such as solar, wind, hydro, and biomass. The Ivory Coast has developed a national strategy and action plan to conserve biodiversity, protect ecosystems, and promote sustainable use of natural resources. The plan includes measures to protect endangered species, establish protected areas, and promote community-based conservation initiatives.

The Ivory Coast has adopted a national forestry policy to promote sustainable forest management, conservation, and restoration to enhance biodiversity, mitigate climate change, and improve livelihoods for forest-dependent communities. The policy includes measures to combat deforestation, promote afforestation and reforestation, and strengthen community-based forest management (Ngondjeb et al., 2020). The Ivory Coast has developed a national policy framework for promoting sustainable water management, conservation, and allocation to enhance water security, support ecosystem health, and improve access to safe drinking water and sanitation services. The policy emphasizes integrated water resources management, water conservation, and efficient water use practices. The Ivory Coast has formulated a national sustainable transport policy to promote sustainable urban mobility, reduce greenhouse gas emissions, and improve air quality. The policy includes measures to promote nonmotorized transport, public transportation systems, and low-carbon technologies to reduce traffic congestion and pollution in urban areas. The Ivory Coast has implemented a national strategy to improve waste management and promote recycling and resource efficiency. The strategy aims to reduce waste generation, increase recycling rates, and minimize environmental pollution. The Ivory Coast's commitment to waste management is demonstrated through initiatives such as the establishment of waste recycling facilities and the promotion of community-based waste management initiatives. These national strategies and initiatives demonstrate the Ivory Coast's commitment to promoting green growth, addressing climate change, and advancing sustainable development objectives in line with international agreements such as the Paris Agreement and the SDGs.

The Ivory Coast stands out in West Africa for embracing green growth as a cornerstone of its economic advancement and sustainable progress. With backing from the Global Green Growth Institute (GGGI), the country is leveraging green growth strategies and climate initiatives to tackle developmental hurdles and mitigate the impacts of climate change. These endeavors are currently manifested through various channels such as the Nationally Determined Contributions, SDGs, the NDP, and the Public Investment Program. Supported by the GGGI, Ivory Coast is actively pursuing several focal areas including green investments, climate action, resilient agriculture, sustainable forestry, and efficient mobility, all aimed at fostering green growth. The GGGI's green growth index serves as a metric to gauge progress toward achieving this vision in Ivory Coast (Figure 1).

Observations from Figure 1 reveal a pattern akin to the broader trend across the West African region, where strides toward achieving green growth are predominantly evident in the realms of efficient resource utilization and the preservation of natural capital. Here, growth indices consistently span from moderate to very high. However, the aspect of social inclusion demonstrates notable underperformance, marked by growth indices ranging from moderate to low. Regarding green economic prospects, green investment demonstrates significant strength with a notably high growth index, whereas green trade exhibits considerable weakness, characterized by a very low growth index. Additionally, the absence of data presents a significant obstacle to assessing the green growth index, particularly in areas such as sustainable land use, green employment, and innovation. Consequently, while the Ivory Coast shows commendable progress in efficient resource utilization and natural capital preservation, challenges persist in fostering social inclusion and capitalizing on green economic opportunities.

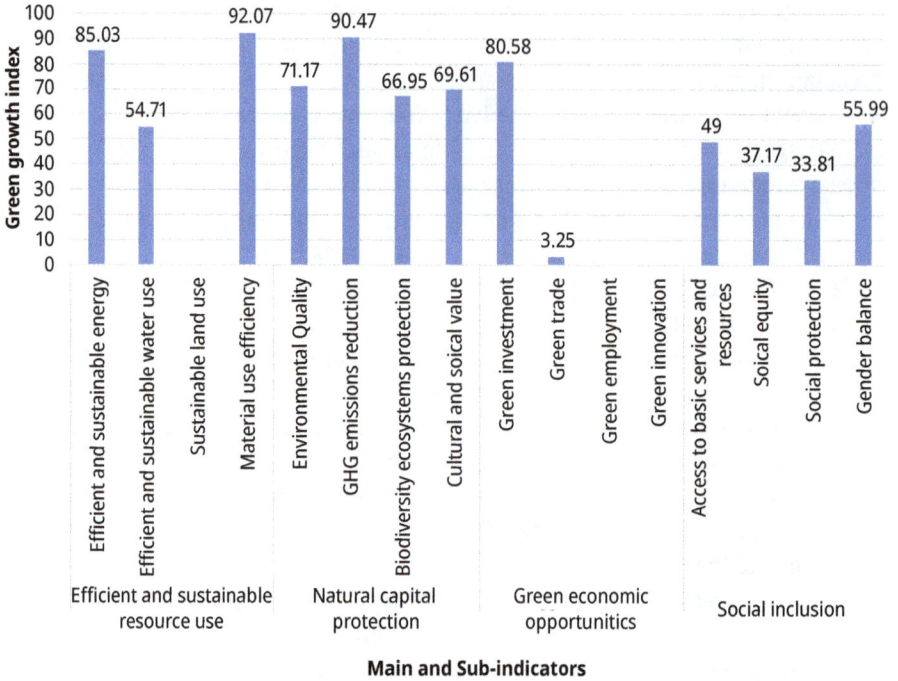

Figure 1: Green growth index for the Ivory Coast (Legend: 1–20 = very low; 20–40 = low; 40–60 = moderate; 60–80 = high; and 80–100 = very high). (source: AfDB, 2023b).

4.10 Sierra Leone

Sierra Leone has initiated several national strategies and policies to foster green growth and sustainable development with key examples including the National Renewable Energy Action Plan (NREAP), NCCP, National Forestry Policy, National Environmental Protection Agency (NEPA), National SDGs Framework, National Waste Management Strategy, National Water Resources Management Policy, and the National Coastal Zone Management Policy (Richards, 2016; Smith and Torrente-Murciano, 2021; Karim and Bah, 2022; Moinina, 2023). Sierra Leone's NREAP outlines strategies and targets for increasing the share of renewable energy in the country's energy mix. It aims to enhance energy access, reduce reliance on fossil fuels, and mitigate greenhouse gas emissions by promoting investments in renewable energy sources such as solar, wind, hydro, and biomass. Sierra Leone has developed a national policy framework to address climate change mitigation, adaptation, and resilience-building. The NCCP outlines sectoral priorities, policy interventions, and action plans to mainstream climate change considerations into national development planning and implementation. Sierra Leone's National Forestry Pol-

icy aims to promote sustainable forest management, conservation, and restoration to enhance biodiversity, mitigate climate change, and improve livelihoods for forest-
dependent communities. The policy includes measures to combat deforestation, promote
afforestation and reforestation, and strengthen community-based forest management.

Sierra Leone has established NEPA to coordinate and regulate environmental protection and management efforts in the country. NEPA plays a crucial role in implementing environmental policies, conducting environmental assessments, and promoting sustainable development practices. Sierra Leone has aligned its national development
priorities with the SDGs to promote inclusive and sustainable development. The government has developed a framework for implementing the SDGs, integrating environmental sustainability, social equity, and economic development objectives into national development planning and policy implementation. Sierra Leone has implemented a
national strategy to improve waste management and promote recycling and resource
efficiency. The strategy aims to reduce waste generation, increase recycling rates, and
minimize environmental pollution. Sierra Leone's commitment to waste management
is demonstrated through initiatives such as the establishment of waste recycling facilities and the promotion of community-based waste management initiatives. Sierra
Leone's National Water Resources Management Policy provides a policy framework for
promoting sustainable water management, conservation, and allocation to enhance
water security, support ecosystem health, and improve access to safe drinking water
and sanitation services. The policy emphasizes integrated water resources management, water conservation, and efficient water use practices. Sierra Leone has developed
a national policy framework for managing and protecting its coastal zones, which are
vulnerable to climate change impacts such as sea-level rise and coastal erosion. The policy aims to promote sustainable coastal development, conserve coastal ecosystems, and
enhance resilience to climate change impacts. These national strategies and policies
demonstrate Sierra Leone's commitment to promoting green growth, addressing climate change, and advancing sustainable development objectives in line with international agreements such as the Paris Agreement and the SDGs.

4.11 Mauritius

Mauritius has implemented several national strategies and initiatives to promote
green growth and sustainable development such as the Mauritius Sustainable Development Plan (SDP), NCCAP, Renewable Energy Roadmap, NBSAP, National Waste Management Strategy, National Water Resources Management Policy, and the National
Coastal Zone Management Policy (Chamroo, 2012; Teeluck et al., 2013; Sultan, 2013;
Jogoo, 2014; Sultan and Harsdorff, 2014; Bhiwajee and Docile, 2019). The Mauritius SDP
outlines the country's long-term vision for achieving sustainable development across
various sectors. It integrates environmental, social, and economic objectives to ensure
inclusive and environmentally sustainable growth. The plan emphasizes the need to

transition to a low-carbon economy, promote renewable energy, and enhance environmental conservation. The NCCAP provides a comprehensive framework for addressing climate change mitigation, adaptation, and resilience-building in Mauritius. It outlines sectoral priorities, policy interventions, and action plans to mainstream climate change considerations into national development planning and implementation. Mauritius has developed a Renewable Energy Roadmap to promote investments in renewable energy technologies and reduce reliance on fossil fuels (Greene-Dewasmes, 2021). The roadmap includes targets and strategies for increasing the share of renewable energy in the country's energy mix, with a focus on solar, wind, and biomass energy sources.

Mauritius has developed an NBSAP to conserve biodiversity, protect ecosystems, and promote sustainable use of natural resources. The plan includes measures to protect endangered species, establish protected areas, and promote community-based conservation initiatives. Mauritius has implemented a national strategy to improve waste management and promote recycling and resource efficiency. The strategy aims to reduce waste generation, increase recycling rates, and minimize environmental pollution. Mauritius' commitment to waste management is demonstrated through initiatives such as the establishment of waste recycling facilities and the promotion of community-based waste management initiatives. Mauritius' National Water Resources Management Policy provides a policy framework for promoting sustainable water management, conservation, and allocation to enhance water security, support ecosystem health, and improve access to safe drinking water and sanitation services. The policy emphasizes integrated water resources management, water conservation, and efficient water use practices. Mauritius has developed a national policy framework for managing and protecting its coastal zones, which are vulnerable to climate change impacts such as sea-level rise and coastal erosion. The policy aims to promote sustainable coastal development, conserve coastal ecosystems, and enhance resilience to climate change impacts. These national strategies and initiatives demonstrate Mauritius' commitment to promoting green growth, addressing climate change, and advancing sustainable development objectives in line with international agreements such as the Paris Agreement and the SDGs.

4.12 Namibia

Namibia has implemented several national strategies and initiatives to promote green growth and sustainable development such as the NCCP, NREP, NBSAP, National Water Policy, National Waste Management Strategy, National Forestry Policy, and the National SDG Framework (Harper, 2008; Kararach et al., 2018; Carver, 2020; Ruppel and Katoole, 2023). Namibia's NCCP provides a comprehensive framework for addressing climate change mitigation, adaptation, and resilience-building. It outlines sectoral priorities, policy interventions, and action plans to mainstream climate change considerations into national development planning and implementation. Namibia has developed an NREP to promote investments in renewable energy technologies and reduce reliance on fossil fuels. The policy includes targets and strategies for increasing the share of renewable energy in the country's energy mix, with a focus on solar, wind, and biomass energy sources (Von Oertzen, 2021; Monteith, 2022). Namibia has developed an NBSAP to conserve biodiversity, protect ecosystems, and promote sustainable use of natural resources. The plan includes measures to protect endangered species, establish protected areas, and promote community-based conservation initiatives. Namibia's National Water Policy provides a policy framework for promoting sustainable water management, conservation, and allocation to enhance water security, support ecosystem health, and improve access to safe drinking water and sanitation services. The policy emphasizes integrated water resources management, water conservation, and efficient water use practices.

Namibia has implemented a national strategy to improve waste management and promote recycling and resource efficiency. The strategy aims to reduce waste generation, increase recycling rates, and minimize environmental pollution. Namibia's commitment to waste management is demonstrated through initiatives such as the establishment of waste recycling facilities and the promotion of community-based waste management initiatives. Namibia has adopted a national forestry policy to promote sustainable forest management, conservation, and restoration to enhance biodiversity, mitigate climate change, and improve livelihoods for forest-dependent communities. The policy includes measures to combat deforestation, promote afforestation and reforestation, and strengthen community-based forest management. Namibia has aligned its national development priorities with the SDGs to promote inclusive and sustainable development. The government has developed a framework for implementing the SDGs, integrating environmental sustainability, social equity, and economic development objectives into national development planning and policy implementation (Wijesinghe and Thorn, 2021). These national strategies and initiatives demonstrate Namibia's commitment to promoting green growth, addressing climate change, and advancing sustainable development objectives in line with international agreements such as the Paris Agreement and the SDGs.

4.13 Botswana

Botswana has initiated several national strategies and policies to promote green growth and sustainable development namely the Botswana Climate Change Policy and Strategy, Botswana Renewable Energy Master Plan (BREMP), NBSAP, National Waste Management Strategy, National Water Policy, National Forestry Policy, and the National SDGs Framework (Kelly, 2010; Barbier, 2016; Mogende and Ramutsindela, 2020; Hikkaduwa Liyanage, 2021; Ofori et al., 2022). Botswana's Climate Change Policy and Strategy outline the country's approach to addressing climate change mitigation, adaptation, and resilience-building. It emphasizes the importance of mainstreaming climate change considerations into national development planning and implementing strategies to reduce greenhouse gas emissions and enhance climate resilience across various sectors. The BREMP aims to increase the share of renewable energy in Botswana's energy mix to enhance energy security, reduce reliance on fossil fuels, and mitigate greenhouse gas emissions. The plan focuses on promoting investments in renewable energy sources such as solar, wind, hydro, and biomass. Botswana has developed an NBSAP to conserve biodiversity, protect ecosystems, and promote sustainable use of natural resources. The plan includes measures to protect endangered species, establish protected areas, and promote community-based conservation initiatives.

Botswana has implemented a national strategy to improve waste management and promote recycling and resource efficiency. The strategy aims to reduce waste generation, increase recycling rates, and minimize environmental pollution. Botswana's commitment to waste management is demonstrated through initiatives such as the establishment of waste recycling facilities and the promotion of community-based waste management initiatives. Botswana's National Water Policy provides a policy framework for promoting sustainable water management, conservation, and allocation to enhance water security, support ecosystem health, and improve access to safe drinking water and sanitation services. The policy emphasizes integrated water resources management, water conservation, and efficient water use practices. Botswana has adopted a national forestry policy to promote sustainable forest management, conservation, and restoration to enhance biodiversity, mitigate climate change, and improve livelihoods for forest-dependent communities. The policy includes measures to combat deforestation, promote afforestation and reforestation, and strengthen community-based forest management. Botswana has aligned its national development priorities with the SDGs to promote inclusive and sustainable development (Banerjee et al., 2020). The government has developed a framework for implementing the SDGs, integrating environmental sustainability, social equity, and economic development objectives into national development planning and policy implementation. These national strategies and initiatives demonstrate Botswana's commitment to promoting green growth, addressing climate change, and advancing sustainable development objectives in line with international agreements such as the Paris Agreement and the SDGs.

4.14 Senegal

Senegal has implemented several national strategies and initiatives to promote green growth and sustainable development including the National Renewable Energy Strategy, National Climate Change Adaptation Plan, the Great Green Wall Initiative (GGWI), Emerging Senegal Plan (Plan Sénégal Émergent – PSE), NBSAP, National Sustainable Energy Strategy (NSES), National Waste Management Strategy, and the National Water Policy (Louveau, 2021; Tawiah et al., 2021; Mentz et al., 2022; Karambiri, 2022; Delay et al., 2022; Macia et al., 2023; Fassinou et al., 2024). Senegal has developed a National Renewable Energy Strategy to increase the share of renewable energy in its energy mix. The strategy aims to enhance energy security, reduce greenhouse gas emissions, and promote sustainable development through investments in renewable energy sources such as solar, wind, hydro, and biomass. Senegal has formulated a National Climate Change Adaptation Plan to address the impacts of climate change and enhance resilience across various sectors. The plan includes measures to promote climate-resilient agriculture, water management, coastal zone protection, and disaster risk reduction. The PSE is Senegal's national development strategy aimed at achieving sustainable and inclusive economic growth. It integrates green growth principles and objectives into its development agenda, focusing on sectors such as agriculture, energy, infrastructure, and urban development.

Senegal has developed an NBSAP to conserve biodiversity, protect ecosystems, and promote sustainable use of natural resources. The plan includes measures to protect endangered species, establish protected areas, and promote community-based conservation initiatives. Senegal's NSES outlines strategies and targets for promoting sustainable energy access, efficiency, and affordability. It aims to expand access to modern energy services, reduce reliance on traditional biomass fuels, and promote clean and efficient energy technologies. Senegal has implemented a national strategy to improve waste management and promote recycling and resource efficiency. The strategy aims to reduce waste generation, increase recycling rates, and minimize environmental pollution through initiatives such as waste sorting, recycling facilities, and public awareness campaigns. Senegal's National Water Policy provides a policy framework for promoting sustainable water management, conservation, and allocation to enhance water security, support ecosystem health, and improve access to safe drinking water and sanitation services. These national strategies and initiatives demonstrate Senegal's commitment to promoting green growth, addressing climate change, and advancing sustainable development objectives in line with international agreements such as the Paris Agreement and the SDGs.

4.15 Cameroon

Cameroon has implemented several national strategies and initiatives to promote green growth and sustainable development such as the NCCP, National Renewable Energy Plan, NBSAP, National Sustainable Development Plan, National Waste Management Strategy, National Water Policy, Operation Green Sahel, National Development Strategy (NDS30), and the National Forestry Policy (Ackah and Kizys, 2015; Kararach et al., 2018; Ndzembanteh, 2020; Ndzembanteh and Eryiğit, 2020; Tawiah et al., 2021). Cameroon's NCCP provides a framework for addressing climate change mitigation, adaptation, and resilience-building. It outlines sectoral priorities, policy interventions, and action plans to mainstream climate change considerations into national development planning and implementation. Cameroon has developed a National Renewable Energy Plan to increase the share of renewable energy in its energy mix. The plan aims to enhance energy security, reduce greenhouse gas emissions, and promote sustainable development through investments in renewable energy sources such as solar, wind, hydro, and biomass. Cameroon has developed an NBSAP to conserve biodiversity, protect ecosystems, and promote sustainable use of natural resources (Schoneveld and Zoomers, 2015). The plan includes measures to protect endangered species, establish protected areas, and promote community-based conservation initiatives.

Cameroon's National Sustainable Development Plan and the National Development Strategy (NDS30) integrate green growth principles and objectives into its development agenda (Awazi et al., 2024b). It aims to achieve sustainable and inclusive economic growth while promoting environmental conservation and social equity across various sectors. Cameroon has implemented a national strategy to improve waste management and promote recycling and resource efficiency. The strategy aims to reduce waste generation, increase recycling rates, and minimize environmental pollution through initiatives such as waste sorting, recycling facilities, and public awareness campaigns (Kimengsi and Fogwe, 2017; Ngwasiri et al., 2022). Cameroon's National Water Policy provides a policy framework for promoting sustainable water management, conservation, and allocation to enhance water security, support ecosystem health, and improve access to safe drinking water and sanitation services. Cameroon has adopted a national forestry policy to promote sustainable forest management, conservation, and restoration. The policy includes measures to combat deforestation, promote afforestation and reforestation, and strengthen community-based forest management (Assoua and Molua, 2018; Kimengsi et al., 2022). These national strategies and initiatives demonstrate Cameroon's commitment to promoting green growth, addressing climate change, and advancing sustainable development objectives in line with international agreements such as the Paris Agreement and the SDGs.

4.16 Common Ground Between African Countries in Their National Strategies to Promote Green Growth

Several common themes emerge across African countries in their national strategies to promote green growth and sustainable development. These common grounds include climate change mitigation and adaptation, renewable energy development, biodiversity conservation, sustainable natural resource management, waste management and pollution reduction, water resource management, sustainable agriculture and land use, and the community engagement and capacity building (Kararach et al., 2018; Bergius and Buseth, 2019; IMF, 2022). Many African countries prioritize climate change mitigation and adaptation in their national strategies. This involves reducing greenhouse gas emissions, enhancing resilience to climate impacts, and promoting low-carbon development pathways. African countries recognize the importance of transitioning to renewable energy sources to enhance energy security, reduce reliance on fossil fuels, and mitigate climate change. Strategies often include promoting investments in solar, wind, hydro, and biomass energy. Protecting biodiversity and ecosystems is a common goal across African countries. Strategies aim to conserve biodiversity, establish protected areas, and promote sustainable management of natural resources to support ecological balance and ecosystem services. African countries emphasize sustainable management of natural resources, including water, forests, fisheries, and minerals, to ensure their long-term availability and contribute to economic development while preserving environmental integrity.

Addressing waste management challenges and reducing pollution are key priorities for many African countries. Strategies focus on waste reduction, recycling, and pollution prevention to minimize environmental degradation and public health risks. Sustainable water management is critical for ensuring water security, supporting ecosystems, and enhancing resilience to climate change. Strategies include promoting integrated water resources management, water conservation, and access to safe drinking water and sanitation services. Promoting sustainable agriculture practices, land use planning, and land management is essential for food security, poverty reduction, and environmental sustainability. Strategies aim to improve agricultural productivity, conserve soil and water resources, and promote climate-smart agriculture. Many African countries emphasize community engagement, stakeholder participation, and capacity building in their strategies to promote green growth. This involves empowering local communities, fostering partnerships, and building institutional and human capacity for sustainable development. The national strategies for green growth in Africa demonstrate a commitment to sustainable development, environmental stewardship, and climate resilience, reflecting the continent's efforts to address pressing socioeconomic and environmental challenges while seizing opportunities for inclusive and sustainable growth. By focusing on these common grounds, African countries can work together to address shared environmental challenges, promote sustainable development, and achieve collective goals for a greener and more resilient future. Collaboration and knowledge sharing among African nations can further enhance the effectiveness of national strategies and contribute to regional and continental efforts toward sustainable development.

5 Conclusion and Perspectives

The transition to green growth in Africa is gaining momentum as governments and stakeholders recognize the pressing need for sustainable development amid mounting environmental challenges and socioeconomic inequality. As Africa grapples with the dual pressures of climate change and rapid population growth, green growth provides a pathway to a more resilient and inclusive economy. Currently, the continent remains heavily dependent on natural resources, with many countries still reliant on carbon-intensive industries. However, national strategies and frameworks are increasingly focusing on decarbonization, sustainable resource management, and fostering green innovation. The rationale for green growth in Africa is multifaceted. Environmental degradation, such as deforestation, soil erosion, and water scarcity, not only threatens the continent's natural resources but also jeopardizes the livelihoods of millions. Green growth seeks to reconcile economic development with environmental preservation, offering opportunities for renewable energy, sustainable agriculture, and circular economies. In the face of global pressure for reduced carbon emissions, Africa's vast renewable energy potential, such as solar and wind, positions it as a key player in the green energy revolution despite its relatively low historical emissions. National strategies, such as Africa's Agenda 2063, emphasize the integration of green growth into development plans. Countries like Kenya, South Africa, and Ethiopia have made significant strides by prioritizing renewable energy, green infrastructure, and climate-smart agriculture. However, these efforts are hindered by challenges including limited financing, weak institutional capacities, and the need for more robust policy frameworks. In perspective, the future of green growth in Africa hinges on overcoming these barriers while ensuring that the transition benefits all sectors of society. Governments must continue to develop and refine policies that incentivize green investments, encourage innovation, and support local communities. International cooperation, technology transfer, and financing mechanisms will also be crucial in facilitating a successful green transition. Africa's green growth agenda offers an opportunity not only to address climate change but also to drive sustainable, inclusive economic growth for future generations.

References

Abernethy, K., Maisels, F., & White, L. J. (2016). Environmental issues in central Africa. *Annual Review of Environment and Resources, 41,* 1–33.

Ackah, I. & Kizys, R. (2015). Green growth in oil producing African countries: A panel data analysis of renewable energy demand. *Renewable and Sustainable Energy Reviews, 50,* 1157–1166.

Adams, B. (2008). *Green Development: Environment and Sustainability in a Developing World.* Routledge.

Addisu, B. (2019). Green Economy: Challenges of Sustainable Consumption and Production in Ethiopia. *Journal of Economics and Sustainable Development.*

Adejumo, A. V. & Asongu, S. A. (2020). *Foreign Direct Investment, Domestic Investment and Green Growth in Nigeria: Any Spillovers?* (pp. 839–861). Springer International Publishing.

Adetiloye, K. A., Babajide, A. A., & Taiwo, J. N. (2022). Powering the Sustainable Development Goals for Green Growth in Nigeria. In *Research Anthology on Measuring and Achieving Sustainable Development Goals* (pp. 650–665). IGI Global.

AfDB. (2023b). *African Economic Outlook 2023: Mobilizing Private Sector Financing for Climate and Green Growth in Africa* (p. 222). The African Development Bank Group. https://www.afdb.org/sites/default/files/documents/publications/afdb23-01_aeo_main_english_0602.pdf.

AfDB. (2024). *African Economic Outlook 2024: Driving Africa's Transformation, the Reform of the Global Financial Architecture* (p. 252). The African Development Bank Group. https://www.afdb.org/en/documents/african-economic-outlook-2024.

Afionis, S., Mkwambisi, D. D., & Dallimer, M. (2020). Lack of cross-sector and cross-level policy coherence and consistency limits urban green infrastructure implementation in Malawi. *Frontiers in Environmental Science, 8,* 558619.

African Development Bank (AfDB). (2016). *Transitioning the African Continent toward Green Growth: An Introductory Guide to Understanding AfDB's Green Growth Framework.* Abidjan: The African Development Bank Group.

African Development Bank (AfDB). (2018). *Climate Change and Green Growth: 2018 Annual Report.* Abidjan: The African Development Bank Group.

African Development Bank (AfDB). (2019). *African Economic Outlook 2019: Integration for Africa's Economic Prosperity.* Abidjan: The African Development Bank Group.

African Development Bank (AfDB). (2021). *NDC Implementation in Africa through Green Investments by Private Sector: A Scoping Study.* Abidjan: The African Development Bank Group.

African Development Bank (AfDB). (2022a). *African Economic Outlook 2022: Supporting Climate Resilience and A Just Energy Transition in Africa.* Abidjan: The African Development Bank Group.

African Development Bank (AfDB). (2022b). *East Africa Economic Outlook 2022: Supporting Climate Resilience and A Just Energy Transition.* Abidjan: The African Development Bank Group.

African Development Bank (AfDB). (2023a). *Africa's Macroeconomic Performance and Outlook* January 2023 edition. Abidjan: The African Development Bank Group.

African Development Bank (AfDB) and GGGI. (2021). Africa Green Growth Readiness Assessment. *African Development Bank Group: Abidjan, and Global Green Growth Institute: Seoul.*

African Natural Resources Management and Investment Centre. (2022). *Debt for Nature Swaps – Feasibility and Policy Significance in Africa's Natural Resources Sector.* African Development Bank. Abidjan, Côte d'Ivoire.

Ainou, F. Z., Ali, M., & Sadiq, M. (2023). Green energy security assessment in Morocco: Green finance as a step toward sustainable energy transition. *Environmental Science and Pollution Research, 30*(22), 61411–61429.

Akanmu, A. O., Akol, A. M., Ndolo, D. O., Kutu, F. R., & Babalola, O. O. (2023). Agroecological techniques: Adoption of safe and sustainable agricultural practices among the smallholder farmers in Africa. *Frontiers in Sustainable Food Systems, 7,* 1143061.

Albagoury, S. (2016). *Inclusive Green Growth in Africa: Ethiopia Case Study.* Germany: University Library of Munich.

Ali, E. B., Anufriev, V. P., & Amfo, B. (2021). Green economy implementation in Ghana as a road map for a sustainable development drive: A review. *Scientific African, 12,* e00756.

Amankwah-Amoah, J. & Sarpong, D. (2016). Historical pathways to a green economy: The evolution and scaling-up of solar PV in Ghana, 1980–2010. *Technological Forecasting and Social Change, 102*, 90–101.

Amin, A., Naidoo, C., & Whitley, S. (2016). Green Growth in Practice: Lessons from Country Experiences. *Mobilising Investments*.

Anabaraonye, B., Ewa, B. O., Anukwonke, C. C., Eni, M., & Anthony, P. C. (2021). The role of green entrepreneurship and opportunities in agripreneurship for sustainable economic growth in Nigeria. *Covenant Journal of Entrepreneurship*.

Andersen, M. M., Ogallo, E., & Diniz Faria, L. G. (2022). Green economic change in Africa–green and circular innovation trends, conditions and dynamics in Kenyan companies. *Innovation and Development, 12*(2), 231–257.

Archer, E., Dziba, L., Mulongoy, K. J., Maoela, M. A., Walters, M., Biggs, R. O., . . . Sitas, N. (2018). The regional assessment report on biodiversity and ecosystem services for Africa: summary for policymakers.

Arimoro, A. E. (2021). Private Sector Investment in Infrastructure in Sub-Saharan Africa Post-COVID-19: The Role of Law. *Public Works Management & Policy, 0*(0), 1–19. https://doi.org/10.1177/1087724X211059531.

Asongu, S. A. & Odhiambo, N. M. (2020). Economic development thresholds for a green economy in sub-Saharan Africa. *Energy Exploration & Exploitation, 38*(1), 3–17.

Assoua, J. E. & Molua, E. L. (2018). Opportunities and Challenges of Sustainable Forest Management for a Green Economy Transition in Cameroon. *Journal of Economics and Sustainable Development, 9*(12), 1–8.

Auktor, G. V. (2020). *Green Industrial Skills for a Sustainable Future. United Nations Industrial Development Organization, Vienna*.

AVCA (2022). *2022 H1 Africa Private Capital Activity Report*, September, African Private Equity and Venture Capital Association, London, UK. Available.

Awazi, N. P., Kimengsi, J. N., Balgah, R. A., Mairomi, H. W., Tume, S. J. P., & Tsufac, A. R. (2024b). Bioeconomy transition for the attainment of Cameroon's National Development Strategy (NDS30) goal of environmental and nature protection: Assessing the all-encompassing contribution of agroforestry. In *Biodiversity and Bioeconomy* (pp. 325–345). Elsevier.

Azzeddine, B. B., Hossaini, F., & Savard, L. (2024). Greenhouse gas emissions and economic growth in Morocco: A decoupling analysis. *Journal of Cleaner Production, 450*, 141857.

Badraoui, M. & Dahan, R. (2011). The Green Morocco Plan in relation to food security and climate change. *Food Security and Climate Change in Dry Areas, 61*.

Banerjee, O., Bagstad, K. J., Cicowiez, M., Dudek, S., Horridge, M., Alavalapati, J. R., . . . Rutebuka, E. (2020). Economic, land use, and ecosystem services impacts of Rwanda's Green Growth Strategy: An application of the IEEM+ ESM platform. *Science of the Total Environment, 729*, 138779.

Barbier, E. B. (2016). Is green growth relevant for poor economies? *Resource and Energy Economics, 45*, 178–191.

Barbier, E. B. (2015). "Is Green Growth Relevant for Poor Economies?", Working Paper 44, FERDI.

Bari, M. & Dessus, S. (2022). "Adapting to natural disasters in Africa: What's in it for the private sector?", IFC Working Paper, November.

Bass, S., Wang, S. S., Ferede, T., & Fikreyesus, D. (2013). Making growth green and inclusive: The case of Ethiopia.

Basubas, D. E. (2023). *A critical examination of the 'green economy' concept as a vehicle for inclusion: a qualitative case study of the Agulhas Plain, South Africa* (Doctoral dissertation, University of Otago).

Bergius, M. & Buseth, J. T. (2019). Towards a green modernization development discourse? The new, green revolution in Africa.

Bergius, M., Benjaminsen, T. A., & Widgren, M. (2018). Green economy, Scandinavian investments and agricultural modernization in Tanzania. *The Journal of Peasant Studies, 45*(4), 825–852.

Bergius, M., Benjaminsen, T. A., Maganga, F., & Buhaug, H. (2020). Green economy, degradation narratives, and land-use conflicts in Tanzania. *World Development, 129*, 104850.

Bhiwajee, S. L. & Docile, R. P. (2019). Adoption of green jobs in Mauritius: Drivers and challenges. *International Journal of Business and Economic Development (IJBED), 7*(2).

Bilame, O. (2023). Inclusive Green Growth and Shared Prosperity: Are they Basic Indictors for Tanzania to Attain an Upper Middle-Income Country? A Theoretical Review. *Tanzania Journal of Development Studies, 21*(1).

Borchgrevink, A. (2014). Ethiopia: Rapid and green growth for all? In *Emerging Economies and Challenges to Sustainability* (pp. 247–260). Routledge.

Borel-Saladin, J. M. & Turok, I. N. (2013). The impact of the green economy on jobs in South Africa: News & views. *South African Journal of Science, 109*(9), 1–4.

Burkolter, P. & Perch, L. (2014). Greening growth in the south: Practice, policies and new frontiers. *South African Journal of International Affairs, 21*(2), 235–259.

Buseth, J. T. (2017). The green economy in Tanzania: From global discourses to institutionalization. *Geoforum, 86*, 42–52.

Carver, R. (2020). Lessons for blue degrowth from Namibia's emerging blue economy. *Sustainability Science, 15*(1), 131–143.

Chamroo, D. (2012). Developing green industry in Mauritius. In *International Trade Forum* (Vol. 4, p. 22). International Trade Centre.

Chandrashekeran, S., Morgan, B., Coetzee, K., & Christoff, P. (2017). Rethinking the green state beyond the Global North: A South African climate change case study. *Wiley Interdisciplinary Reviews: Climate Change, 8*(6), e473.

Chavula, P. & Turyasingura, B. (2022). Critical thinking on green economy for sustainable development in Africa. *International Journal of Academic and Applied Research, 6*(8), 181–188.

Chirambo, D. (2018). Towards the achievement of SDG 7 in sub-Saharan Africa: Creating synergies between Power Africa, Sustainable Energy for All and climate finance in-order to achieve universal energy access before 2030. *Renewable and Sustainable Energy Reviews, 94*, 600–608.

Chisholm, J. M., Zamani, R., Negm, A. M., Said, N., Abdel Daiem, M. M., Dibaj, M., & Akrami, M. (2021). Sustainable waste management of medical waste in African developing countries: A narrative review. *Waste Management & Research, 39*(9), 1149–1163.

Chuku, C. & Ajayi, V. (2024). Growing green: Enablers and barriers for Africa. *Journal of Productivity Analysis, 61*(3), 195–214.

Chukwuemeka, N. S., Ugonna, A. P., Ugochukwu, O. B., Immaculata, E. N., Chiziterem, E. K., & Chukwubuikem, O. P. (2023). The Challenges and Opportunities of Energy Transition across Africa. *International Journal of Environment and Climate Change, 13*(10), 4312–4339.

CPI. (2022a). *Landscape of Climate Finance in Africa*. San Francisco: Climate Policy Initiative (CPI).

CPI. (2022b). *The State of Climate Finance in Africa: Climate Finance Needs of African Countries*. San Francisco: Climate Policy Initiative (CPI).

CPI. (2022c). *Climate Finance Innovation for Africa*. San Francisco: Climate Policy Initiative (CPI).

Death, C. (2014). The green economy in South Africa: Global discourses and local politics. *Politikon, 41*(1), 1–22.

Degbedji, D. F., Akpa, A. F., Chabossou, A. F., & Osabohien, R. (2024). Institutional quality and green economic growth in West African economic and monetary union. *Innovation and Green Development, 3*(1), 100108.

Degila, J., Assogbadjo, A., Avakoudjo, H., Souand, T. A. H. I., & Houetohossou, A. C. (2022). Accelerating inclusive green growth through agri-based innovation in Western Africa (AGriDI)'.

Delay, E., Ka, A., Niang, K., Touré, I., & Goffner, D. (2022). Coming back to a Commons approach to construct the Great Green Wall in Senegal. *Land Use Policy, 115*, 106000.

Dovie, D. B. (2017). A communication framework for climatic risk and enhanced green growth in the eastern coast of Ghana. *Land Use Policy, 62*, 326–336.

Dupar, M., Henriette, E., & Hubbard, E. (2023). Nature-based green infrastructure: A review of African experience and potential.

Durán-Díaz, P. (2023). Sustainable land governance for water–energy–food systems: A framework for rural and peri-urban revitalisation. *Land, 12*(10), 1828.

EIB. (2021). *Finance in Africa: For Green, Smart and Inclusive Private Sector Development*. Luxembourg: European Investment Bank.

Elmoukhtar, M., Touhami, F., Taouabit, O., & Mouhtat, I. (2022). Promoting Green Entrepreneurship in Morocco as a Roadmap to Sustainable Development: A literature Review. *International Journal of Accounting, Finance, Auditing, Management and Economics, 3*(4), 174–190.

Elum, Z. A. & Mjimba, V. (2020). Potential and challenges of renewable energy development in promoting a green economy in Nigeria. *Africa Review, 12*(2), 172–191.

Fadly, D. (2019). Low-carbon transition: Private sector investment in renewable energy projects in developing countries. *World Development, 122*(October), 552–569. https://doi.org/10.1016/j.worlddev.2019.06.015.

Fassinou, F. J. C., Cesaro, J. D., Nungi-Pambu, M., Fensholt, R., Brandt, M., Akodewou, A., . . . Taugourdeau, S. (2024). Quantifying the impact of Great Green Wall and Corporate plantations on tree density and biomass in Sahelian Senegal. *Trees, Forests and People, 16*, 100569.

Fonta, W. M., Ayuk, E. T., & Van Huysen, T. (2018). Africa and the Green Climate Fund: Current challenges and future opportunities. *Climate Policy, 18*(9), 1210–1225.

Freire, M. E. (2013). Urbanization and green growth in Africa. *The Growth Dialogue, 1*, 1–38.

GCA and CPI (2021). *Financing Innovation for Climate Adaptation in Africa*.

Greene-Dewasmes, G. S. E. (2021). *A Comparative Exploration of Green Energy Transition Outcomes across Small Island Developing States: Jamaica, Barbados, and Mauritius* Doctoral dissertation. University of York).

Gu, J., Renwick, N., & Xue, L. (2018). The BRICS and Africa's search for green growth, clean energy and sustainable development. *Energy Policy, 120*, 675–683.

Günay, C., Haddad, C., Gharib, S., Jamea, E. M., Zejli, D., & Komendantova, N. (2018). Green growth and its global-local meanings – Insights from Morocco.

Harper, S. A. (2008). *Towards the Development of a "green" Worldview, and Criteria to Assess the "green-ness" of a Text: Namibia Vision 2030 as Example* (Doctoral dissertation, University of Pretoria).

Heldt, S. (1977). *Towards the attainment of self-sustaining growth: Ghana and the Ivory Coast* (No. 57). Kiel Working Paper.

Hickel, J. & Kallis, G. (2020). Is green growth possible? *New Political Economy, 25*(4), 469–486.

Hikkaduwa Liyanage, S. I. (2021). *A framework for Greening Universities in Knowledge Based Economy, Botswana* (Doctoral dissertation, North-West University (South Africa)).

Hudani, S. E. (2020). The green masterplan: Crisis, state transition and urban transformation in post-genocide Rwanda. *International Journal of Urban and Regional Research, 44*(4), 673–690.

IEA (2022). *Africa Energy Outlook 2022*, Paris, France.

Ike, A. N. (2019). Green Growth and Sustainable Economic Development in Nigeria: Benefits and Challenges. *International Journal, 6*(1).

IMF. (2022). *Regional Economic Outlook. SSA: Living on the Edge* October edition. Washington, DC: International Monetary Fund.

IMF, OECD, UN and the World Bank (2015). "Options for Low Income Countries' Effective and Efficient Use of Tax Incentives for Investment", A Report to The G-20 Development Working Group.

Jabik, B. B. & Bawakyillenuo, S. (2016). Green entrepreneurship for sustainable development in Ghana: A review. *GHANA SOCIAL SCIENCE, 13*(2), 96.

Jogoo, V. K. (2014). The Political Economy of Transitioning to a Green Economy in Mauritius. *Transitioning to a Green Economy: Political Economy of Approaches in Small States*, 128–153.

Kakou, A. E. V. (2023). *Institutional Framework for Green Hydrogen Project in West Africa: The Case Study of Cote d'Ivoire* Doctoral dissertation. WASCAL).

Karambiri, M. (2022). Analysis of policies relevant to the Great Green Wall Initiative in Senegal.

Kararach, G., Nhamo, G., Mubila, M., Nhamo, S., Nhemachena, C., & Babu, S. (2018). Reflections on the Green Growth Index for developing countries: A focus of selected African countries. *Development Policy Review, 36*, O432–O454.

Karim, S. & Bah, C. J. (2022). An analysis of the impacts of climate change on green economy of Sierra Leone.

Kassa, H., Seyoum, Y., Tolera, M., Tadesse, W., Tesfaye, Y., & Asfaw, Z. (2015). Enhancing the Role of the Forestry Sector in Building Climate Resilient Green Economy in Ethiopia.

Kedir, A. M. (2014). Debating critical issues of green growth and energy in Africa: Thinking beyond our lifetimes. *Managing Africa's Natural Resources: Capacities for Development*, 185–205.

Kelly, S. (2010). Rethinking Socio-economic and Political Institutions in Botswana in Light of HIV, the" Green Movement" and Globalization. *International Journal of Interdisciplinary Social Sciences, 5*(8).

Khisa, K. (2016). *Development of an Industrial Ecology Model for the Athi River Special Economic Zone: Policy Implications for Green Growth in Kenya*. Doctoral dissertation, University of Nairobi.

Kimengsi, J. N. & Fogwe, Z. N. (2017). Urban green development planning opportunities and challenges in sub-saharan Africa: Lessons from Bamenda city, Cameroon. *International Journal of Global Sustainability, 1*(1), 1–17.

Kimengsi, J. N., Forje, G. W., & Awazi, N. P. (2022). Non-timber forest products and bioeconomy transitioning in Cameroon: Potentials and challenges. *The Bioeconomy and Non-timber Forest Products*, 109–127.

King-Okumu, C. (2015). Inclusive green growth in Kenya: Opportunities in the dryland water and rangeland sectors: Study in support of the Danish Green Growth and Employment Programme in Kenya 2015–2020. In *Inclusive Green Growth in Kenya: Opportunities in the Dryland Water and Rangeland Sectors: Study in Support of the Danish Green Growth and Employment Programme in Kenya 2015–2020: King-Okumu, Caroline*. London, UK: International Institute for Environment and Development.

Lambert, E. & Deyganto, K. O. (2024). The Impact of Green Legacy on Climate Change in Ethiopia. *Green and Low-Carbon Economy, 2*(2), 97–105.

Louveau, F. (2021). The Spirits of the Great Green Wall in Senegal: Spirituality, Ecology, and Secularization. *Journal for the Study of Religion, Nature & Culture, 15*(3).

Macia, E., Allouche, J., Sagna, M., Diallo, A. H., Boëtsch, G., Guisse, A., . . . Duboz, P. (2023). The Great Green Wall in Senegal: Questioning the idea of acceleration through the conflicting temporalities of politics and nature among the Sahelian populations.

Maconachie, R. (2019). Green grabs and rural development: How sustainable is biofuel production in post-war Sierra Leone? *Land Use Policy, 81*, 871–877.

Makate, C., Makate, M., Mango, N., & Siziba, S. (2019). Increasing resilience of smallholder farmers to climate change through multiple adoption of proven climate-smart agriculture innovations. Lessons from Southern Africa. *Journal of Environmental Management, 231*, 858–868.

Martinelli, L. T. (2022). Debt Relief as an Instrument for Promoting Green Growth–The Case of Ghana.

Mathez, A. & Loftus, A. (2023). Endless modernisation: Power and knowledge in the Green Morocco Plan. *Environment and Planning E: Nature and Space, 6*(1), 87–112.

McKinsey & Co. (2021). *Solving Africa's Infrastructure paradox.*

Menkeh, M. I. (2021). Cocoa sustainability in Ghana and Ivory Coast: The role of green financing. *Global Journal of Business and Integral Security, 1*(2).

Mentz, S., Karambiri, M., & Smith Dumont, E. (2022). The Great Green Wall Initiative in Senegal-Country Review.

Mihayo, I. Z. (2020). Role of policies in the sustainability of fish species in Lake Victoria: A pathway to the green economy in Tanzania. *International Journal of Agricultural Science, 5.*

Mihayo, I. Z. & Swai, R. M. (2019). Green economy in Tanzania: Is it foreseeable. *Journal of Applied and Advanced Research, 4*(4), 112–118.

Mogende, E. & Ramutsindela, M. (2020). Political leadership and non-state actors in the greening of Botswana. *Review of African Political Economy, 47*(165), 399–415.

Mohamed, N. & Montmasson-Clair, G. (2018). Policies for sustainability transformations in South Africa: A critical review. *Sustainability Transitions in South Africa,* 80–100.

Mohsin, M., Taghizadeh-Hesary, F., Iqbal, N., & Saydaliev, H. B. (2022). The role of technological progress and renewable energy deployment in green economic growth. *Renewable Energy, 190,* 777–787.

Moinina, V. (2023). Understanding the Green Economy of African Politics and Democracy-A Case of Sierra Leone 2018–2023. *Disiplinlerarası Afrika Çalışmaları, 1*(2), 205–227.

Monteith, W. (2022). Critical cartographies of the green hydrogen rush in Namibia.

Moore, L. (2019). Ambitions for Greening Solid Waste Management: Perspectives from Urban (ising) Africa. *JSTOR Sustainability Collection.*

Muhammad, A., Ibitomi, T., Amos, D., Idris, M., & Ahmad Ishaq, A. (2023). Comparative Analysis of sustainable finance initiatives in Asia and Africa: A Path towards Global Sustainability. *Glob. Sustain. Res, 2,* 33–51.

Mulugetta, Y. (2016). ETHIOPIA – Green Economy in Policy-Making Processes. *The Rise of the Green Economies,* 165.

Mungai, E. M., Ndiritu, S. W., & Da Silva, I. (2022). Unlocking climate finance potential and policy barriers – A case of renewable energy and energy efficiency in Sub-Saharan Africa. *Resources, Environment and Sustainability, 7,* 100043.

Musango, J. K., Brent, A. C., & Bassi, A. M. (2014). Modelling the transition towards a green economy in South Africa. *Technological Forecasting and Social Change, 87,* 257–273.

Musvoto, C., Nortje, K., De Wet, B., Mahumani, B. K., & Nahman, A. (2015). Imperatives for an agricultural green economy in South Africa. *South African Journal of Science, , 111*(1–2), 01–08.

Muthama, N. J. (2021). Transitioning to green growth in Kenya: The Horticulture Productivity, Fuel Consumption and Short-Lived Climate Pollutants nexus. In *East African Journal of Science, Technology and Innovation.* Vol. 2.

Ndzembanteh, A. N. (2020). *The Role of Financial Development, Human Capital, and Economic Growth on Environmental Sustainability: An Empirical Analysis of Cameroon.* Turkey: Doctoral dissertation, Bursa Uludag University.

Ndzembanteh, A. N. & Eryiğit, K. Y. (2020). The role of financial development, education and economic growth on environmental quality in Cameroon. *Journal of Business Economics and Finance, 9*(3), 232–244.

Neves, J. L., Rocha, V., & Rocha, D. K. (2022). The Importance of Nature-Based Solutions to Enhance Cabo Verde's Environment. In *Enhancing Environmental Education Through Nature-Based Solutions* (pp. 63–81). Cham: Springer International Publishing.

Ngondjeb, D. Y., Atewamba, C., & Macalou, M. (2020). Insights on Africa's future in its transition to the green economy. *Inclusive Green Growth: Challenges and Opportunities for Green Business in Rural Africa,* 309–326.

Ngwasiri, P. N., Ambindei, W. A., Adanmengwi, V. A., Ngwi, P., Mah, A. T., Ngangmou, N. T., . . . Aba, E. R. (2022). A Review Paper on Agro-food Waste and Food by-Product Valorization into Value Added

Products for Application in the Food Industry: Opportunities and Challenges for Cameroon Bioeconomy. *Asian Journal of Biotechnology and Bioresource Technology, 8*(3), 32–61.

Nhamo, S. & Nhamo, G. (2014). Mainstreaming green economy into sustainable development policy frameworks in SADC. *Environmental Economics, 5*(2), 55–65.

Niyigaba, J., Ya Sun, J., Peng, D., & Uwimbabazi, C. (2020). Agriculture and Green Economy for Environmental Kuznets Curve Adoption in Developing Countries: Insights from Rwanda. *Sustainability, 12*(24), 10381.

Nwokolo, S. C., Singh, R., Khan, S., Kumar, A., & Luthra, S. (2023). Remedies to the Challenges of Renewable Energy Deployment in Africa. In *Africa's Path to Net-Zero: Exploring Scenarios for a Sustainable Energy Transition* (pp. 59–74). Cham: Springer Nature Switzerland.

Nyika, J. M. (2021). Green energy technologies as the road map to sustainable economic growth in Kenya. In *Eco-Friendly Energy Processes and Technologies for Achieving Sustainable Development* (pp. 167–184). IGI Global.

OECD. (2020). *Blended Finance in the Least Developed Countries 2020 Supporting a Resilient COVID-19 Recovery.* Paris: OECD Publishing.

OECD (2021). Scaling up green, social, sustainability and sustainability-linked bond issuances in developing countries.

OECD. (2023). Private finance mobilised by official development finance interventions. In *Development Co-operation Directorate*. Paris: OECD Publishing.

OECD, World Bank and UN Environment. (2018). *Financing Climate Futures: Rethinking Infrastructure.* Paris: OECD Publishing. https://doi.org/10.1787/9789264308114-en.

Ofori, I. K., Gbolonyo, E. Y., & Ojong, N. (2022). Towards Inclusive Green Growth in Africa: Critical energy efficiency synergies and governance thresholds. *Journal of Cleaner Production, 369*, 132917.

Okereke, C., Coke, A., Geebreyesus, M., Ginbo, T., Wakeford, J. J., & Mulugetta, Y. (2019). Governing green industrialisation in Africa: Assessing key parameters for a sustainable socio-technical transition in the context of Ethiopia. *World Development, 115*, 279–290.

Okoh, A. I. S. (2014). Green Economy Vs Green Growth: A Quest for an Ecologically Sustainable Polity in Nigeria. *SCSR Journal of Development, 1*(4), 22–40.

Onyeji-Nwogu, I. (2017). Harnessing and integrating Africa's renewable energy resources. In *Renewable Energy Integration* (pp. 27–38). Academic Press.

Owino, T., Kamphof, R., Kuneman, E., Van Tilburg, X., Van Schaik, L., & Rawlins, J. (2016). *Towards a 'Green' trajectory of Economic Growth and Energy Security in Kenya?* (pp. 7). Clingendael Institute.

Oyebanji, I. J., Bamidele, A., Khobai, H., & Le Roux, P. (2017). Green growth and environmental sustainability in Nigeria. *International Journal of Energy Economics and Policy, 7*(4), 216–223.

Paul, C. J. & Weinthal, E. (2019). The development of Ethiopia's Climate Resilient Green Economy 2011–2014: Implications for rural adaptation. *Climate and Development, 11*(3), 193–202.

Pausata, F. S., Gaetani, M., Messori, G., Berg, A., De Souza, D. M., Sage, R. F., & DeMenocal, P. B. (2020). The greening of the Sahara: Past changes and future implications. *One Earth, 2*(3), 235–250.

Pedercini, M., Zuellich, G., Dianati, K., & Arquitt, S. (2018). Toward achieving sustainable development goals in Ivory Coast: Simulating pathways to sustainable development. *Sustainable Development, 26*(6), 588–595.

Perry, W. (2020). Social sustainability and the argan boom as green development in Morocco. *World Development Perspectives, 20*, 100238.

Popp, D. (2012). The role of technological change in green growth.

Raoux, J. (2024). *Greening the Economy in Ethiopia: A Critical Discourse Analysis of Ethiopia's Climate.* Resilient Green Economy Strategy.

Resnick, D., Tarp, F., & Thurlow, J. (2012). The political economy of green growth: Cases from Southern Africa. *Public Administration and Development, 32*(3), 215–228.

Richards, P. (2016). Toward an African green revolution? An anthropology of rice research in Sierra Leone. In *Ecology of Practice* (pp. 201–252). Routledge.

Rihab, B. (2019). Sustainable Development in Morocco: The Green Future. *Economic and Social Development: Book of Proceedings*, 285–293.

Ruppel, O. C. & Katoole, M. (2023). A regulatory green hydrogen framework for Namibia.

Satgar, V. (2014). South Africa's Emergent 'Green Developmental State'? In *The End of the Developmental State?* (pp. 126–153) Routledge.

Scherr, S. J., Milder, J. C., Buck, L. E., Hart, A. K., & Shames, S. A. (2013). *A Vision for Agriculture Green Growth in the Southern Agricultural Growth Corridor of Tanzania (SAGCOT): Overview. Dar es Salaam: SAGCOT Centre*.

Schoneveld, G. & Zoomers, A. (2015). Natural resource privatisation in Sub-Saharan Africa and the challenges for inclusive green growth. *International Development Planning Review*, 37(1), 95–118.

Schwerhoff, G. & Sy, M. (2017). Financing renewable energy in Africa – Key challenge of the sustainable development goals. *Renewable and Sustainable Energy Reviews*, 75, 393–401.

Seidu, D. (2020). Green Recovery and Green Jobs in Africa: The Case of Ghana. *South African Institute of International Affairs*.

Shilomboleni, H. (2017). A sustainability assessment framework for the African green revolution and food sovereignty models in southern Africa. *Cogent Food & Agriculture*, 3(1), 1328150.

Simbanegavi, W. (2019). Expediting growth and development: Policy challenges confronting Africa. *Journal of Development Perspectives*, 3(1–2), 46–79.

Smith, C. & Torrente-Murciano, L. (2021). The potential of green ammonia for agricultural and economic development in Sierra Leone. *One Earth*, 4(1), 104–113.

Steiner, A. (2008). Africa's natural resources key to powering prosperity. *Environment and Poverty Times*, 5.

Sultan, R. M. (2013). A green industry for sustainable trade strategies: The case of the manufacturing sector in Mauritius. *International Journal of Green Economics*, 7(2), 162–180.

Sultan, R. & Harsdorff, M. (2014). *Green Jobs Assessment: Mauritius*. ILO.

Swilling, M., Musango, J. K., & Wakeford, J. (Eds.). (2016). Greening the South African Economy: Scoping the issues, challenges and opportunities.

Swilling, M., Musango, J., & Wakeford, J. (2016). Developmental states and sustainability transitions: Prospects of a just transition in South Africa. *Journal of Environmental Policy & Planning*, 18(5), 650–672.

Tawiah, V., Zakari, A., & Adedoyin, F. F. (2021). Determinants of green growth in developed and developing countries. *Environmental Science and Pollution Research*, 28(29), 39227–39242.

Tearfund (2022). *Dying to adapt: A comparison of African healthcare spending and climate adaptation costs*, United Kingdom.

Teeluck, S., Pudaruth, S., & Kishnah, S. (2013). How Green is Mauritius? *International Journal of Computer Applications*, 74(19).

The Africa CEO Forum (2022). "Six Key Recommendations on Climate Finance for African Growth: What can public and private decision-makers do to move faster?"

Theodory, T. F. (2022). Framing the forests future: A transition to green growth among the forests dependent communities in Mafinga Town Council, Tanzania. *African Journal of Economic and Sustainable Development*, 9(1), 68–85.

Thorn, J. P. R., Hejnowicz, A. P., Marchant, R., Ajala, O. A., Delgado, G., Shackleton, S., . . . Cinderby, S. (2021). Dryland nature-based solutions for informal settlement upgrading schemes in Africa.

Tiamgne, X. T., Kalaba, F. K., & Nyirenda, V. R. (2022). Mining and socio-ecological systems: A systematic review of Sub-Saharan Africa. *Resources Policy*, 78, 102947.

Tsegay, B. (2023). *Green Economy for Climate Change Mitigation and Poverty Reduction in Sub-Saharan Africa: A Critical Analysis of Carbon Finance in Ethiopia*. Doctoral dissertation, SOAS University of London.

Tunji-Olayeni, P. F., Mosaku, T. O., Oyeyipo, O. O., & Afolabi, A. O. (2018). Sustainability strategies in the construction industry: Implications on Green Growth in Nigeria. In *IOP Conference Series: Earth and Environmental Science*. (Vol. 146, No. 1, p. 012004). IOP Publishing.

UN (2015). Addis Ababa Action Agenda of the Third International Conference on Financing for Development (Addis Ababa Action Agenda).

UNECA (2020). *Economic Report on Africa 2020: Innovative Finance for Private Sector Development in Africa Development in Africa*, Addis Ababa, Ethiopia.

UNFCCC (2015). *Paris Agreement*.

Vazquez-Brust, D., Smith, A. M., & Sarkis, J. (2014). Managing the transition to critical green growth: The 'Green Growth State'. *Futures*, *64*, 38–50.

Von Oertzen, D. (2021). Issues, challenges and opportunities to develop green hydrogen in Namibia.

Wijesinghe, A. & Thorn, J. P. (2021). Governance of urban green infrastructure in informal settlements of Windhoek, Namibia. *Sustainability*, *13*(16), 8937.

Wodajo, B. T. (2021). "Decarbonisation of transport in Africa: A Transport Planning Perspective", Summary report of IAP-NASAC workshop November 15–17, 2021.

World Bank (2019). The Role of the Public Sector in Mobilizing Commercial Finance for Grid-Connected Solar Projects: Lessons Learned and Case Studies.

World Bank et al. (2022). *Off-Grid Solar Market Trends Report 2022: State of the Sector*, Washington, DC: The World Bank.

Yeboah, O. A., Amoah, N. M., Fuseini, S., & Sugri, I. (2023). The impact of the local green economy of Ghana: A general equilibrium analysis. *Sustainability*, *15*(23), 16358.

Zaatari, R. (2022). The green economy of Morocco (Doctoral dissertation, University of Glasgow).

Zahir, K., Nechad, A., & Kasbaoui, T. (2022). Green Economy: Challenges and Opportunities in Morocco. Economic and Social Development: Book of Proceedings, 228–235.

Devansh Jain, Harshita Jain, Suresh Singh Kushwah,
Vijay Singh Solanki, Laxmi Narayan Malviya, and Dungar Singh*

Enhancing Sustainability Through Plastic Waste Utilization in Construction: A Review

Abstract: The global consumption of plastics has steadily increased due to their widespread and unavoidable use, resulting in tons of plastic waste. This waste is nonbiodegradable, and its increasing volume poses significant environmental challenges. To mitigate these harmful effects, plastic waste can be recycled and reused as an alternative construction material. Plastics are strong, durable, waterproof, lightweight, and flexible, making them suitable for various construction applications. This chapter examines numerous studies that explore the use of plastic waste in construction materials. The focus is on recycling plastic waste and its application in road construction, concrete, cementitious composites, bricks, and blocks. The chapter is divided into three sections, each addressing the utilization of plastic waste in these specific areas. The findings indicate that incorporating plastic waste into construction materials significantly enhances environmental sustainability and offers economic benefits. The researcher highlighted the optimum inclusion of plastic waste improved compressive strength in concrete by up to 17 MPa, reduced water absorption below 0.7%, and decreased bitumen usage in asphalt by 10%. Moreover, recycled plastics prove to be reliable materials for construction, contributing to the development of sustainable and durable infrastructure.

Keywords: Construction materials, plastic, plastic waste, recycling, polyethylene terephthalate

1 Introduction

Plastic is one of the most widely used fine materials in the world and brings comfort and convenience to human life. The production of plastic has grown unusually fast in recent decades due to the use of other man-made materials. This event generates 300 million tons of plastic waste every year (Wang et al., 2020), of which only 8% are incinerated, 7% are recycled, and the rest are dumped (Nyika and Dinka, 2022). Plastics have become a major environmental concern, with approximately 8,300 million

---

*Corresponding author: Dungar Singh, L. N. Malviya Infra Projects Pvt. Ltd., Bhopal 462023, Madhya
Pradesh, India, e-mail: dsdudi97@gmail.com
Devansh Jain, Vijay Singh Solanki, Laxmi Narayan Malviya, L. N. Malviya Infra Projects Pvt. Ltd.,
Bhopal 462023, India
Harshita Jain, Jai Narain College of Technology, Bhopal 462038, India
Suresh Singh Kushwah, Rajiv Gandhi Proudyogiki Vishwavidyalaya, Bhopal 462033, India

https://doi.org/10.1515/9783111563046-006

metric tons of virgin plastics produced to date. As of 2015, around 9% of the 6,300 million metric tons of plastic waste generated had been recycled, 12% incinerated, and 79% accumulated in landfills or the natural environment. If current trends continue, an estimated 12,000 million metric tons of plastic waste will be in landfills or the natural environment by 2050 (Geyer et al., 2017). The recent evaluation in India produces plastic waste of around 26,000 tons every day if these tons of waste will not be collected and will produce garbage (de Hita et al., 2018). Plastic waste disposal has become a major environmental issue because of its degradability. Since plastic is not biodegradable and the quantity of waste plastic in our environment is gradually increasing. Plastic waste is often the most unwanted waste and can be seen in landfills without decomposing for several months (Taghavi et al., 2021, Rochman, 2013). Remaining plastic waste is discarded in our surroundings or in large bodies of water, resulting as negative environmental effects, such as soil pollution due to the release of toxic chemicals, impeded groundwater movement and pollution, marine pollution, unbalanced water and aquatic life survival, and a low aesthetic value for the environment (Kamaruddin et al., 2017). The COVID-19 pandemic has worsened the situation by encouraging the use of plastic in the production of hand sanitizer dispensers, masks, gloves, and personal protective equipment (PPE) kits, resulting in the pandemic of plastic pollution (Nyika and Dinka, 2022). Although, certain varieties of plastic toxins can also be released into the atmosphere, which means that plastic waste dumping in landfills is not a sustainable solution. The incineration process entirely removes these plastic wastes and can be a source of energy, carbon dioxide, and other toxic chemicals, creating toxic bottom and fly ash, which are typically delivered by this method (Altieri et al., 2021). Plastic waste is often improperly discarded whether dumped in landfills or littered in the environment leading to ecological harm. A mainstream of life cycle assessment (LCA) studies associating recycling of plastic with different waste management choices have concluded that there are usually optimistic outcomes, including an important modification of environmental influences. Also, one of the possible approaches to plastic waste recycling is incorporating it into a building material (del Rey Castillo et al., 2020).

The possibility of using plastic waste as a building material will decrease the entire environmental risk of this plastic production. Additionally, the potential application of plastic waste in building materials will enable the production industry to accomplish its sustainability goals (Meng et al., 2016). Reducing the amount of recycled and newly produced plastics translates into a notable reduction in energy dissipation and CO_2 emissions from the reuse of plastic waste. As plastic waste has proven its effectiveness, the packaging industry has investigated different ways to recycle this waste, but its uses in the construction industry are limited (Wong et al., 2020). Building material manufacturing is a promising industry in which plastic waste can be used as a variety of building materials, as it is the largest industry and consumption of raw materials in numerous countries. In the construction sector, plastic waste is used as aggregate in cement and asphalt composites, fillers, insulation materials for bricks and pavement construction, etc. Despite the magnificent potential for utilizing plastic waste as a building material, its growth and usage are still very inadequate (Kabirifar et al., 2020). This study reviews existing literature and critically evaluates the role of recycled plastics in various construction materials, highlighting material behavior, performance metrics, and environmental benefits. The uniquely integrated recent findings across multiple construction applications (asphalt, concrete, and bricks) highlighted the performance gaps, and identified future directions for durability, LCA, and large-scale applicability.

2 Types of Plastic

Plastic is a mixture of a wide variety of materials, and the mixing proportions of such materials are created to provide different products and meet the needs of the industry. Plastics are classified into two key groups, such as thermoplastics and thermosetting plastics (Möllnitz et al., 2020). For example, vinyl esters, silicones, phenol-formaldehyde, epoxies, and polyurethane are heated to change their chemical characteristics, occurring in a three-dimensional network from which many types of plastics are made. However, while some plastics cannot be reshaped after heating, thermoplastics are capable of being reformed. Thermoplastics can melt when heated and cured with repeated vulnerability to cold (Wang, 2021). There are approximately 50 varieties of plastic, including hundreds of different varieties. The American Plastics Industry Association has developed a standard identification code. Thus, the ranking can be easily done when ranking. Seven groups of plastics are marked and classified as follows: polypropylene (PP), polystyrene (PS), low-density polyethylene (LDPE), polyethylene terephthalate (PET), vinyl/polyvinyl chloride (PVC), high-density polyethylene (HDPE), and various kinds of plastics (Kamaruddin et al., 2017). These plastics have many good characteristics, which include versatility, lightness, hardness, and resistance to chemicals, water, and impact. Polymer plastic can be used for many things, like cups, bags, CD cases, packaging, clothing, water bottles, and so on. Types of plastic with their applications, symbol, type of recycling, and properties are given in Table 1.

One effective method for minimizing plastic waste involves recycling plastics into building materials. Not only does this approach mitigate environmental issues associated with plastic disposal, but it also enhances the structural integrity of building materials (Boucedra et al., 2020).

Table 1: Types of plastics and their properties.

Plastic	Code	Symbol	Recyclable or not	Few applications	Types of recycling	Properties
PET (Fakirov, 2021; Çepelioğullar and Pütün, 2013)	1	△ 1 PET	Yes	Water bottles, soft drink bottles, food jars, films, sheets, furniture, carpets, and paneling	Converted back to polymer and used for making apparel	Clear, tough, solvent resistant, barrier to as and moisture, softens at 80 °C
HDPE (Demets et al., 2020; Anuar Sharuddin et al., 2016)	2	△ 2 HDPE	Yes	Milk pouches, bottles, carry bags, recycling bins, and base cups	Converted to pellets and used to produce new HDPE	Resistant to chemicals and moisture, easily processed, hard to semiflexible, colored, and formed
PVC (Ragaert, 2020; Wang, 2019)	3	△ 3 PVC	Yes	Pipes, hoses, sheets, wire cable insulations, multilayer tubes, window profile, fencing, and lawn chairs	Pyrolysis, hydrolysis and heating are used to convert PVC waste into calcium chloride, hydrocarbon products and heavy metals. These are used to produce new PVC or other manufacturing processes or as fuel for energy recovery	Tough, strong, softens at 80 °C, can be solvent welded
LDPE (Mohammed et al., 2020; Wang, 2019)	4	△ 4 LDPE	Yes	Plastic bags, various containers, dispensing bottles, and wash bottles	Converted to pellets and used to produce new LDPE	Tough, flexible, good moisture barrier properties

Table 1 (continued)

Plastic	Code	Symbol	Recyclable or not	Few applications	Types of recycling	Properties
PP (Kerdlap et al., 2020; Wang, 2019)	5		Yes	Disposable cups, bottle caps, straws, auto parts, and industrial fibers	Converted to pellets and used to produce new PP	Hard, strong, flexible, and excellent chemical resistance
PS (Awasthi and Majumder, 2017; Anuar Sharuddin et al., 2016)	6		No	Disposable cups, glasses, plates, spoons, trays, CD covers, cassette boxes, and foams	Not recyclable	Rigid, hard, brittle, and high clarity
Others (Gaytán et al., 2021)	7		No	Thermoset plastics, multilayer and laminates, nylon SMC, FRP, CD, melamine plates, helmets, and shoe soles	Not recyclable – however, multilayer packaging could be crushed and turned into sheets and boards for roofing, using adhesives	Other polymers have wide range of uses; these are identified with number 7

3 World Plastic Scenario

3.1 Production of Plastic

Plastics are manufactured in both developed countries like the United States and Europe, as well as in developing countries, such as India and China.

Plastics are flexible materials with a wide variety of industrial and daily applications. Currently, the production of plastics uses 6% of the world's oil output, and if current trends continue, that percentage will increase to 20% by 2050 (Wang, 2019). Plastic waste is expanding globally and is generally hazardous to the environment, causing pollution that harms aquatic animals and people. According to Geyer et al. (2017), approximately 8.4% of total plastic waste generated was recycled, while 75.8% ended up in landfills and the environment. Figure 1 illustrates the historical trends in cumulative plastic waste management and disposal from 1950 to 2015, with projections through 2050.

The majority of plastic products are manufactured in China, accounting for approximately 32% of global production. Meanwhile, the rest of Asia accounted for an additional 19% while North America contributes around 17% of worldwide output, with Europe following closely at 14% (Plastics Europe and EPRO, 2016). The extreme impacts of plastic and the awareness among people worldwide within the last 3 years have brought about the reduction in the production of plastic. The total worldwide production of plastics surpassed 400 million metric tons in 2022. Figure 2 illustrates the percentage of annual global production of plastic in 2022 (Plastics Europe and EPRO, 2016).

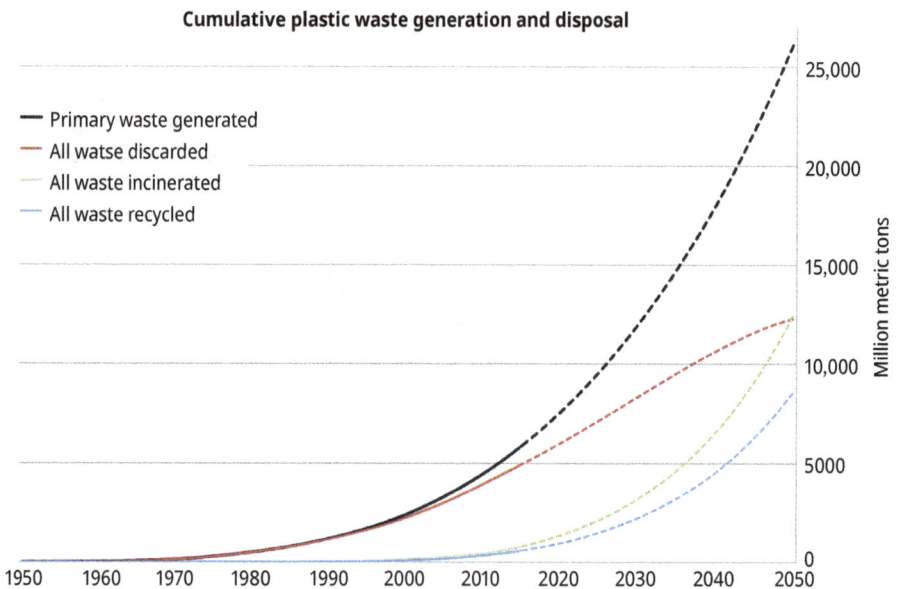

Cumulative plastic waste generation and disposal

Legend:
- Primary waste generated
- All watse discarded
- All waste incinerated
- All waste recycled

Figure 1: Accumulated generation and disposal of plastic waste (in million metric tons).

3.2 Consumption of Plastic

Plastic consumption is escalating worldwide, driven by extensive usage across various industries. In Europe, for example, PP, LDPE, and HDPE dominate, collectively constituting 49.1% of plastic usage as shown in Figure 3. These materials are integral to multiple sectors globally, prominently in packaging (39.90%), building and construction (19.70%), automotive (10%), electronics, household goods, leisure, and support indus-

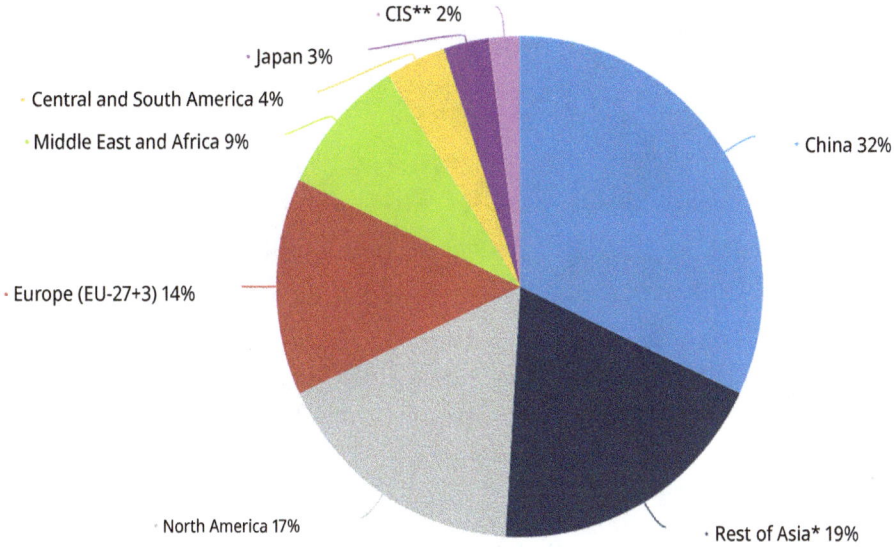

CIS** 2%
Japan 3%
Central and South America 4%
Middle East and Africa 9%
China 32%
Europe (EU-27+3) 14%
North America 17%
Rest of Asia* 19%

Figure 2: Worldwide annual production of plastic by area.

tries as shown in Figure 4. This widespread application underscores the pervasive role of plastics in modern economies and daily life (PEMRG, 2017).

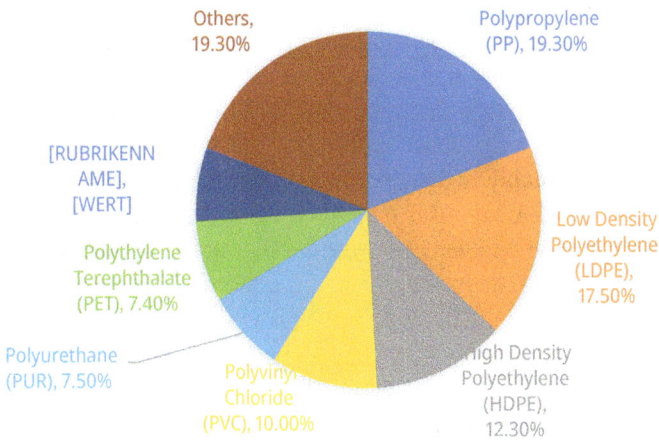

Others, 19.30%
Polypropylene (PP), 19.30%
[RUBRIKENN AME], [WERT]
Polythylene Terephthalate (PET), 7.40%
Low Density Polyethylene (LDPE), 17.50%
Polyurethane (PUR), 7.50%
Polyvinyl Chloride (PVC), 10.00%
High Density Polyethylene (HDPE), 12.30%

Figure 3: Demand of plastic type.

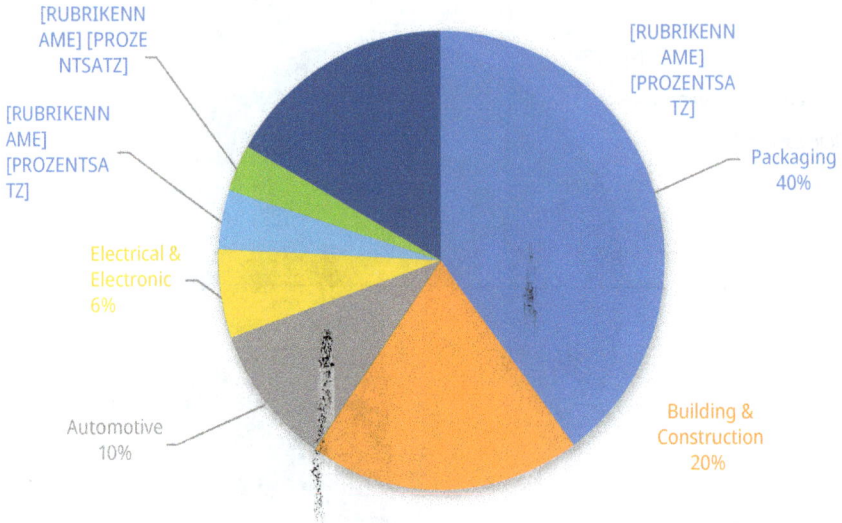

Figure 4: Utilization of plastic by different industries.

4 Plastic Waste

The per capita generation of plastic waste worldwide has shown consistent growth across various regions over the years. Specifically, in 1980, 2005, and 2015, this trend is evident across all areas. Notably, the North American Free Trade Agreement (NAFTA) region has recorded the highest per capita generation of plastic waste. Over the period from 2005 to 2015, countries in Central Europe and the Commonwealth of Independent States (CIS) witnessed a significant increase from 24 to 48 kg per person. Industrialized nations exhibit notably higher per capita plastic consumption compared to other regions, with NAFTA countries topping at 139 kg per person. In contrast, Asia's per capita consumption remains below the global average. By 2015, global plastic waste accumulation had reached a staggering 5.8 billion metric tons (Geyer et al., 2017).

5 Recycling of Plastic Wastes

The recycling of plastic waste begins with collection from various sources like households and businesses. Next, the collected plastics undergo sorting based on type and color to ensure purity. They are then processed through shredding or melting into small pellets or flakes. These materials are used to manufacture new products, such

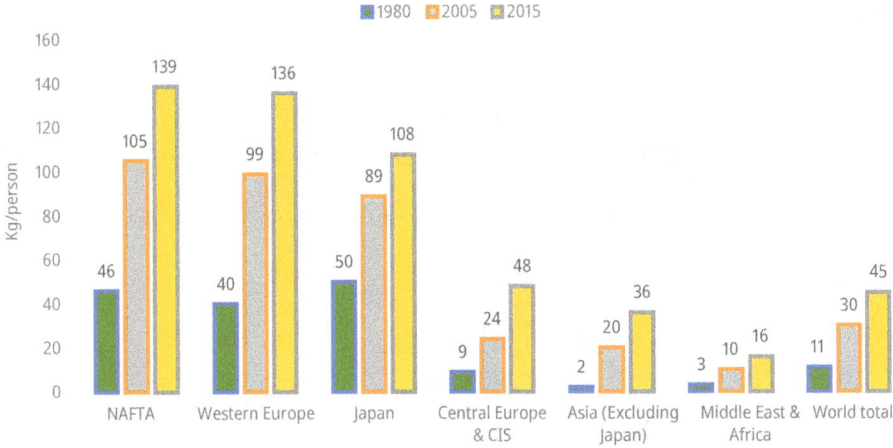

Figure 5: Per capita generation of plastic waste worldwide.

as packaging, textiles, and construction materials. Finally, the recycled products are distributed to markets, completing the continuous cycle of plastic recycling.

Due to its widespread production and relatively low cost, plastic is extensively used in everyday consumer goods. However, of the 830 million metric tons produced between 1950 and 2015, only 500 million metric tons were recycled. Figure 6 visually depicts the trends in plastic production, recycling, and disposal during this period (Geyer et al., 2017).

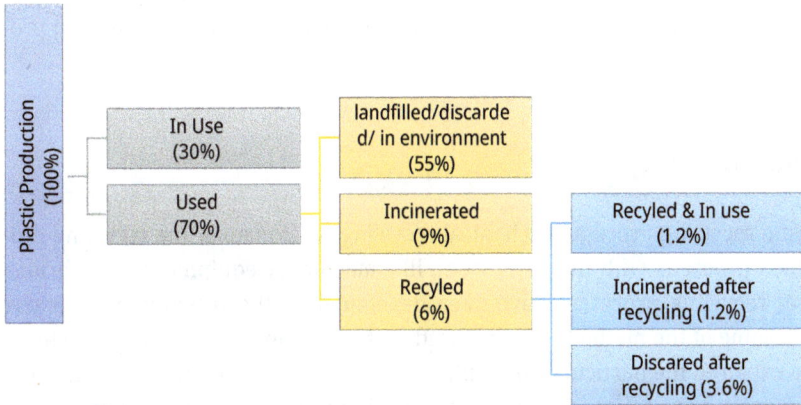

Figure 6: Amount of plastic produced, recycled, and disposed.

In the world, different kinds of plastics are utilized for many applications, for example, food packaging and water bottles. This plastic is very harmful to human lives and environmental conditions. So, the recycling of plastics is a necessary requirement in day-to-day life. Hence, different types of plastics and types of recycling methods are described in this section.

The recovery of plastic waste into functional and efficient products is known as "plastic recycling" (Faraca and Astrup, 2019). Recycling of this plastic waste is most significant for developing appropriate waste management policies as well as for preventing plastic pollution, especially in the oceans. Currently, around 60% of plastic waste in India is recycled. Plastic waste recycling can be done by the methods shown in Figure 7 (Jang et al., 2020).

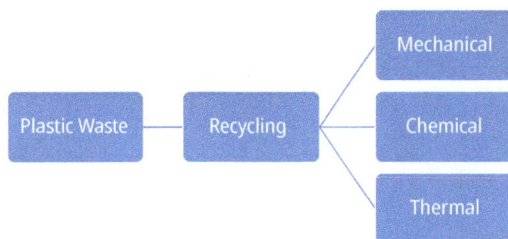

Figure 7: Block diagram of plastic recycling process.

A recycling method is a technological approach made for plastic waste disposal. Fundamental recycling of plastic waste is processed through a mechanical recycling method, which was the most reasonable way to recycle waste. The first step is to cut or shred, which includes using scissors or a saw to turn the plastic waste into small sections that are easier to work with. Pieces of paper, dirt, and small particles are divided from the plastic supplied by a hurricane separator during the disinfection process of detachment (Aryan et al., 2019).

5.1 Mechanical Recycling

In the plastic recycling process, mechanical recycling is also one of the recycling processes, which proceeds with the equipment, like machinery equipment. The various mechanical recycling processes have been developed by the researchers (Ragaert et al., 2017). Some of the methods are reviewed in this section. Current sorting technologies have explored the particular difficulties of mechanical recycling and the inevitability of polymer composites, such as thermomechanical or endurance degradation. These comprise PET recycling, such as electric (on) IC devices, or end-of-life vehicles, and solid plastic waste (SPW) from consumer post-packaging. A separate section is devoted to the relationship between design and recycling, maintaining the role of concepts, such as design recycling. Faraca et al. (2019) examined the place of recycling plants, different kinds of energy supply, and recycled material use. In their research, authors found the final ranking of cumulative prices and potential global warming. That standard reveals that contamination can have a variety of outcomes for mechanical processing, from the efficiency of reduced processing, to decreased mechanical properties of recycling, to the incorporation of unwanted (and potentially hazardous) chemicals into the product matrix. Plastic waste cleanliness is an essential factor for the quality and economic value of recycled waste and, in turn, depends on the type and number of sorting and recycling stages. Research shows that obtaining high-quality recycled plastics can lead to environmental savings and financial returns. Maris et al. (2018) presented a mechanical recycling process that leads to the recovery of plastic solid waste. Therefore, recycled plastic waste can be converted into new material, as an alternative to original polymers. This method worsens the material properties of recycling; it results in the decomposition of the polymers, the plastic waste diversity, and the low molecular weight compounds (additives, degradable materials, and impurities). This means that there is a necessity to promote relevant technology to enhance the characteristics of waste products and adapt them to new uses.

5.2 Chemical Recycling

The plastic waste is chemically reduced to its original monomeric form, so that, it can eventually be recycled and converted into new material for further use. Meys et al. (2020) have accomplished whether chemical recycling gives environmental gains, the method for estimating the supreme environmental benefit of LCA-based chemical recycling technology leads to the ecological probable of chemical recycling. Feghali et al. (2020) have utilized the chemical recycling method to reduce the impact of global warming and the use of fossil resources using sorted plastic wrapping waste, which is processed in the incineration of municipal waste. Deng et al. (2021) have explained various types of plastic recycling methods employed to create plastics, which split down polymers into their constituent monomers. This can be reused at refineries or used in petrochemical and chemical production. Chemical modification and depolymerization are the two processes of recycling. Kumar (2020) have explained the chemical treatment process, which is the best approach to treating carbon fiber because of the comfort of fiber separation and the operating parameters of the environment. This method facilitates the reclamation of degraded polymer matrix commodities and their reuse in the manufacture of new resin plastics. Recycled carbon fibers are employed less for structural purposes because they require longer fibers. Besides, fiber processing is economical and proven.

5.3 Thermal Recycling

The plastic waste can be recycled thermally using a pyrolysis-gasification process for energy production. Awoyera and Adesina (2020) have presented the thermal recycling of plastic waste. It involves heating plastic waste at a high molten temperature, followed by splashing it into the mold to produce a novel material. The recycling potential of different hard plastic products, such as PET, PP, PS, PVC, and HDPE, was investigated. This investigation also revealed that the particular properties of plastics influence their recyclability. Moreover, this investigation presented a comparison of the sorting, treatment, and recycling possibilities of different types of plastics. The plastic reprocessing and sorting capacity in this group is over 50%. But PET plastic waste shows low recycling potential when compared to others. This suggests that earlier recycling will be more advantageous in terms of energy consumption and its administration. Turukmane et al. (2018) have analyzed the techniques for recycling polyester fiber acquired in different textile industries throughout the world. This is the most reliable technique for polyester recycling, in which waste PET bottles are initially shredded into small pieces, washed, and melted through a spinneret to obtain the desired polyester fibers, which are further changed into fabric and clothing. Remelting of the polymer lowers the crystallinity and strength of the fibers.

6 Processes Involved in Recycling Plastics

Recycling plastics requires a number of operations, including the collection of trash at the place of production or disposal, its sorting, compression, crushing, and pelletization into raw materials. The following thermal, chemical, and mechanical processes produce the final product. As a result, plastic waste recycling is complex and less desirable than recycling other materials, such as aluminium, glass, ceramics, and paper (Hahladakis and Iacovidou, 2019). Recycling plastic trash chemically, thermally, or mechanically begins with their automated sorting employing techniques, such as spectroscopy, infrared, fluorescence, flotation, and electrostatics (Wang et al., 2020). During mechanical recycling, plastic waste is processed by shredding and grinding (Singh et al., 2017). However, the process is not preferred if the waste combination is complicated; instead, incineration is preferred (Khoo, 2019). During the chemical recycling of plastic trash, it is broken down into monomers or smaller chains and chemically changed to produce virgin raw materials that are utilized to manufacture new goods. Thermal processes function by melting plastic trash at high temperatures and reusing the resulting mold to produce new goods. The steps involved in the recycling of plastics are summarized in Figure 8.

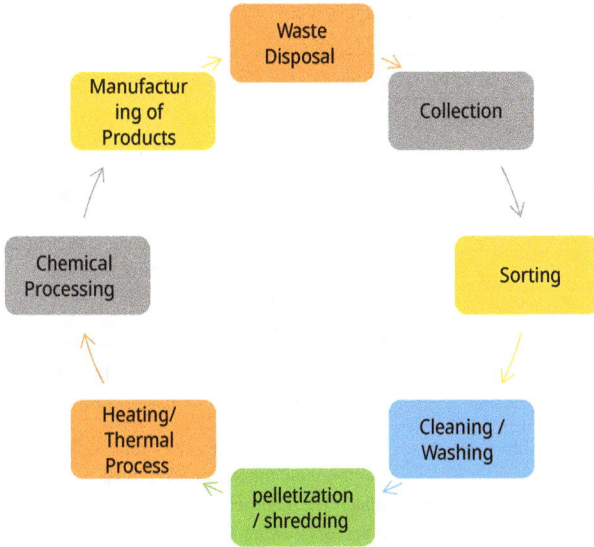

Figure 8: Process involved in the recycling of plastic waste (Nyika and Dinka, 2022).

7 Types of Recyclable Plastic Wastes

Plastics are generally classified as either thermosets or thermoplastics, depending on their ability to retain their original shape when exposed to heat (Ogundairo et al., 2019). However, certain types of hard plastic waste, such as PVC, PS, PP, PET, and HDPE, have been successfully repurposed (Wang et al., 2020). Research highlighted in the work of Faraca and Astrup (2019) compared the recyclability potential of these plastics and found that their properties significantly influence their recycling feasibility. PVC was identified as the easiest material to separate, whereas PET posed the greatest challenge. PET, PS, HDPE, and PP demonstrated higher reprocessing capabilities than PVC, with PET showing the lowest recycling potential among them. A detailed breakdown of the characteristics and applications of recyclable plastic waste, is shown in Table 2 (Kamaruddin et al., 2017; Ogundairo et al., 2019).

8 Utilization of Plastic Waste as Construction Material

Plastic is one of the most widely used fine materials in the world and brings comfort and convenience to human life. The production of plastic has grown unusually fast in

Table 2: The characteristics of recyclable plastics and general applications of their wastes.

Plastic type	Properties	Application of plastics' waste
PVC (Thomas and Moosvi, 2020)	Elastic, transparent, and flexible	Factory floors
PS (Mondal et al., 2019)	Glossy, transparent, stiff, and brittle	Manufacture of hangers, pegs, coat hangers, and insulation materials
PP (Bhogayata and Arora, 2017)	Flexible and hard	Manufacture of crates and bins. Used as aggregates in asphalt mixtures
PET (Hameed and Ahmed, 2019)	Fiber-like, transparent solid	Making of rain jackets, carpet fibers, soft packaging, and wrapping items. They are used as fibers in cement-based composites
HDPE (Awoyera et al., 2020)	Colored and sometimes white plastics	Making of soap bottles, bins, and crates. Components for chairs and tables
LDPE (Pooja et al., 2019)	Mostly white, flexible, and soft	Making of wrappers, cling films, and nursery bags for growing seedlings. Used in making blocks and bricks
PE (Lamba et al., 2021)	Tough, translucent, water proof, semirigid, and chemical resistance	Manufacture of cling film wrappers
PA (Jassim, 2017)	Nylons	Packaging and wrapping items
Polymethyl methacrylate(PMMA) (Záleská, 2018)	Transparent, durable, chemical resistant, and with high UV light	Manufacture of shampoo bottles, bins, and crates

recent decades due to the use of other man-made materials. Factors, such as the increase in the population, low-cost production, and a broad range of applications, have led to an increase in plastics production (Janani and Kaveri, 2020). Certain varieties of plastic toxins can also be released into the atmosphere, which means that plastic waste dumping in landfills is not a sustainable solution. The incineration process entirely removes these plastic wastes and can be a source of energy, carbon dioxide, and other toxic chemicals, creating toxic bottom and fly ash, which are typically delivered by this method (Altieri et al., 2021). Therefore, it is necessary to recycle the disposed of plastic waste. A mainstream of LCA studies associating recycling of plastic with different waste management choices have concluded that there are usually optimistic outcomes, including an important modification of environmental influences. Also, one of the possible approaches to plastic waste recycling is incorporating it into a building material (del Rey Castillo et al., 2020). The utilization of plastic waste as a different building material has cost-effective and methodological advantages for resolving the disposal of large amounts of waste plastic.

Exploiting this possibility of plastic waste as a building material will decrease the entire environmental risk of this plastic production. Additionally, the potential application of plastic waste in building materials will enable the production industry to accomplish its sustainability goals (Meng et al., 2016). Reducing the amount of recycled and newly produced plastics translates into a notable reduction in energy dissipation and CO_2 emissions from the reuse of plastic waste. As plastic waste has proven its effectiveness, the packaging industry has investigated different ways to recycle this waste, but its uses in the construction industry are limited at the building material manufacturing is a promising industry in which plastic waste can be used as a variety of building materials as it is the largest industry and consumption of raw materials in numerous countries (de Hita et al., 2018; Nur 2022). In the construction sector, plastic waste is used as aggregate in cement and asphalt composites, fillers, insulation materials for bricks and pavement construction, etc. Despite the magnificent potential for utilizing plastic waste as a building material, its growth and usage are still very inadequate (Kabirifar et al., 2020).

Construction manufacturers are the backbone of each country and are a significant contributor to its economy. As a result, the potential application of waste increases the longevity of building methods and procedures. The continued use of plastic waste in construction also brings economic benefits (Aldahdooh et al., 2018). Using plastic waste in construction in a sustainable and creative way will greatly reduce the amount of plastic waste that ends up in water and give the building industry access to new materials that meet their needs. The reuse of plastic waste in the manufacturing of construction materials is an effective solution, which has a positive impact on the building industry and simultaneously preserves natural resources (Biswas et al., 2020). The recycled plastics are utilized in different kinds of building materials.

8.1 Plastic Waste in Construction of Road

Pavements are generally classified into two types: flexible and rigid. However, many roads experience premature deterioration due to the inadequate performance of asphalt materials. To improve asphalt mix behavior, plastic waste is now widely used as a bitumen modifier, offering enhanced performance and durability. Recycled plastic blended with bitumen raises the melting point and helps retain elasticity during colder temperatures, acting as a binding agent that strengthens the asphalt and reduces wear (Veropalumbo et al., 2021). When plastic is incorporated into bitumen, high-temperature mixing becomes necessary. Typically, shredded plastic is coated onto aggregates and then mixed with hot bitumen. Commonly used plastics include HDPE, LDPE, and PP, which help reduce bitumen usage by up to 10% while increasing strength and resistance to rutting and pothole formation (Kumar and Khan, 2020; Tulashie et al., 2020). In addition to structural benefits, titanium dioxide has been suggested as a supplementary material to reduce vehicular emissions, acting as a smoke absorber that suits India's hot, humid climate and urban air quality challenges

(Martin-Alfonso et al., 2019). Plastic waste can be incorporated into road construction through two methods: the dry method, which mixes shredded plastic with aggregates, and the wet method, which blends melted plastic directly with bitumen. These approaches can increase pavement durability by up to 40% and are effective in lowering bitumen consumption (Ponnada and Krishna, 2020). In the dry process, plastic is heated to around 170 °C to coat aggregates, which are then mixed with bitumen at 160 °C. IS-code-compliant aggregates are selected to ensure optimal porosity, strength, and water absorption (Vishnu and Singh, 2021, I. Journal).

In the wet method, powdered plastic is mixed with hot bitumen at 155–165 °C, followed by the addition of coated aggregates. The typical plastic content ranges from 6% to 8%. This method significantly boosts the Marshall stability of the mix – up to two or three times that of conventional bitumen roads (Duggal et al., 2020, Nizamuddin et al., 2020). Among various plastics, recycled linear low-density polyethylene (LLDPE) is notable for its flexibility and durability, making it especially suitable for road construction applications (Nizamuddin et al., 2020). Research has further confirmed the effectiveness of plastic waste in asphalt binders. For instance, El-Naga and Ragab (2019) showed enhanced pavement life and reduced thickness when using plastic-modified mixes. Movilla-Quesada et al. (2019) explored the replacement of bitumen with fine and coarse plastic scraps, aiming to lower binder content while maintaining performance. Leng et al. (2018) studied PET-based compounds, which significantly improved asphalt durability. Collectively, these studies report improvements, such as a minimum 15% increase in rutting resistance and up to 60% improvement in fatigue cracking resistance. PET replacement of bitumen at 1%, 3%, and 5% had minimal effect on optimum binder content while maintaining high performance on parameters like Marshall flow, stability, bulk density, and VTM. Badejo (2017) and Padhan and Gupta (2018) introduced an efficient method to produce polymer-modified bitumen using PET, presenting a promising approach for PET recycling in road construction. Plastic waste is also used as a partial or total substitute for aggregates in base and subbase layers, where it enhances stiffness, shear strength, and load-bearing capacity (Wang et al., 2020; Nyika and Dinka, 2022; Ogundairo et al., 2019). However, the smooth texture of some plastic materials may reduce their performance in terms of resilient modulus and stiffness when compared to traditional materials (Wang et al., 2020). Still, incorporating plastic waste improves the mechanical and volumetric properties of pavement mixtures, extends service life, and contributes to sustainable infrastructure (Abdullah, 2018). Moreover, quantitative improvements, such as "pavement durability increased by up to 40%" and "compressive strength increased by 8–25%." Table 3 shows different types of plastics used in road construction.

8.2 Plastic Waste in Concrete and Cementitious Composites

Plastic waste is increasingly being integrated into concrete and cementitious composites to promote sustainability, reduce dependency on natural resources, and address

Table 3: Plastics used in road construction.

References	Plastic waste	Role of the plastic waste	Plastic content (%)
Hameed and Ahmed (2019)	PET	Replacement of sand	1–10
Jaivignesh and Sofi (2017)	PET	Fine aggregate	0–50
Vanitha et al. (2015)	HDPE	Aggregate	0–10
Jain et al. (2019)	Shredded plastic bags		0–5
Bhogayata and Arora (2017)	PP	Coarse aggregate	0–2
Jain et al. (2019)	Fine plastic waste	Partial substitute for aggregate	0–12.5
Bansal et al. (2017)	Recycled plastic aggregate	Substitute	25, 50, 77, and 100
Olofinnade et al. (2021)	LDPE	Partial substitute to sand	0–50
Dawale (2016)	Thermoset plastic waste	Modifier	–
Gavhane et al. (2016)	Electronic plastic waste	Aggregate	0–30
Lamba et al. (2021)	PE	Binder	5–11
Awoyera et al. (2020)	PET, HDPE, LDPE	Component of asphalt	–
Almeshal et al. (2020)	HDPE	Component of subbase and base	–

disposal challenges. It serves either as a partial replacement for natural aggregates or as a strengthening agent. Studies show that plastics, such as PET, HDPE, LDPE, PP, and resins (e.g., PVC, PLA, and PS), can be added without significantly affecting the hydration process of concrete (Gu and Ozbakkaloglu, 2016). When mixed with supplementary materials like fly ash, silica fume, or reclaimed concrete, plastics help form durable products, such as bricks, tiles, and blocks (Awoyera and Adesina, 2020; Jain et al., 2019). The approach known as direct volume replacement – substituting natural aggregates with plastics of equal volume – has shown positive results in reducing deadweight, although excessive plastic content can compromise mechanical strength (Silva et al., 2013; Ismail and AL-Hashmi, 2008).

Research findings vary depending on the type of plastic, particle size, and mix design. While some studies indicate reductions in compressive and tensile strength, others demonstrate that, up to a threshold (typically 5–10%), plastic waste can improve impact resistance, flexibility, and fresh concrete properties like workability and air content (Hama and Hilal, 2017; Hama, 2022; Steyn et al., 2021). For example, mixes using LDPE, PET flakes, and crumb rubber showed reduced workability but increased durability in harsh environments. Hameed and Ahmed (2019) found that PET flakes at 1–10% replacement by weight of cement enhanced compressive, flexural, and tensile strength. Similarly, Vasanthi et al. (2020) demonstrated that combining M-sand with LDPE milk covers or PET bottles improved concrete strength in both slabs and cubes. Agyeman et al. (2019) further showed that plastic waste used in paving blocks offered better water absorption and rapid curing – making it suitable for waterlogged areas and lightweight construction.

However, as plastic content increases – especially with 100% recycled plastic aggregates (RPA) properties like modulus of elasticity, adhesive strength, and flexural strength tend to decline, highlighting a performance trade-off (Basha et al., 2020). Despite this, several studies confirm the potential of plastic waste in producing eco-friendly, functional concrete, especially for nonstructural and light-duty applications (Gesoglu et al., 2017). Overall, optimizing plastic type, content, and mix proportions is a key to balancing sustainability with mechanical performance. The utilization of plastic waste can improve the strength of concrete. Experimental outcomes conclude that replacing natural aggregates with plastic waste in concrete shows promise for mass concrete production by reducing aggregate deformation. Moreover, using waste plastics contributes to cost-effective construction and supports environmental sustainability by lowering the demand for cement and sand (Bharathi et al., 2020). Table 4 lists the types of plastics used in making cementitious composites and concrete.

Table 4: Plastics used in making of cementitious composites/concrete.

References	Plastic type	Material made	Type and quantity of replacement ratio
Needhidasan and Sai (2020)	E-plastic waste	Concrete	Coarse aggregate up to 16% by volume
Suleman and Needhidasan (2020)	E-plastic waste	Concrete	Coarse aggregate up to 16.5% by volume
Mustafa et al. (2019)	Polycarbonate	Concrete	Fine aggregate up to 20% by volume
Punitha et al. (2020)	HDPE	Concrete	Fine aggregate up to 30% by volume
Mohammed and Rahim (2020)	PET	Concrete	Fine aggregate up to, 15% by volume
Olofinnade et al. (2020)	HIPs and LDPE	Concrete	Fine aggregate up to 50% by weight
Vanitha et al. (2015)	HDPE	Concrete	Coarse aggregate up to 10% by volume
Almeshal et al. (2020)	PET	Concrete	Fine aggregate up to 50% by volume
Thomas and Moosvi (2020)	PVC	Concrete	Fine or coarse aggregate 30% by weight

Table 4 (continued)

References	Plastic type	Material made	Type and quantity of replacement ratio
Shahjalal et al. (2021)	PET	Concrete	Fine aggregate 0.8% by volume
Sule (2017)	PP	Concrete	Coarse aggregate up to 30% by volume
Pooja et al. (2019)	LDPE	Concrete	Fine aggregate up to 30% by volume
Remadnia et al. (2009)	PET and PS	Concrete	Sand 10–20% by volume
Sule (2017)	PET	Mortar	Fine aggregates 30–70% by volume
Hannawi et al. (2010)	LDPE	Lightweight concrete	Fine aggregates 5%, 10%, 15%, 20%, and 30% by volume
Hama and Hilal (2019)	PET and PC	Mortar	Fine aggregates 3%, 10%, 20%, and 50% by volume
Haghighatnejad et al. (2016)	HDPE and PET	Concrete	Fine aggregates 10%, 15%, 20%, 25%, 30%, 35%, and 40% by volume
Haghighatnejad et al. (2016)	PVC	Mortar	Sand 20%, 30%, 40%, and 50%
Záleská (2018)	Plastic waste mixture	Concrete	Fine aggregate 10% by weight
Jassim (2017)	PP	Lightweight concrete	Sand 10%, 20%, 30%, 40%, and 50% by weight
Chaudhary et al. (2014)	HDPE	Plastic cement	Sand 15–80% by volume
Kennedy (1993)	LDPE	Concrete	Fine aggregates 0.4–1% total weight

8.3 Plastic Waste in Bricks and Blocks

Recycling plastic waste in brickmaking offers multiple advantages, including enhanced strength, durability, and thermal insulation, while simultaneously reducing environmental hazards like landfill accumulation and incineration (Uvarajan et al., 2022; Lamba et al., 2021). This sustainable practice supports a circular economy by converting plastic waste into lightweight, resilient building materials. One study demonstrated that recycled bricks incorporating mixed plastic waste and foundry sand achieved a compressive strength of 8.16 MPa at 70% PET content, with reduced water absorption and tensile strength approximately 1.35 times higher than that of traditional clay bricks (Lamba et al., 2021). Additionally, materials such as aerated concrete and recycled plastic bricks can improve thermal insulation by up to 17.4% (Jonnala et al., 2024).

Further research involving fly ash and waste plastics reported increased hardness and water absorption below 0.7%, indicating superior durability (Hait et al., 2024). In paver blocks, plastic waste improves both compressive strength and heat resistance, broadening its applicability while also contributing to environmental protection (Rajan et al., 2024). Various laboratory and field tests – including abrasion, soundness, compressive strength, hardness, and flexural strength – have validated the structural performance of plastic-infused bricks. These bricks are widely favored in construction due to their strength, stress resistance, quality consistency, and cost-efficiency.

Using plastic waste in brick manufacturing not only reduces overall weight and enhances water absorption capacity but also lowers production costs by minimizing landfill expenses and preserving natural clay resources. The specific plastic waste compositions and their functional roles in brick production are summarized in Table 5.

Table 5: Plastic wastes used in making bricks/blocks.

References	Plastic waste	Added materials	Role of the plastic waste	Plastic content (%)
Geyer et al. (2017)	PET	Foundry sand	Aggregate	20–40
	PET	Recycled glass	Aggregate	20–40
Akinwumi et al. (2019)	PET	Cement and clayey sand	Aggregate	0–7
Limami et al. (2021)	PET and HDPE	Clay	Additives	0–20
Jnr et al. (2018)	LDPE	Sand	Aggregate	–
Ansori (2015)	PP	Fly ash and river sand	Aggregate	20–30
Mondal et al. (2019)	PP and PET	Bitumen and quarry dust	Aggregate	65–80
Mondal et al. (2019)	PS and polycarbonate (PC)	Regular cement, ash, and sand	Aggregate	0–10
Kognole et al. (2019)	Nylon, PE, and PET	Stone crush, river sand, and red soil	Aggregate	–
Sonone et al. (2017)	PE	Fly ash	Aggregate	40–45
Sellakutty (2016)	HDPE and PE	Sand	Aggregate	Waste plastic to sand ratio (1:2, 1:3, 1:4, 1:5, and 1:6)

Table 5 (continued)

References	Plastic waste	Added materials	Role of the plastic waste	Plastic content (%)
Manjarekar (2017)	LDPE, HDPE, and PET	Stone dirt	Aggregate	–
Seghiri et al. (2017)	HDPE	Sand	Aggregate	30–80
Safinia and Alkalbani (2016)	PET bottles	–	Substitute	–
Mokhtar (2016)	PET	Sand	Alternate	–
Limami et al. (2021)	HDPE and PET	Earth clay	Additive	–

To address concerns regarding durability and microplastic release, several studies have evaluated the long-term performance of plastic waste materials under environmental stressors, such as wear, UV exposure, moisture, and temperature variations. Recycled plastics, particularly PET, HDPE, and LDPE, have demonstrated considerable resistance to weathering, making them suitable for construction applications. When used in roads, bricks, or concrete, these plastics are typically encapsulated within a matrix of bitumen, cement, or aggregate, which significantly reduces direct environmental exposure and limits microplastic leaching. Under UV radiation, certain plastics may degrade over extended periods, but this process is notably slower when the material is shielded within structural elements. Furthermore, the addition of stabilizers or coatings can enhance UV resistance. In mechanical durability assessments, plastic-modified materials have shown improved flexural and compressive strength, reduced water permeability, and increased resistance to abrasion and fatigue, thereby extending their service life. While microplastic release remains a theoretical concern, the inert embedding of plastic waste in dense construction matrices has been shown to minimize fragmentation and limit environmental dispersion, especially when end-of-life recycling or proper disposal protocols are followed.

9 Conclusion

This chapter provides valuable insights into the simultaneous utilization of waste plastics for construction materials, with promising properties for practical applications in construction and landscaping. This chapter also indicates that extensive efforts have been made to recycle plastic waste from municipal solid waste and other sources, utilizing it as a fundamental component in the production of roads, concrete, bricks, mortar, and more. Most researchers have used different forms of plastic (LDPE, PET, HDPE, etc.) as substitutes for coarse/fine aggregates in making concrete, mortar, and other construction materials like plastic-dust bricks, sand-plastic bricks, and paver blocks. LDPE and PET are majorly used in concrete. Shredded plastic is used as fiber to reinforce concrete and as a replacement for sand and stone in concrete and mortar. Heated polystyrene foams are also used in concrete as aggregates to enhance its properties. LDPE and soft plastics like water sachets are melted to make sand-plastic bricks and paver blocks, and melted plastic also combines with aggregates to make flexible pavement. Researchers have also developed mortar using resin and plastic.

Bituminous plastic roads containing waste plastic in various proportions show better results than traditional flexible bituminous roads, requiring less maintenance, performing better, and having a longer life cycle. Plastic mix decreases the bitumen content by 10%, reinforcing the road surface and increasing its strength while reducing ruts and potholes. Plastic bricks and blocks have the compressive strength of second-class bricks, making them efficient for construction. The tensile and compressive strengths of plastic bricks are greater compared to conventional bricks. For instance, injection molding of plastic waste composites has demonstrated compressive strengths up to 17 MPa, depending on the plastic type and processing temperature. Due to their high polymer content, plastic bricks have good thermal insulation properties and absorb less water than conventional bricks, improving the mechanical strength. Using plastic waste as an aggregate replacement in concrete reduces the unit weight of the concrete. However, it can decrease the compressive strength, density, and tensile strength of concrete. The importance of the water-cement ratio on strength increase is not significant due to plastic aggregates diminishing the adhesive strength of concrete. This results in disintegration because of the weak bond between the cement paste and the plastic aggregate. Adding recycled plastic to concrete increases its flexibility, enhancing its ability to deform before breaking, which is beneficial in extreme weather conditions. Future research on recycled plastics in construction should focus on enhancing material formulations to improve mechanical properties, and extreme weather, LCA over decade conducting LCAs to evaluate environmental impacts and implementing long-term field studies to assess real-world performance. We should focus on evaluating long-term durability (e.g., freeze-thaw resistance and UV degradation) and mechanical properties, such as tensile and flexural strength under field conditions to validate lab-scale results. Investigating novel applications and developing industry standards will ensure safety and quality, while studying the behavior of plastic-based materials under extreme conditions can provide valuable insights. Additionally, understanding public perception and improving recycling processes will promote acceptance and availability. Exploring integration with smart technologies and analyzing the economic implications will further support the sustainable use of recycled plastics in the construction industry.

References

Abdullah, M. E. *et al*. (2018). Effects of Kaolin Clay on the Mechanical Properties of Asphaltic Concrete AC14. *IOP Conf. Ser. Earth Environ. Sci, 140*(1). doi: 10.1088/1755-1315/140/1/012121.

Agyeman, S., Obeng-Ahenkora, N. K., Assiamah, S., & Twumasi, G. (2019). Exploiting recycled plastic waste as an alternative binder for paving blocks production. *Case Stud. Constr. Mater, 11*, 0–7. doi: 10.1016/j.cscm.2019.e00246.

Akinwumi, I. I., Domo-Spiff, A. H., & Salami, A. (2019). Marine plastic pollution and affordable housing challenge: Shredded waste plastic stabilized soil for producing compressed earth bricks. *Case Stud. Constr. Mater, 11*, e00241. doi: 10.1016/j.cscm.2019.e00241.

Aldahdooh, M. A. A., Jamrah, A., Alnuaimi, A., Martini, M. I., Ahmed, M. S. R., & Ahmed, A. S. R. (2018). Influence of various plastics-waste aggregates on properties of normal concrete. *J. Build. Eng, 17*, 13–22. doi: 10.1016/j.jobe.2018.01.014.

Almeshal, I., Tayeh, B. A., Alyousef, R., Alabduljabbar, H., & Mohamed, A. M. (2020). Eco-friendly concrete containing recycled plastic as partial replacement for sand. *J. Mater. Res. Technol, 9*(3), 4631–4643. doi: 10.1016/j.jmrt.2020.02.090.

Altieri, V. G., De Sanctis, M., Sgherza, D., Pentassuglia, S., Barca, E., & Di Iaconi, C. (2021). Treating and reusing wastewater generated by the washing operations in the non-hazardous plastic solid waste recycling process: Advanced method vs. conventional method. *J. Environ. Manage, 284* (December 2020), 112011. doi: 10.1016/j.jenvman.2021.112011.

Ansori, M. (2015). Utilization of polypropylene plastic waste with fly ash and river sand in the production of bricks. Towards a Media History Documentation, 3(April), 49–58.

Anuar Sharuddin, S. D., Abnisa, F., Wan Daud, W. M. A., & Aroua, M. K. (2016). A review on pyrolysis of plastic wastes. *Energy Convers. Manag, 115*, 308–326. doi: 10.1016/j.enconman.2016.02.037.

Aryan, Y., Yadav, P., & Samadder, S. R. (2019). Life Cycle Assessment of the existing and proposed plastic waste management options in India: A case study. *J. Clean. Prod, 211*, 1268–1283. doi: 10.1016/j.jclepro.2018.11.236.

Awasthi, A. K. & Majumder, S. (2017). *Potential Use of Plastic Waste as Construction Materials: Recent Progress and Future Prospect Potential Use of Plastic Waste as Construction Materials: Recent Progress and Future Prospect*. doi: 10.1088/1757-899X/267/1/012011.

Awoyera, P. O. & Adesina, A. (2020). Plastic wastes to construction products: Status, limitations and future perspective. *Case Stud. Constr. Mater, 12*, e00330. doi: 10.1016/j.cscm.2020.e00330.

Awoyera, P., Onoja, E., & Adesina, A. (2020). Fire resistance and thermal insulation properties of foamed concrete incorporating pulverized ceramics and mineral admixtures. *Asian J. Civ. Eng, 21*(1), 147–156. doi: 10.1007/s42107-019-00203-4.

Badejo, A. A. *et al*. (2017). Plastic waste as strength modifiers in asphalt for a sustainable environment. *African J. Sci. Technol. Innov. Dev, 9*(2), 173–177. doi: 10.1080/20421338.2017.1302681.

Bansal, S., Misra, A. K., & Bajpai, P. (2017). Evaluation of modified bituminous concrete mix developed using rubber and plastic waste materials. *Int. J. Sustain. Built Environ, 6*(2), 442–448. doi: 10.1016/j.ijsbe.2017.07.009.

Basha, S. I., Ali, M. R., Al-Dulaijan, S. U., & Maslehuddin, M. (2020). Mechanical and thermal properties of lightweight recycled plastic aggregate concrete. *J. Build. Eng, 32*, 101710. doi: 10.1016/j.jobe.2020.101710.

Bharathi, S. M. L., Johnpaul, V., Kumar, R. P., Surya, R., & Kumar, T. V. (2020). Experimental investigation on compressive behaviour of plastic brick using M sand as fine aggregate. *Mater. Today Proc, xxxx*, 1–5. doi: 10.1016/j.matpr.2020.10.252.

Bhogayata, A. C. & Arora, N. K. (2017). Fresh and strength properties of concrete reinforced with metalized plastic waste fibers. *Constr. Build. Mater, 146*, 455–463. doi: 10.1016/j.conbuildmat.2017.04.095.

Biswas, A., Goel, A., & Potnis, S. (2020). Performance comparison of waste plastic modified versus conventional bituminous roads in Pune city: A case study. *Case Stud. Constr. Mater*, *13*, e00411. doi: 10.1016/j.cscm.2020.e00411.

Boucedra, A., Bederina, M., & Ghernouti, Y. (2020). Study of the acoustical and thermo-mechanical properties of dune and river sand concretes containing recycled plastic aggregates. *Constr. Build. Mater*, *256*, 119447. doi: 10.1016/j.conbuildmat.2020.119447.

Çepelioğullar, Ö. & Pütün, A. E. (2013). Thermal and kinetic behaviors of biomass and plastic wastes in co-pyrolysis. *Energy Convers. Manag*, *75*, 263–270. doi: 10.1016/j.enconman.2013.06.036.

Chaudhary, M., Srivastava, V., & Agarwal, V. C. (2014). Effect of Waste Low Density Polyethylene on Mechanical Properties of Concrete Utilization of waste materials in concrete View project Blending of concrete with innovative materials View project Effect of Waste Low Density Polyethylene on Mechanical Proper. *J. Acad. Ind. Res*, *3*(3), 123.

Dawale, S. A. (2016), "Prof. Dawale S.A Use of waste plastic coated aggregates in bituminous road construction," pp. 118–126.

de Hita, P. R., Pérez-Gálvez, F., Morales-Conde, M. J., & Pedreño-Rojas, M. A. (2018). Reuse of plastic waste of mixed polypropylene as aggregate in mortars for the manufacture of pieces for restoring jack arch floors with timber beams. *J. Clean. Prod*, *198*, 1515–1525. doi: 10.1016/j.jclepro.2018.07.065.

Del Rey Castillo, E., Almesfer, N., Saggi, O., & Ingham, J. M. (2020). Light-weight concrete with artificial aggregate manufactured from plastic waste. *Constr. Build. Mater*, *265*(December). doi: 10.1016/j.conbuildmat.2020.120199.

Demets, R., Roosen, M., Vandermeersch, L., Ragaert, K., Walgraeve, C., & De Meester, S. (2020). Development and application of an analytical method to quantify odour removal in plastic waste recycling processes. *Resour. Conserv. Recycl*, *161*(January). doi: 10.1016/j.resconrec.2020.104907.

Deng, Y., Dewil, R., Appels, L., Ansart, R., Baeyens, J., & Kang, Q. (2021). Reviewing the thermo-chemical recycling of waste polyurethane foam. *J. Environ. Manage*, *278*(October). doi: 10.1016/j.jenvman.2020.111527.

Duggal, P., Shisodia, A. S., Havelia, S., & Jolly, K. (2020). *Use of Waste Plastic in Wearing Course of Flexible Pavement*. Vol. 38, Springer Singapore.

El-Naga, I. A. & Ragab, M. (2019). Benefits of utilization the recycle polyethylene terephthalate waste plastic materials as a modifier to asphalt mixtures. *Constr. Build. Mater*, *219*, 81–90. doi: 10.1016/j.conbuildmat.2019.05.172.

Fakirov, S. (2021). Advanced Industrial and Engineering Polymer Research A new approach to plastic recycling via the concept of micro fi brillar composites. *Adv. Ind. Eng. Polym. Res*, *4*(xxxx).

Faraca, G. & Astrup, T. (2019). Plastic waste from recycling centres: Characterisation and evaluation of plastic recyclability. *Waste Manag*, *95*, 388–398. doi: 10.1016/j.wasman.2019.06.038.

Faraca, G., Martinez-Sanchez, V., & Astrup, T. F. (2019). Environmental life cycle cost assessment: Recycling of hard plastic waste collected at Danish recycling centres. *Resour. Conserv. Recycl*, *143*(June 2018), 299–309. doi: 10.1016/j.resconrec.2019.01.014.

Feghali, E., Tauk, L., Ortiz, P., Vanbroekhoven, K., & Eevers, W. (2020). Catalytic chemical recycling of biodegradable polyesters. *Polym. Degrad. Stab*, *179*, 109241. doi: 10.1016/j.polymdegradstab.2020.109241.

Gavhane, M. A., Sutar, M. D., Soni, M. S., & Patil, M. P. (2016). Utilisation of E – Plastic Waste in Concrete. *Int. J. Eng. Res*, *V5*(02), 594–601. doi: 10.17577/ijertv5is020538.

Gaytán, I., Burelo, M., & Loza-Tavera, H. (2021). Current status on the biodegradability of acrylic polymers: Microorganisms, enzymes and metabolic pathways involved. *Appl. Microbiol. Biotechnol*, *105*(3), 991–1006. doi: 10.1007/s00253-020-11073-1.

Gesoglu, M., Güneyisi, E., Hansu, O., Etli, S., & Alhassan, M. (2017). Mechanical and fracture characteristics of self-compacting concretes containing different percentage of plastic waste powder. *Constr. Build. Mater*, *140*, 562–569. doi: 10.1016/j.conbuildmat.2017.02.139.

Geyer, R., Jambeck, J. R., & Law, K. L. (2017). Production, use, and fate of all plastics ever made. *Sci. Adv, 3* (7), 3–8. doi: 10.1126/sciadv.1700782.

Gu, L. & Ozbakkaloglu, T. (2016). Use of recycled plastics in concrete: A critical review. *Waste Manag, 51*, 19–42. doi: 10.1016/j.wasman.2016.03.005.

Hahladakis, J. N. & Iacovidou, E. (2019). An overview of the challenges and trade-offs in closing the loop of post-consumer plastic waste (PCPW): Focus on recycling. *J. Hazard. Mater, 380*(July), 120887. doi: 10.1016/j.jhazmat.2019.120887.

Hait, P., Dhara, D., Ghanta, I., Biswas, C., & Basu, P. (2024). Simultaneous Utilization of Fly Ash and Waste Plastics for Making Bricks and Paver Blocks. *J. Inst. Eng. Ser. D*, 1–8. doi: 10.1007/s40033-024-00639-2.

Hama, S. M. & Hilal, N. N. (2017). Fresh properties of self-compacting concrete with plastic waste as partial replacement of sand. *Int. J. Sustain. Built Environ, 6*(2), 299–308. doi: 10.1016/j.ijsbe.2017.01.001.

Hama, S. M. & Hilal, N. N. (2019). *Fresh Properties of Concrete Containing Plastic Aggregate*. Elsevier Ltd.

Haghighatnejad, N., Mousavi, S. Y., Khaleghi, S. J., Tabarsa, A., & Yousefi, S. (2016). Properties of recycled PVC aggregate concrete under different curing conditions. *Constr. Build. Mater, 126*, 943–950. doi: 10.1016/j.conbuildmat.2016.09.047.

Hama, S. M. (2022). Behavior of concrete incorporating waste plastic as fine aggregate subjected to compression, impact load and bond resistance. *Eur. J. Environ. Civ. Eng, 26*(8), 3372–3386. doi: 10.1080/19648189.2020.1798287.

Hameed, A. M. & Ahmed, B. A. F. (2019). Employment the plastic waste to produce the light weight concrete. *Energy Procedia, 157*(2018), 30–38. doi: 10.1016/j.egypro.2018.11.160.

Hannawi, K., Kamali-Bernard, S., & Prince, W. (2010). Physical and mechanical properties of mortars containing PET and PC waste aggregates. *Waste Manag, 30*(11), 2312–2320. doi: 10.1016/j.wasman.2010.03.028.

I. Journal. IOSR Journal of Engineering (IOSR-. *3*(9). 9–19.

Ismail, Z. Z. & AL-Hashmi, E. A. (2008). Reuse of waste iron as a partial replacement of sand in concrete. *Waste Manag, 28*(11), 2048–2053. doi: 10.1016/j.wasman.2007.07.009.

Jain, A., Siddique, S., Gupta, T., Jain, S., Sharma, R. K., & Chaudhary, S. (2019). Fresh, Strength, Durability and Microstructural Properties of Shredded Waste Plastic Concrete. *Iran. J. Sci. Technol. – Trans. Civ. Eng, 43*, 455–465. doi: 10.1007/s40996-018-0178-0.

Jaivignesh, B. & Sofi, A. (2017). Study on Mechanical Properties of Concrete Using Plastic Waste as an Aggregate. *IOP Conf. Ser. Earth Environ. Sci, 80*(1). doi: 10.1088/1755-1315/80/1/012016.

Janani, R. & Kaveri, V. (2020). A critical literature review on reuse and recycling of construction waste in construction industry. *Mater. Today Proc, 37*(Part 2), 3077–3081. doi: 10.1016/j.matpr.2020.09.015.

Jang, Y. C., Lee, G., Kwon, Y., Hong Lim, J., & Hyun Jeong, J. (2020). Recycling and management practices of plastic packaging waste towards a circular economy in South Korea. *Resour. Conserv. Recycl, 158* (December 2019). doi: 10.1016/j.resconrec.2020.104798.

Jassim, A. K. (2017). Recycling of Polyethylene Waste to Produce Plastic Cement. *Procedia Manuf, 8* (October 2016), 635–642. doi: 10.1016/j.promfg.2017.02.081.

Jnr, A. K., Yunana, D., Kamsouloum, P., Webster, M., Wilson, D. C., & Cheeseman, C. (2018). Recycling waste plastics in developing countries: Use of low-density polyethylene water sachets to form plastic bonded sand blocks. *Waste Manag, 80*, 112–118. doi: 10.1016/j.wasman.2018.09.003.

Jonnala, S. N., Gogoi, D., Devi, S., Kumar, M., & Kumar, C. (2024). A comprehensive study of building materials and bricks for residential construction. *Constr. Build. Mater, 425*(April), 135931. doi: 10.1016/j.conbuildmat.2024.135931.

Kabirifar, K., Mojtahedi, M., Wang, C., & Tam, V. W. Y. (2020). Construction and demolition waste management contributing factors coupled with reduce, reuse, and recycle strategies for effective waste management: A review. *J. Clean. Prod, 263*(April). doi: 10.1016/j.jclepro.2020.121265.

Kamaruddin, M. A., Abdullah, M. M. A., Zawawi, M. H., & Zainol, M. R. R. A. (2017). Potential use of plastic waste as construction materials: Recent progress and future prospect. *IOP Conf. Ser. Mater. Sci. Eng*, *267*(1). doi: 10.1088/1757-899X/267/1/012011.

Kennedy, J. F. (1993). to 1989. *30*(3), 1993.

Kerdlap, P., Purnama, A. R., Low, J. S. C., Tan, D. Z. L., Barlow, C. Y., & Ramakrishna, S. (2020). Environmental evaluation of distributed versus centralized plastic waste recycling: Integrating life cycle assessment and agent-based modeling. *Procedia CIRP*, *90*, 689–694. doi: 10.1016/j. procir.2020.01.083.

Khoo, H. H. (2019). LCA of plastic waste recovery into recycled materials, energy and fuels in Singapore. Resour. Conserv. Recycl, 145(February), 67–77. doi: 10.1016/j.resconrec.2019.02.010.

Kognole, R. S., Shipkule, K., S., K., P., M., P., L., & Survase, U. (2019). Utilization of Plastic waste for Making Plastic Bricks. *Int. J. Trend Sci. Res. Dev*, *3*(4), 878–880. doi: 10.31142/ijtsrd23938.

Kumar, R. & Khan, M. A. (2020). Use of plastic waste along with bitumen in construction of flexible pavements. *Int. J. Eng. Res*, *V9*(03), 115–124. doi: 10.17577/ijertv9is030069.

Kumar, R. (2020). Tertiary and quaternary recycling of thermoplastics by additive manufacturing approach for thermal sustainability. *Mater. Today Proc*, *37*(Part 2), 2382–2386. doi: 10.1016/j.matpr.2020.08.183.

Lamba, P., Kaur, D. P., Raj, S., & Sorout, J. (2021). Recycling/reuse of plastic waste as construction material for sustainable development: A review. *Environ. Sci. Pollut. Res*, 0123456789. doi: 10.1007/s11356-021-16980-y.

Leng, Z., Sreeram, A., Padhan, R. K., & Tan, Z. (2018). Value-added application of waste PET based additives in bituminous mixtures containing high percentage of reclaimed asphalt pavement (RAP). *J. Clean. Prod*, *196*, 615–625. doi: 10.1016/j.jclepro.2018.06.119.

Limami, H., Manssouri, I., Cherkaoui, K., & Khaldoun, A. (2021). Mechanical and physicochemical performances of reinforced unfired clay bricks with recycled Typha-fibers waste as a construction material additive. *Clean. Eng. Technol*, *2*(July 2020), 100037. doi: 10.1016/j.clet.2020.100037.

Manjarekar, P. A. S. (2017). Utilization of Plastic Waste in Foundry Sand Bricks. *Int. J. Res. Appl. Sci. Eng. Technol*, *V*(IV), 977–982. doi: 10.22214/ijraset.2017.4178.

Maris, J., Bourdon, S., Brossard, J. M., Cauret, L., Fontaine, L., & Montembault, V. (2018). Mechanical recycling: Compatibilization of mixed thermoplastic wastes. *Polym. Degrad. Stab*, *147*, 245–266. doi: 10.1016/j.polymdegradstab.2017.11.001.

Martin-Alfonso, J. E., Cuadri, A. A., Torres, J., Hidalgo, M. E., & Partal, P. (2019). Use of plastic wastes from greenhouse in asphalt mixes manufactured by dry process. *Road Mater. Pavement Des*, *20*(sup1), S265–S281. doi: 10.1080/14680629.2019.1588776.

Meng, T., Klepacka, A. M., Florkowski, W. J., & Braman, K. (2016). Determinants of recycling common types of plastic product waste in environmental horticulture industry: The case of Georgia. *Waste Manag*, *48*(July 2020), 81–88. doi: 10.1016/j.wasman.2015.11.013.

Meys, R., Frick, F., Westhues, S., Sternberg, A., Klankermayer, J., & Bardow, A. (2020). Towards a circular economy for plastic packaging wastes – The environmental potential of chemical recycling. *Resour. Conserv. Recycl*, *162*(May). doi: 10.1016/j.resconrec.2020.105010.

Mohammed, A. A. & Rahim, A. A. F. (2020). Experimental behavior and analysis of high strength concrete beams reinforced with PET waste fiber. *Constr. Build. Mater*, *244*, 118350. doi: 10.1016/j. conbuildmat.2020.118350.

Mohammed, A. S., Ali, K. M., Rajab, N. A., & Hilal, N. (2020). Mechanical Properties of Concrete and Mortar Containing Low Density Polyethylene Waste Particles as Fine Aggregate. *J. Mater. Eng. Struct*, *7*, 57–72.

Mokhtar, M. *et al.* (2016). Application of plastic bottle as a wall structure for green house. *ARPN J. Eng. Appl. Sci*, *11*(12), 7617–7621.

Möllnitz, S., Khodier, K., Pomberger, R., & Sarc, R. (2020). Grain size dependent distribution of different plastic types in coarse shredded mixed commercial and municipal waste. *Waste Manag*, *103*, 388–398. doi: 10.1016/j.wasman.2019.12.037.

Mondal, M. K., Bose, B. P., & Bansal, P. (2019). Recycling waste thermoplastic for energy efficient construction materials: An experimental investigation. *J. Environ. Manage*, *240*(February), 119–125. doi: 10.1016/j.jenvman.2019.03.016.

Mondal, M. K., Bose, B. P., & Bansal, P. (2019). Recycling waste thermoplastic for energy efficient construction materials: An experimental investigation. *J. Environ. Manage*, *240*(November 2020), 119–125. doi: 10.1016/j.jenvman.2019.03.016.

Movilla-Quesada, D., Raposeiras, A. C., Silva-Klein, L. T., Lastra-González, P., & Castro-Fresno, D. (2019). Use of plastic scrap in asphalt mixtures added by dry method as a partial substitute for bitumen. *Waste Manag*, *87*, 751–760. doi: 10.1016/j.wasman.2019.03.018.

Mustafa, M. A. T., Hanafi, I., Mahmoud, R., & Tayeh, B. A. (2019). Effect of partial replacement of sand by plastic waste on impact resistance of concrete: Experiment and simulation. *Structures*, *20*(April), 519–526. doi: 10.1016/j.istruc.2019.06.008.

Needhidasan, S. & Sai, P. (2020). Demonstration on the limited substitution of coarse aggregate with the E-waste plastics in high strength concrete. *Mater. Today Proc*, *22*(xxxx), 1004–1009. doi: 10.1016/j.matpr.2019.11.255.

Nizamuddin, S., Jamal, M., Gravina, R., & Giustozzi, F. (2020). Recycled plastic as bitumen modifier: The role of recycled linear low-density polyethylene in the modification of physical, chemical and rheological properties of bitumen. *J. Clean. Prod*, *266*, 121988. doi: 10.1016/j.jclepro.2020.121988.

Nyika, J. & Dinka, M. (2022). Recycling plastic waste materials for building and construction Materials: A minireview. *Mater. Today Proc*, *62*(April), 3257–3262. doi: 10.1016/j.matpr.2022.04.226.

Ogundairo, T. O., Adegoke, D. D., Akinwumi, I. I., & Olofinnade, O. M. (2019). Sustainable use of recycled waste glass as an alternative material for building construction – A review. *IOP Conf. Ser. Mater. Sci. Eng*, *640*(1). doi: 10.1088/1757-899X/640/1/012073.

Olofinnade, O., Chandra, S., & Chakraborty, P. (2020). Recycling of high impact polystyrene and low-density polyethylene plastic wastes in lightweight based concrete for sustainable construction. *Mater. Today Proc*, *38*(xxxx), 2151–2156. doi: 10.1016/j.matpr.2020.05.176.

Olofinnade, O., Morawo, A., Okedairo, O., & Kim, B. (2021). Solid waste management in developing countries: Reusing of steel slag aggregate in eco-friendly interlocking concrete paving blocks production. *Case Stud. Constr. Mater*, *14*. doi: 10.1016/j.cscm.2021.e00532.

Padhan, R. K. & Gupta, A. A. (2018). Preparation and evaluation of waste PET derived polyurethane polymer modified bitumen through in situ polymerization reaction. *Constr. Build. Mater*, *158*, 337–345. doi: 10.1016/j.conbuildmat.2017.09.147.

Plastics Europe and EPRO. (2016). Plastics – The Facts 2016. *Plast. – Facts 2016*, 37.

Plastics Europe and Plastics Europe Market Research Group (PEMRG) / Consultic Marketing & Industrieberatung GmbH, "Plastics – the Facts 2017," p. 16, 2017, doi: 10.1016/j.marpolbul.2013.01.015.

Ponnada, S. & Krishna, K. V. (2020). Experimental investigation on modification of rheological parameters of bitumen by using waste plastic bottles. *Mater. Today Proc*, *32*(xxxx), 692–697. doi: 10.1016/j.matpr.2020.03.243.

Pooja, P., Vaitla, M., Sravan, G., Reddy, M. P., & Bhagyawati, M. (2019). Study on behavior of concrete with partial replacement of fine aggregate with waste plastics. *Mater. Today Proc*, *8*, 182–187. doi: 10.1016/j.matpr.2019.02.098.

Punitha, V., Sakthieswaran, N., & Babu, O. G. (2020). Experimental investigation of concrete incorporating HDPE plastic waste and metakaolin. *Mater. Today Proc*, *37*(Part 2), 1032–1035. doi: 10.1016/j.matpr.2020.06.288.

Ragaert, K. *et al*. (2020). Design from recycling: A complex mixed plastic waste case study. *Resour. Conserv. Recycl*, *155*(March 2019), 104646. doi: 10.1016/j.resconrec.2019.104646.

Ragaert, K., Delva, L., & Van Geem, K. (2017). Mechanical and chemical recycling of solid plastic waste. *Waste Manag*, *69*, 24–58. doi: 10.1016/j.wasman.2017.07.044.

Rajan, M. S., Manoj, K. M., Dharun, M., Navanithakrishnan, S., & Shiyamala,. (2024). Comparative Study on Various Properties of Paver Block Produced from Municipal Plastic Waste. *E3S Web Conf, 529*. doi: 10.1051/e3sconf/202452901020.

Remadnia, A., Dheilly, R. M., Laidoudi, B., & Quéneudec, M. (2009). Use of animal proteins as foaming agent in cementitious concrete composites manufactured with recycled PET aggregates. *Constr. Build. Mater, 23*(10), 3118–3123. doi: 10.1016/j.conbuildmat.2009.06.027.

Rochman, C. M. *et al.* (2013). Policy: Classify plastic waste as hazardous. *Nature, 494*(7436), 169–170. doi: 10.1038/494169a.

Safinia, S. & Alkalbani, A. (2016). Use of Recycled Plastic Water Bottles in Concrete Blocks. *Procedia Eng, 164* (June), 214–221. doi: 10.1016/j.proeng.2016.11.612.

Seghiri, M., Boutoutaou, D., Kriker, A., & Hachani, M. I. (2017). The Possibility of Making a Composite Material from Waste Plastic. *Energy Procedia, 119*(November), 163–169. doi: 10.1016/j.egypro.2017.07.065.

Sellakutty, D., UTILISATION OF WASTE PLASTIC IN MANUFACTURING OF BRICKS AND PAVER. No January, 2016.

Shahjalal, M., Islam, K., Rahman, J., Ahmed, K. S., Karim, M. R., & Billah, A. M. (2021). Flexural response of fiber reinforced concrete beams with waste tires rubber and recycled aggregate. *J. Clean. Prod, 278*, 123842. doi: 10.1016/j.jclepro.2020.123842.

Silva, R. V., De Brito, J., & Saikia, N. (2013). Influence of curing conditions on the durability-related performance of concrete made with selected plastic waste aggregates. *Cem. Concr. Compos, 35*(1), 23–31. doi: 10.1016/j.cemconcomp.2012.08.017.

Singh, N., Hui, D., Singh, R., Ahuja, I. P. S., Feo, L., & Fraternali, F. (2017). Recycling of plastic solid waste: A state of art review and future applications. *Compos. Part B Eng, 115*, 409–422. doi: 10.1016/j.compositesb.2016.09.013.

Sonone, P., Devalkar, R., & Student, P. G. (2017). *Green Sustainable Bricks Made of Fly Ash and Discarded Polyethylene Waste*. 6509–6516. doi: 10.15680/IJIRSET.2017.0604063.

Steyn, Z. C., Babafemi, A. J., Fataar, H., & Combrinck, R. (2021). Concrete containing waste recycled glass, plastic and rubber as sand replacement. *Constr. Build. Mater, 269*(xxxx), 121242. doi: 10.1016/j.conbuildmat.2020.121242.

Sule, J. *et al.* (2017). Use of Waste Plastics in Cement-Based Composite for Lightweight Concrete Production. *Int. J. Res. Eng. Technol, 2*(5), 44–54.

Suleman, S. & Needhidasan, S. (2020). Utilization of manufactured sand as fine aggregates in electronic plastic waste concrete of M30 mix. *Mater. Today Proc, 33*(xxxx), 1192–1197. doi: 10.1016/j.matpr.2020.08.043.

Taghavi, N., Singhal, N., Zhuang, W. Q., & Baroutian, S. (2021). Degradation of plastic waste using stimulated and naturally occurring microbial strains. *Chemosphere, 263*, 127975. doi: 10.1016/j.chemosphere.2020.127975.

Thomas, L. M. & Moosvi, S. A. (2020). Hardened properties of binary cement concrete with recycled PET bottle fiber: An experimental study. *Mater. Today Proc, 32*(xxxx), 632–637. doi: 10.1016/j.matpr.2020.03.025.

Tulashie, S. K., Boadu, E. K., Kotoka, F., & Mensah, D. (2020). Plastic wastes to pavement blocks: A significant alternative way to reducing plastic wastes generation and accumulation in Ghana. *Constr. Build. Mater, 241*, 118044. doi: 10.1016/j.conbuildmat.2020.118044.

Turukmane, R. N., Daberao, A., & Gulhane, S. S. (2018). Recycling of PET Clothes and Bottles. *Int. J. Res. Sci. Innov, V*(May), 295–296.

Uvarajan, T., Gani, P., Ng, C. C., & Zulkernain, N. H. (2022). Reusing plastic waste in the production of bricks and paving blocks: A review. *Eur. J. Environ. Civ. Eng, 26*(14), 1–21.

Vanitha, S., Natarajan, V., & Praba, M. (2015). Utilisation of waste plastics as a partial replacement of coarse aggregate in concrete blocks. *Indian J. Sci. Technol, 8*(12). doi: 10.17485/ijst/2015/v8i12/54462.

Vasanthi, P., Devaraju, A., Haripriya, P., Prasanth, E., & Vaishnavi, T. S. (2020). Performance of concrete by using milk cover and pet flakes replacement in concrete constituent. *Mater. Today Proc*, *39*(xxxx), 459–466. doi: 10.1016/j.matpr.2020.07.720.

Veropalumbo, R., Russo, F., Viscione, N., Biancardo, S. A., & Oreto, C. (2021). Investigating the rheological properties of hot bituminous mastics made up using plastic waste materials as filler. *Constr. Build. Mater*, *270*(xxxx), 121394. doi: 10.1016/j.conbuildmat.2020.121394.

Vishnu, T. B. & Singh, K. L. (2021). A study on the suitability of solid waste materials in pavement construction: A review. *Int. J. Pavement Res. Technol*, *14*(5), 625–637. doi: 10.1007/s42947-020-0273-z.

Wang, C., Zhao, L., Lim, M. K., Chen, W. Q., & Sutherland, J. W. (2020). Structure of the global plastic waste trade network and the impact of China's import Ban. *Resour. Conserv. Recycl*, *153*. doi: 10.1016/j.resconrec.2019.104591.

Wang, L. *et al*. (2021). Bacterial and fungal assemblages and functions associated with biofilms differ between diverse types of plastic debris in a freshwater system. *Environ. Res*, *196*(September). doi: 10.1016/j.envres.2020.110371.

Wang, W. *et al*. (2019). Current influence of China's ban on plastic waste imports. *Waste Dispos. Sustain. Energy*, *1*(1), 67–78. doi: 10.1007/s42768-019-00005-z.

Wong, Y. C., Perera, S., Zhang, Z., Arulrajah, A., & Mohammadinia, A. (2020). Field study on concrete footpath with recycled plastic and crushed glass as filler materials. *Constr. Build. Mater*, *243*, 118277. doi: 10.1016/j.conbuildmat.2020.118277.

Záleská, M. *et al*. (2018). Biomass ash-based mineral admixture prepared from municipal sewage sludge and its application in cement composites. *Clean Technol. Environ. Policy*, *20*(1), 159–171. doi: 10.1007/s10098-017-1465-3.

Shashikant Nishant Sharma*, Kavita Dehalwar, Gopal Kumar,
Krishna Yadav, and Devraj Verma

Utilization of Advanced Materials for Permeable Paving, Biocrete, and Piezoelectric Materials in Walkways to Transit Stations

Abstract: Transit stations are critical nodes in urban transportation networks and require innovative, resilient pedestrian infrastructure to support high footfall and enhance commuter experience. This study explores the application of advanced materials – specifically permeable paving, biocrete, and piezoelectric surfaces – for footpaths and walkways within transit-oriented development zones. The exploration was conducted through a comprehensive literature review of peer-reviewed publications, technical reports, and pilot project evaluations.

Permeable paving improves stormwater management and reduces urban runoff, contributing to pedestrian safety and environmental resilience. Biocrete, a self-healing and eco-friendly material, offers reduced maintenance costs and enhanced structural longevity. Piezoelectric materials, which convert pedestrian kinetic energy into electrical energy, present new opportunities for powering station lighting and smart sensors, promoting energy efficiency in dense transit hubs. Case studies from cities, such as Portland, Delft, and New Delhi, illustrate the feasibility and benefits of these technologies in real-world settings.

The findings suggest that integrating such materials can significantly improve durability, safety, comfort, and long-term cost-efficiency in pedestrian infrastructure. The study recommends targeted pilot implementation in high-traffic transit zones, supported by performance monitoring and life-cycle cost analysis. Future research should focus on optimizing material properties, addressing cost barriers, and developing region-specific design guidelines. Ultimately, the study underscores the importance of interdisciplinary collaboration to advance smart, sustainable, and resilient transit infrastructure in urban environments.

---

*Corresponding author: **Shashikant Nishant Sharma**, Maulana Azad National Technology, Bhopal, Madhya Pradesh, India, e-mail: urp2025@gmail.com
Kavita Dehalwar, Department of Architecture and Planning, Maualana Azad National Institute of Technology, Bhopal, India, email: kdehalwar@manit.ac.in
Gopal Kumar, Department of Architecture and Planning, Maualana Azad National Institute of Technology, Bhopal, India, email: gopalkumar594@gmail.com
Krishna Yadav, Department of Architecture and Planning, Maualana Azad National Institute of Technology, Bhopal, India, email: kkyadav96@gmail.com
Devraj Verma, Department of Architecture and Planning, Maualana Azad National Institute of Technology, Bhopal, India, email: dvnitb@gmail.com

https://doi.org/10.1515/9783111563046-007

Keywords: Permeable paving, biocrete, piezoelectric materials, sustainable pedestrian infrastructure, transit station walkways

1 Introduction

Transit stations serve as vital nodes in urban transportation networks, facilitating seamless movement for large volumes of pedestrians (Sharma et al., 2024). However, traditional footpaths and walkways often face significant challenges, including water-logging, wear and tear, and inefficiency in energy utilization. These issues not only reduce the lifespan of pedestrian infrastructure but also compromise safety and accessibility (Saltarin-Molino et al., 2023). Addressing these concerns through innovative material solutions can enhance the functionality and sustainability of pedestrian facilities in transit stations. This chapter explores the potential benefits and applicability of permeable paving, biocrete, and piezoelectric materials in the design and construction of pedestrian infrastructure, offering a sustainable and technologically advanced approach to urban mobility.

Permeable paving is an effective solution for mitigating waterlogging, a common issue in high-traffic transit areas (Patel et al., 2024). By allowing water to infiltrate through its surface and drain into underlying layers, this material reduces surface runoff and prevents the formation of hazardous puddles (Mayer et al., 2024). Additionally, permeable pavements can integrate filtration systems to remove pollutants, thereby improving urban water management (Kayhanian et al., 2019). The use of such materials in transit station walkways ensures enhanced durability while promoting environmental sustainability (De Graaf-van Dinther et al., 2021).

Biocrete, a bioengineered concrete infused with self-healing bacteria, addresses the issue of wear and tear in pedestrian infrastructure (De Souza & Sanchez, 2023). Traditional concrete walkways often develop cracks due to constant pressure and environmental stressors, leading to costly maintenance and potential safety hazards. Biocrete, however, possesses self-repairing properties, wherein bacteria embedded within the concrete activate upon exposure to moisture and fill the cracks with calcite deposits (Jeon et al., 2022). This innovation extends the lifespan of pedestrian pathways, reducing maintenance costs and minimizing disruptions in transit areas.

Piezoelectric materials introduce an energy-efficient approach by harnessing kinetic energy from pedestrian movement (Zheng et al., 2024). These materials generate electricity when subjected to mechanical stress, offering a renewable energy source that can be used to power transit station lighting, electronic displays, and other essential facilities (Chen et al., 2019). Implementing piezoelectric flooring in high-footfall areas of transit stations transforms pedestrian traffic into a valuable energy resource, contributing to overall energy efficiency and sustainability (Li et al., 2023). The growing emphasis on sustainable and smart urban infrastructure has necessitated the development of innovative materials for pedestrian pathways. Walkways leading to transit stations play a critical role in last-mile connectivity, influencing both commuter experience and environmental impact. This research investigates how advanced materials – permeable paving, biocrete, and piezoelectric materials – can be utilized to improve sustainability, durability, and energy efficiency in pedestrian infrastructure.

Thus, the research question arises as follows: "How can the utilization of advanced materials, such as permeable paving, biocrete, and piezoelectric materials enhance the sustainability, durability, and functionality of walkways leading to transit stations?" The integration of these advanced materials in transit station infrastructure presents a forward-thinking solution to the challenges faced by traditional pedestrian facilities. By combining permeability, self-healing capabilities, and energy generation, transit authorities can create resilient, low-maintenance, and eco-friendly walkways. This approach not only enhances pedestrian comfort and safety but also aligns with global sustainability goals, reinforcing the role of innovative materials in shaping the future of urban mobility.

2 Methodology

This study employs a comprehensive research methodology based on a literature review and a review of case studies from around the world to assess the utilization of advanced materials – permeable paving, biocrete, and piezoelectric materials – in walkways leading to transit stations. The methodology is structured to systematically analyze existing knowledge, identify best practices, and evaluate the feasibility and effectiveness of these materials in improving pedestrian infrastructure.

This chapter is based on the above conceptual smart materials integration into the walkways to transit stations, which get the maximum number of footfalls.

2.1 Literature Review

A comprehensive literature review of published academic articles, technical reports to understand the potential application of advanced materials – permeable paving, biocrete, and piezoelectric surfaces – in pedestrian infrastructure at transit stations. The review focused on identifying material properties, functional benefits, limitations, and implementation outcomes in contexts aligned with ttransit-oriented development (TOD) principles. Particular attention was paid to studies that addressed pedestrian safety, environmental resilience, cost-efficiency, and energy harvesting potential as depicted in the Figure 1. This review provided the foundational understanding needed to assess the practical viability of these materials in real-world transit environments. A thorough literature review published papers in last 10 years has

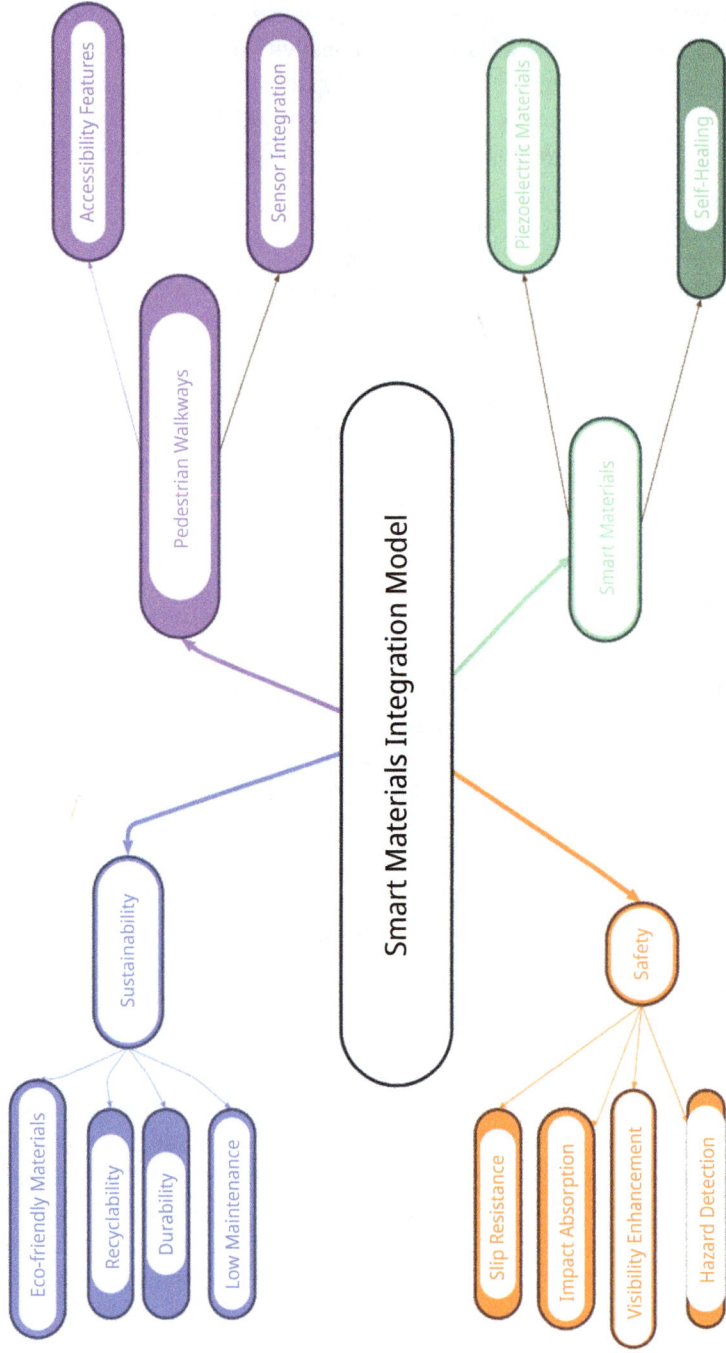

Figure 1: Conceptual diagram to visualize the suggested integration model.

been conducted to establish a foundational understanding of the properties, advantages, limitations, and practical applications of permeable paving, biocrete, and piezoelectric materials. The review will focus on academic research papers, technical reports, government publications, and industry guidelines related to advanced materials in urban transportation infrastructure. Key topics will include:

1. **Permeable paving:** Studies on material composition, permeability rates, structural durability, and environmental benefits, such as stormwater management and reduction in urban heat island (UHI) effects (Saadeh et al., 2019).
2. **Biocrete:** Research on self-healing concrete, microbial-induced calcite precipitation, durability, maintenance cost savings, and its potential for improving walkway longevity (Sharma et al., 2024).
3. **Piezoelectric materials:** Analysis of the principles of piezoelectric energy harvesting, case studies of its implementation in pedestrian pathways, and its contribution to energy-efficient transit infrastructure (Zheng et al., 2024).

The literature review also included a comparative analysis of different material compositions, technological advancements, and implementation challenges. The findings will help determine the suitability of these materials for transit station walkways and their potential role in promoting sustainability.

PRISMA (Preferred Reporting Items for Systematic Reviews and Meta-Analyses) framework for the research methodology shown in Table 1 outlines the selection process of studies and case studies included in the review.

Table 1 shows the PRISMA framework that ensures a transparent and systematic approach to identifying and selecting relevant studies and case studies, improving the reliability and comprehensiveness of the research.

2.2 Review of Case Studies

To supplement the literature review, case studies of cities and transit systems that have implemented these materials will be analyzed. The selection of case studies will be based on their relevance, geographic diversity, and availability of performance data. The review will focus on:

1. **Successful applications:** Examining real-world implementations of permeable paving (Iqbal et al., 2022), biocrete (Sharma et al., 2024), and piezoelectric materials in walkways across different transit hubs, with an emphasis on material performance, maintenance, and cost-effectiveness.
2. **Challenges and lessons learned:** Identifying technical, financial, and policy challenges encountered during implementation and how they were addressed.
3. **Comparative performance assessment:** Evaluating the long-term benefits of these materials in terms of durability, environmental impact, and pedestrian safety based on reported data and feedback from transit authorities.

Table 1: PRISMA (Preferred Reporting Items for Systematic Reviews and Meta-Analyses).

Stage	Description
Identification	– Systematic search conducted in academic databases (e.g., Scopus, Web of Science, and Google Scholar) and institutional reports on permeable paving, biocrete, and piezoelectric materials – Grey literature (government reports, industry white papers, and policy documents) included to capture real-world applications – Keywords used: *permeable paving in transit stations, biocrete for walkways, piezoelectric materials in pedestrian infrastructure, sustainable pedestrian pathways* – Case studies identified from urban transport agencies, infrastructure projects, and international best practices
Screening	– Initial review based on relevance to the studies focus on transit station walkways – Exclusion of duplicate studies and sources unrelated to pedestrian infrastructure – Title and abstract screening to retain studies that specifically address material application, performance assessment, and sustainability impacts
Eligibility	– Full-text articles, reports, and case studies assessed for methodological rigor and data availability – Studies excluded if they lack empirical data, fail to discuss real-world implementation, or are not peer-reviewed – Preference given to studies published in the last 10 years to ensure contemporary relevance
Inclusion	– Final selection includes studies with experimental results, field implementations, cost-benefit analyses, and policy recommendations – Case studies covering multiple geographic regions to ensure diversity in climatic conditions, regulatory environments, and urban planning strategies – Integration of findings to compare the effectiveness and feasibility of advanced materials in walkways leading to transit stations

By integrating insights from both literature and case studies, this methodology will provide a holistic understanding of the feasibility and potential benefits of utilizing advanced materials in pedestrian pathways leading to transit stations. The findings will serve as a foundation for recommending best practices and guiding future implementation strategies.

3 Findings

3.1 Sustainability

Sustainable pedestrian infrastructure relies heavily on the use of eco-friendly materials that minimize environmental impact throughout their life cycle (Ginting & Mardianta, 2023). Materials such as permeable concrete and biocrete support sustainability through high recyclability, reduced reliance on virgin aggregates, and lower embodied carbon (Giriselvam et al., 2018). Biocrete, in particular, enhances environmental performance by incorporating self-healing bacteria, which extend structural life and reduce maintenance frequency and associated resource use (Jeon et al., 2022). These materials also demonstrate high durability under varying climatic conditions, making them suitable for long-term use in TOD zones (Hattrick-Simpers et al., 2023). Their low maintenance requirements translate into cost savings and lower environmental degradation over time, reinforcing their role in building resilient and sustainable urban transport ecosystems (Wijaksono & Wibowo, 2023).

Permeable paving is an innovative solution that allows water to infiltrate through its surface, effectively reducing surface runoff and mitigating urban flooding (Saadeh et al., 2019). By enabling natural water absorption, it enhances stormwater management, preventing water accumulation in high-footfall areas, such as transit stations. One of the documented environmental benefits of permeable pavement is its contribution to reducing the UHI effect (Mistry & Mehrotra, 2023). The permeable pavements can lower surface temperatures by up to 2–3 °C compared to conventional asphalt surfaces, particularly during peak summer conditions (Wang et al., 2018). This reduction depends on factors, such as material albedo, underlying subbase composition, moisture content, and regional climate. Field experiments conducted in California and Arizona have measured sustained lower surface and near-surface air temperatures in areas with permeable installations (Shiflett et al., 2017). Additionally, permeable paving contributes to groundwater recharge by facilitating the infiltration of rainwater into the soil, replenishing underground water sources and promoting ecological balance (Antunes et al., 2020).

Beyond environmental advantages, permeable paving also enhances pedestrian safety by providing improved skid resistance (Guo et al., 2021), reducing the likelihood of slips and falls. Various materials are used for permeable paving, including porous asphalt, permeable concrete, and interlocking pavers, each offering unique benefits in terms of durability, permeability, and maintenance requirements. Case studies from transit stations worldwide demonstrate the effectiveness of permeable paving in managing water runoff, maintaining structural integrity under heavy foot traffic, and reducing maintenance costs (Liu et al., 2021; Patel et al., 2024). These studies highlight key considerations, such as material selection, climate suitability, and long-term performance, reinforcing the role of permeable paving as a sustainable and practical solution for pedestrian infrastructure in urban transit hubs.

3.2 Safety

Safety is a fundamental aspect of pedestrian infrastructure, especially within high-traffic TOD zones. One of the critical factors in ensuring pedestrian safety is **slip resistance** (Guo et al., 2021). Materials, such as porous concrete and permeable pavers offer enhanced friction due to their textured surfaces, significantly reducing the risk of slips and falls during wet or icy conditions (Antunes et al., 2020). Research has shown that permeable pavements can improve skid resistance by **30–50%** compared to traditional asphalt, thereby providing safer walking surfaces, particularly near transit stations where passenger turnover is high and weather conditions vary (Fwa, 2021).

Another essential safety feature is **impact absorption**, which reduces injuries in the event of trips or falls. Rubberized permeable materials or composites mixed with recycled elastomers can offer shock-absorbing qualities, reducing the severity of impact-related injuries (Mahesh et al., 2021). These materials are particularly useful in zones frequented by elderly individuals or children, such as around metro entrances or bus terminals (Chen et al., 2020). Moreover, surfaces with moderate elasticity help alleviate foot fatigue during long commutes or transfers, thereby improving comfort and reducing minor physical strain over time.

Visibility enhancement plays a pivotal role in night-time or low-light conditions, where the risk of pedestrian accidents increases (Omar et al., 2022). Materials with integrated photoluminescent additives or embedded Light Emitting Diode systems powered by piezoelectric energy harvesting can improve visibility along walkways (Zheng et al., 2024). These solutions not only illuminate paths but also create guidance systems for visually impaired users, enhancing safety and accessibility. In addition, high-contrast surface finishes or color-coded pavements can be used to demarcate movement zones and alert pedestrians to level changes, curbs, or approaching crossings (Jiang et al., 2024).

Finally, advancements in **hazard detection** technologies are increasingly being integrated into smart urban materials (Shen et al., 2025). Walkway systems equipped with embedded sensors can detect obstructions, water pooling, or ice formation in real-time, triggering alerts through connected mobile apps or visual indicators (Menz et al., 2004). This proactive approach allows pedestrians and maintenance teams to respond to hazards before accidents occur. Such features are particularly relevant in TOD environments where pedestrian flow is constant and uninterrupted access is essential for transit efficiency and commuter safety.

3.3 Smart Materials

3.3.1 Biocrete: A Self-Healing and Sustainable Concrete

Biocrete, or bio-concrete, is an innovative construction material that integrates bacteria-based self-healing properties, significantly enhancing the durability and sustainability of pedestrian infrastructure (Sharma et al., 2024). Its primary advantage lies in its ability to self-repair cracks through bacterial calcite precipitation, a process in which embedded bacteria activate upon exposure to moisture and produce limestone, sealing the cracks naturally (Giriselvam et al., 2018). This section evaluates biocrete's practical application in transit environments and explores its integration with other sustainable materials, such as permeable paving and piezoelectric surfaces. By combining self-healing properties with permeability and energy-generating capabilities, transit stations can develop more resilient and eco-friendly pedestrian pathways, enhancing both functionality and sustainability in urban mobility infrastructure.

This self-repairing capability reduces maintenance costs and minimizes material degradation, making biocrete a cost-effective solution for high-footfall areas, such as transit stations.

Improvements in **skid resistance** have also been observed with certain porous pavement materials (Guo et al., 2021). A study by Ferguson (2005) reported that porous asphalt exhibited a **30–50% higher skid resistance** compared to dense-graded asphalt due to increased surface texture and water drainage capability, particularly under wet conditions. Enhanced friction coefficients contribute directly to improved pedestrian safety (Ferguson, 2005). Biocrete has shown self-healing crack capabilities of up to 0.5 mm within 28 days of exposure to water and air, as demonstrated in experimental studies. Additionally, its production can reduce carbon emissions by approximately 20–30% compared to traditional concrete (Wiktor & Jonkers, 2011). However, challenges remain in scaling production and ensuring bacterial viability across diverse climates

Beyond durability, biocrete also extends the lifespan of footpaths and walkways, ensuring long-term structural integrity despite continuous pedestrian loads. Additionally, it contributes to environmental sustainability by absorbing CO_2 during the calcite precipitation process, mitigating the carbon footprint associated with conventional concrete (Antunes et al., 2020). These characteristics make biocrete a promising alternative for transit station walkways, where frequent wear and tear necessitates robust and low-maintenance materials.

3.3.2 Piezoelectric Materials for Energy Harvesting

Piezoelectric materials offer a cutting-edge solution for energy-efficient pedestrian infrastructure by converting mechanical energy from footsteps into electrical energy (Covaci & Gontean, 2020). This technology has significant applications in transit stations, where high foot traffic can be harnessed to generate sustainable power. One of the primary uses of piezoelectric walkways is powering LED lighting and smart signage, ensuring that transit hubs remain well-lit and information displays remain operational without excessive reliance on external power sources. Additionally, these materials support real-time data collection by monitoring footfall, enabling transit authorities to analyze pedestrian movement patterns and optimize facility management (Bai et al., 2023).

Pilot implementations of **piezoelectric flooring systems** in transport hubs and public spaces have shown that each footstep can generate approximately **2–5 J** of energy (Li & Strezov, 2014). In high-footfall environments like metro stations, this can translate into several kilowatt-hours per day, enough to power LED lighting or low-power sensors. However, installation costs remain high – typically around **$250–500 per m²**, limiting widespread adoption (Sabzpoushan & Woias, 2024).

By reducing dependency on conventional energy supplies, piezoelectric systems contribute to sustainability goals and operational cost savings. Recent advancements in piezoelectric technology have focused on improving material efficiency, durability, and energy conversion rates, making them more viable for large-scale applications. Material selection is crucial, with piezoelectric ceramics, polymers, and composite structures offering varying degrees of flexibility, resilience, and output efficiency.

3.4 Pedestrian Walkaways – Integration of Sensors

This section explores the feasibility of integrating piezoelectric materials into transit station walkways through a cost-benefit analysis, assessing installation expenses, maintenance requirements, and long-term energy savings. By leveraging this technology, transit stations can enhance sustainability while improving the overall passenger experience through innovative, self-sustaining infrastructure solutions.

The key benefits, implementation outcomes, and practical challenges associated with biocrete applications, particularly in pedestrian infrastructure, are summarized in Table 2. Similar summaries for permeable paving and piezoelectric surfaces are included in the respective sections.

The case studies (as shown in Table 2) highlight that advanced materials in pedestrian infrastructure not only offer significant sustainability benefits but also come with challenges. Permeable paving, as seen in Rotterdam and New York, effectively mitigates urban flooding but requires frequent maintenance to prevent clogging. Biocrete, used in London, enhances durability and carbon sequestration, though large-scale adoption remains limited due to cost and awareness barriers. Piezoelectric materials in Tokyo and Paris demonstrate potential for energy harvesting, yet high installation and maintenance costs remain obstacles. These findings underscore the need for context-specific material selection, ongoing maintenance strategies, and supportive policies to maximize the benefits of these innovative walkway solutions.

Table 2: Case studies around the world to assess the applicability of the advanced materials.

Location	Materials used	Benefits	Issues	Source
Tokyo, Japan	Piezoelectric walkways	Generated electricity from pedestrian foot traffic, reducing station energy costs	Initial installation cost is approximately **20–30% higher** than conventional concrete (e.g., biocrete sidewalk project, Tokyo Metropolitan Government, 2021), due to bacterial additive and specialized curing requirements	Smart Cities Japan Report 2020 (Fermoso, 2008)
Rotterdam, Netherlands	Permeable paving	Reduced urban flooding and improved groundwater recharge	Requires frequent cleaning to prevent clogging	Veldkamp et al. (2022) and De Graaf-van Dinther et al. (2021)
London, UK	Biocrete pavements	Enhanced durability and carbon sequestration	Limited large-scale implementation and public awareness	Sustainable Materials Research, 2021
Paris, France	Piezoelectric pavers	Used for street lighting in transit areas, promoting energy efficiency	High wear and tear in high-traffic zones	Green Infrastructure Reports, 2022
New York, USA	Permeable concrete sidewalks	Improved pedestrian safety by reducing surface water accumulation	High cost compared to traditional concrete	*Journal of Urban Planning*, 2023

3.5 Comparative Study of the Applicability of Advanced Materials

This detailed comparative table evaluates the utilization of **permeable paving, biocrete, and piezoelectric materials** in walkways leading to transit stations based on key parameters.

Table 3 provides a comprehensive comparison of the three advanced materials, highlighting their potential applications, challenges, and opportunities for future research. The characteristics, benefits, and limitations of each advanced material reviewed are summarized in Table 3. This table provides a comparative overview to aid planners and policymakers in evaluating their applicability to TOD contexts.

The comparative analysis of permeable paving, biocrete, and piezoelectric materials in transit station walkways highlights their unique benefits and challenges. Permeable paving proves highly effective in stormwater management, reducing urban

Table 3: Utilization of permeable paving, biocrete, and piezoelectric materials.

Criteria	Permeable paving	Biocrete	Piezoelectric materials
Primary function	Allows water infiltration to reduce surface runoff and urban flooding	Self-healing material that repairs cracks using bacteria-based calcite precipitation	Converts mechanical energy from footsteps into electrical energy (Zheng et al., 2024)
Key benefits	– Stormwater management – Reduces urban heat island effect – Improves groundwater recharge. – Enhances skid resistance for pedestrians.	– Reduces maintenance costs – Increases walkway lifespan – Absorbs up to 0.5–1.2 kg of CO_2/m^2 of surface area over its life cycle (Wang et al., 2018), depending on environmental exposure and porosity levels – Enhances structural durability	– Generates renewable energy – Supports real-time footfall monitoring – Reduces dependency on external power sources – Can power LED lighting and smart signage
Common materials used	Porous asphalt, permeable concrete, and interlocking pavers	Concrete mixed with self-healing bacteria (e.g., *Bacillus* species)	Piezoelectric ceramics, polymers, or composite materials
Durability and maintenance	Long-lasting but requires periodic cleaning to prevent clogging of pores (Saltarin-Molino et al., 2023)	Highly durable with self-repairing properties, reducing long-term maintenance needs	Durable but requires protection against extreme weather and mechanical damage
Implementation challenges	– Higher initial cost compared to conventional pavement – Requires specialized installation and maintenance	– Higher material and production costs – Limited large-scale implementation experience	– High upfront investment – Lower energy output per square meter – Integration challenges with existing infrastructure
Environmental impact	– Improves water absorption and reduces flooding (Kayhanian et al., 2019) – Lowers urban heat island effect	– Reduces CO_2 emissions – Minimizes concrete degradation and material waste	– Contributes to energy sustainability – Reduces reliance on fossil-fuel-based electricity (Li et al., 2023)

Cost considerations	Moderate to high installation cost but lower maintenance expenses over time	Higher initial cost but long-term savings due to self-repairing properties	High initial investment with long-term potential for energy cost savings (Chen et al., 2019)
Best use cases	– High-footfall transit station walkways – Areas prone to flooding or excessive water runoff	– High-traffic pedestrian zones – Locations requiring durable and low-maintenance pathways	– High-footfall transit stations where energy can be efficiently harvested
Case studies	– Tokyo, Japan: use of permeable pavements in transit stations for flood control – Portland, USA: green street projects integrating permeable pavements	– The Netherlands: Research on biocrete's application in road and walkway infrastructure – Spain: Experimental use of biocrete in urban spaces	– London, UK: Pavegen walkway installation for energy generation in a train station – Toulouse, France: Piezoelectric pavement used for smart city initiatives.
Future research directions	– Improving permeability and structural integrity for high-traffic areas – Reducing maintenance needs through self-cleaning coatings	– Enhancing bacterial efficiency for faster crack repair – Exploring new bio-based materials for increased sustainability	– Increasing energy efficiency and output – Integrating with smart grids for enhanced energy storage and distribution

flooding and the heat island effect, though it requires periodic maintenance to prevent clogging. Biocrete stands out for its self-healing properties, reducing long-term maintenance costs and extending walkway lifespan, but its high initial cost and limited large-scale applications pose challenges. Piezoelectric materials offer an innovative approach to energy generation, powering smart infrastructure and reducing reliance on external electricity sources, though they face hurdles in energy efficiency and high installation costs. Each material presents environmental benefits, from CO_2 absorption in biocrete to energy sustainability in piezoelectric walkways. While all three materials offer long-term advantages, their large-scale adoption depends on technological advancements, cost reductions, and policy incentives. Future research should focus on hybrid applications that combine these materials to create resilient, sustainable pedestrian infrastructure in transit stations.

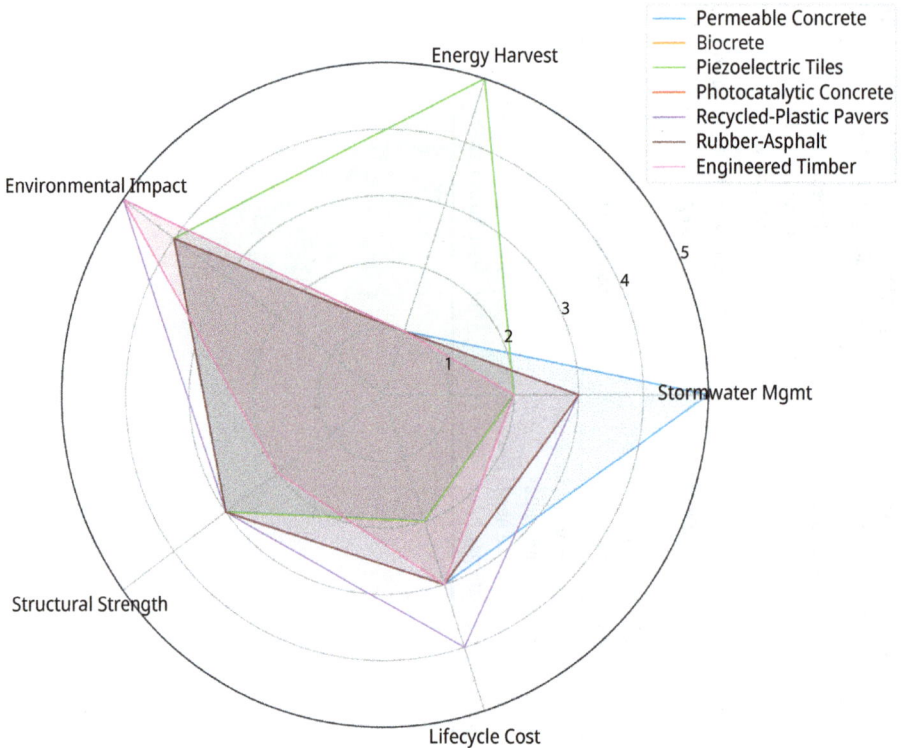

Figure 2: Qualitative performance radar for smart materials.

This visualization (Figure 2) should help you decide which material – or hybrid of materials – best fits a given project priority like green-infrastructure credits versus smart-city energy harvesting.

Modern pedestrian walkways in TOD zones are evolving to prioritize universal accessibility, ensuring that infrastructure serves all users, including the elderly, persons with disabilities, and caregivers with strollers. Accessibility features, such as tactile paving, curb ramps, anti-slip coatings, and appropriate gradient transitions are essential for safe and independent mobility. Materials like biocrete and permeable tiles can be cast or molded to incorporate raised surfaces and textured guides that align with accessibility norms. Additionally, consistent surface smoothness and joint stability reduce tripping hazards and accommodate assistive devices, such as wheelchairs and walking aids.

The integration of sensors into walkway infrastructure represents a significant advancement in creating intelligent pedestrian environments. Embedded sensors can monitor foot traffic patterns, detect obstructions, and measure environmental conditions, such as temperature or surface moisture. These data can be used in real-time to optimize lighting, inform maintenance schedules, or enhance safety by alerting pedestrians and authorities to hazards like spills or ice. In high-density TOD areas, sensors also support crowd management and facilitate better transit coordination by providing dynamic input into station operations.

Beyond functionality, sensor-integrated walkways can also contribute to sustainability and user engagement. For instance, when paired with piezoelectric materials, sensors can help capture and quantify kinetic energy from footfalls, which can then be used to power low-energy devices like LED pathway lights or public information screens. These smart features not only enhance operational efficiency but also foster a responsive, inclusive, and technologically enabled pedestrian experience aligned with the goals of sustainable and resilient urban mobility.

This screener (Table 4) can be used to quickly match project goals (e.g., stormwater control vs. energy harvesting) against the material most suited for your walkway application.

4 Discussion

Materials such as permeable concrete, biocrete, and piezoelectric surfaces not only offer technical advantages like high permeability or energy harvesting but also directly support TOD principles by improving pedestrian experience. For instance, permeable concrete reduces surface water accumulation, which enhances pedestrian safety and walkability by minimizing slip hazards and controlling urban runoff. Similarly, biocrete's self-healing properties reduce maintenance frequency and associated costs, making it a more sustainable and cost-efficient choice for high-footfall TOD zones.

Table 4: Screener/summary innovative pedestrian-walkway materials.

Material type	Key characteristics (headline metrics or traits)	Primary benefits for pedestrian walkways	Key challenges/limitations
Permeable (pervious) concrete	– Void content: 15–25% – Infiltration rate: 200–1,200 mm/h – Compressive strength: 17–28 MPa	– Rapid stormwater infiltration → reduced surface runoff and puddling – Mitigates urban heat island effect (cooler surface) – Can earn LID/green infrastructure credits	– Lower structural strength than conventional concrete (not for heavy vehicular loads) – Prone to clogging if not vacuum-swept periodically – Higher initial cost
Biocrete/bio-based concrete (e.g., bacteria-induced calcite and hempcrete blocks)	– Self-healing microcracks via calcite-precipitating bacteria – Density: 300–900 kg/m^3 (lighter than OPC) – Thermal conductivity: 0.1–0.2 W/m K	– Extends pavement life by autonomously sealing hairline cracks – Lower embodied CO_2; uses agricultural waste or microbial processes – Good insulation for adjacent underground utilities	– Slower strength gain, compressive strength: 3–15 MPa (only for foot traffic) – Needs moisture/nutrients for bacterial viability – Limited long-term field data
Piezoelectric pavement tiles	– Energy output: 2–5 W per footstep (peak), 1–10 kWh/m^2· year (real-world) – Uses PZT ceramics or PVDF films embedded in elastomer tiles	– Harvests kinetic energy to power LED lighting/sensors – Generates real-time pedestrian-flow analytics – High-tech appeal for smart-city branding	– High capital cost ($800–2,000 m^2) – Output sensitive to foot-traffic density and load profile – Electronics require waterproofing and periodic replacement
Photocatalytic (TiO$_2$-doped) concrete/pavers	– NO$_x$ removal rate: 20–40% over 8-h sun exposure – Surface self-cleaning (hydrophilic)	– Improves air quality along busy corridors – Algae and grime resistance keeps walkways brighter, safer	– Performance drops without UV light (shaded streets) – Incremental cost for TiO$_2$ additive
Recycled-plastic composite pavers	– Made from 95% post-consumer HDPE and PP	– Diverts plastic waste from landfill	– Thermal expansion higher than concrete (requires spacing)

Material	Properties	Advantages	Disadvantages
	– Density: 1,000 kg/m³; flexural strength: 15–25 MPa	– High slip resistance, colorfast without paint – Lightweight ⇒ quick installation and modular repairs	– Potential microplastic wear over decades
Rubber asphalt (crumb rubber) walkway surfacing	– Up to 20% scrap-tire rubber blended with asphalt binder – Elastic modulus lower than standard asphalt ⇒ cushioned feel	– Excellent shock absorption, ADA-friendly for joints – Noise reduction of ~3–5 dB	– Temperature-dependent softness, may scuff in hot climates – Specialized mixing equipment needed
Engineered timber decking (acetylated/thermally modified wood)	– Moisture uptake ↓: 65% versus untreated wood – Service life >25 years outdoors	– Warm aesthetic; carbon-negative material – Nontoxic, splinter-free surface after modification	– Higher upfront $/m² than pressure-treated pine – Requires concealed-fastener systems to avoid splitting

Furthermore, materials like piezoelectric surfaces actively contribute to the energy efficiency of TOD hubs by converting pedestrian footfall into usable electrical energy, potentially powering low-energy lighting or sensor-based infrastructure. These functional benefits align with TOD goals of promoting walkable, comfortable, and environmentally resilient public spaces. By integrating the performance outcomes of materials with TOD objectives – such as enhanced safety, thermal comfort, and long-term economic viability – the discussion now offers a more robust and policy-relevant evaluation of material suitability for sustainable pedestrian infrastructure.

For instance, **permeable concrete** has been successfully implemented in **Portland, Oregon's Green Streets Program**, where it is used to manage stormwater in pedestrian-friendly urban corridors. This pilot project demonstrated measurable reductions in surface runoff and improved walkability during rainfall events. Similarly, **biocrete** has been piloted in **the Netherlands** through the work of the Delft University of Technology, where self-healing bacterial concrete was applied to pedestrian infrastructure to reduce maintenance cycles and enhance long-term durability.

Locally, **piezoelectric flooring** was tested in **Select City Walk Mall, New Delhi**, and in **Western Railway stations in Mumbai**, where the material generated small amounts of electricity from foot traffic – illustrating its potential in high-footfall TOD zones. These references underline that while some of these materials are still emerging, they have already shown promising results in context-specific applications relevant to TOD objectives, such as sustainability, safety, and cost-efficiency. These examples have now been cited to ground the discussion in practical, real-world settings.

The successful application of advanced materials, such as permeable paving, biocrete, and piezoelectric surfaces in transit station walkways necessitates a multidisciplinary approach involving urban planners, engineers, and policymakers. Effective integration requires careful consideration of design parameters, cost efficiency, and policy frameworks to ensure long-term sustainability and feasibility.

A critical aspect of implementation is design considerations and material compatibility. The selection of appropriate materials depends on factors, such as load-bearing capacity, environmental conditions, and pedestrian flow. For example, permeable paving must be designed with sufficient structural integrity to withstand high foot traffic while maintaining optimal permeability (De Graaf-van Dinther et al., 2021). Similarly, biocrete should be assessed for its self-healing capabilities in varying climate conditions, and piezoelectric materials must be positioned in areas with maximum foot impact for efficient energy generation.

Cost-effectiveness and life cycle analysis play a crucial role in determining the viability of these materials. While initial investment costs may be higher than traditional materials, long-term benefits such as reduced maintenance expenses, extended durability, and energy savings justify their adoption. Conducting life cycle assessments can help urban planners and decision-makers understand the return on investment and environmental impact over time (Sharma et al., 2024).

Despite the advantages, challenges in large-scale adoption remain significant. These include high upfront costs, lack of standardized guidelines for new materials, and technical constraints in integrating different materials seamlessly. Additionally, public awareness and acceptance of these innovations need to be addressed to ensure widespread implementation.

To facilitate adoption, policy recommendations and future research directions should focus on incentivizing sustainable infrastructure through government subsidies, updating building codes to accommodate innovative materials, and promoting pilot projects to assess real-world feasibility. Further research is required to enhance material performance, optimize cost efficiency, and explore hybrid solutions that combine multiple technologies for maximum impact. Through collaborative efforts, transit stations can evolve into smarter, more sustainable urban spaces.

5 Conclusion

The incorporation of advanced materials, such as permeable paving, biocrete, and piezoelectric surfaces in transit station footpaths offers a promising shift toward sustainable, efficient, and user-centered urban infrastructure. These materials not only provide environmental benefits – such as improved stormwater infiltration, carbon sequestration, and energy harvesting – but also contribute to pedestrian safety through enhanced slip resistance, impact absorption, and visibility. Their inherent durability and low maintenance requirements extend the operational life of walkways, reducing life cycle costs in the long term. Moreover, when strategically applied within TOD zones, they support broader urban goals, such as walkability, inclusivity, and climate resilience.

Despite their potential, several challenges must be addressed for widespread adoption. These include high upfront material and installation costs, technical challenges in integrating new materials into existing infrastructure, and the lack of standardized design guidelines or performance benchmarks. In particular, the implementation of piezoelectric technology at scale remains constrained by efficiency limitations and infrastructure compatibility. Additionally, gaps in empirical data – especially from low- and middle-income countries – highlight the need for localized case studies and real-world testing. Nonetheless, emerging evidence from pilot projects in cities, such as Tokyo, Amsterdam, and London, indicates that, with proper support, these innovative solutions can be both practical and impactful.

5.1 Recommendations

1. **Invest in pilot projects and field testing**: Governments and urban authorities should fund demonstration projects in diverse urban contexts to evaluate the technical, social, and economic performance of these materials. This will help generate evidence-based guidelines and best practices for implementation in TOD zones.
2. **Develop standards and certification protocols**: Urban planning bodies and engineering institutes should work collaboratively to create standardized performance metrics and quality assurance frameworks for advanced paving materials, facilitating easier procurement and adoption by public agencies.
3. **Promote cross-sector collaboration**: Effective integration of these materials requires coordination between engineers, architects, urban planners, environmental scientists, and policymakers. Interdisciplinary working groups or task forces can be established at the municipal or regional level to align infrastructure goals and streamline decision-making.
4. **Encourage policy and financial incentives**: Governments should provide tax breaks, grants, or green infrastructure subsidies for projects that incorporate sustainable walkway materials. Policies that mandate the inclusion of climate-resilient infrastructure in new TOD developments can also accelerate adoption.
5. **Advance research on hybrid materials and life cycle performance**: Future studies should explore material combinations (e.g., biocrete with embedded sensors or piezoelectric layers under permeable paving) that maximize functionality while minimizing cost. Long-term monitoring of maintenance cycles, energy yields, and pedestrian outcomes will be crucial to refining these systems for broader application.

By aligning technological innovation with strategic urban planning, cities can leverage these material solutions to build safer, smarter, and more sustainable transit infrastructure for the future.

Declaration

There is no conflict of interest. There is no funding to be declared.
AI tools like Grammarly have been used to refine the language.

References

Antunes, L. N., Ghisi, E., & Severis, R. M. (2020). Environmental assessment of a permeable pavement system used to harvest stormwater for non-potable water uses in a building. *Science of the Total Environment, 746*, 141087. https://doi.org/10.1016/j.scitotenv.2020.141087.

Bai, D., Deng, S., Li, Y., & Li, H. (2023). A novel inchworm piezoelectric actuator with rhombic amplification mechanism. *Sensors and Actuators A: Physical, 360*, 114515. https://doi.org/10.1016/j.sna.2023.114515.

Chen, J., Qiu, Q., Han, Y., & Lau, D. (2019). Piezoelectric materials for sustainable building structures: Fundamentals and applications. *Renewable and Sustainable Energy Reviews, 101*, 14–25. https://doi.org/10.1016/j.rser.2018.09.038.

Chen, X., Ji, Q., Wei, J., Tan, H., Yu, J., Zhang, P., Laude, V., & Kadic, M. (2020). Light-weight shell-lattice metamaterials for mechanical shock absorption. *International Journal of Mechanical Sciences, 169*, 105288. https://doi.org/10.1016/j.ijmecsci.2019.105288.

Covaci, C. & Gontean, A. (2020). Piezoelectric Energy Harvesting Solutions: A Review. *Sensors, 20*(12), Article 12. https://doi.org/10.3390/s20123512.

De Graaf-van Dinther, R., Leskens, A., Veldkamp, T., Kluck, J., & Boogaard, F. (2021). From Pilot Projects to Transformative Infrastructures, Exploring Market Receptivity for Permeable Pavement in The Netherlands. *Sustainability, 13*(9), 4925. https://doi.org/10.3390/su13094925.

De Souza, D. J. & Sanchez, L. F. M. (2023). Understanding the efficiency of autogenous and autonomous self-healing of conventional concrete mixtures through mechanical and microscopical analysis. *Cement and Concrete Research, 172*, 107219. https://doi.org/10.1016/j.cemconres.2023.107219.

Ferguson, B. (2005). *Porous Pavements* (0 ed.). CRC Press. https://doi.org/10.1201/9781420038439.

Fermoso, J. (2008, December 17). *Power Generating Floor in Train Stations Light Up Holiday Displays.* https://www.wired.com/2008/12/power-generatin/

Fwa, T. F. (2021). Determination and prediction of pavement skid resistance–connecting research and practice. *Journal of Road Engineering, 1*, 43–62. https://doi.org/10.1016/j.jreng.2021.12.001.

Ginting, S. W. & Mardianta, A. V. (2023). Green-Walkable Infrastructure for Sustainable Development. *IOP Conference Series: Earth and Environmental Science, 1188*(1), 012014. https://doi.org/10.1088/1755-1315/1188/1/012014.

Giriselvam, M. G., Poornima, V., Venkatasubramani, R., & Sreevidya, V. (2018). Enhancement of crack healing efficiency and performance of SAP in biocrete. *IOP Conference Series: Materials Science and Engineering, 310*, 012061. https://doi.org/10.1088/1757-899X/310/1/012061.

Guo, F., Pei, J., Zhang, J., Li, R., Zhou, B., & Chen, Z. (2021). Study on the skid resistance of asphalt pavement: A state-of-the-art review and future prospective. *Construction and Building Materials, 303*, 124411. https://doi.org/10.1016/j.conbuildmat.2021.124411.

Hattrick-Simpers, J., Li, K., Greenwood, M., Black, R., Witt, J., Kozdras, M., Pang, X., & Ozcan, O. (2023). Designing durable, sustainable, high-performance materials for clean energy infrastructure. *Cell Reports Physical Science, 4*(1), 101200. https://doi.org/10.1016/j.xcrp.2022.101200.

Iqbal, A., Rahman, M. M., & Beecham, S. (2022). Permeable Pavements for Flood Control in Australia: Spatial Analysis of Pavement Design Considering Rainfall and Soil Data. *Sustainability, 14*(9), 4970. https://doi.org/10.3390/su14094970.

Jeon, S., Hossain, M. S., Han, S., Choi, P., & Yun, -K.-K. (2022). Self-healing characteristics of cement concrete containing expansive agent. *Case Studies in Construction Materials, 17*, e01609. https://doi.org/10.1016/j.cscm.2022.e01609.

Jiang, S., Weng, Z., Wu, D., Du, Y., Liu, C., & Lin, Y. (2024). Pavement compactness estimation based on 3D pavement texture features. *Case Studies in Construction Materials, 21*, e03768. https://doi.org/10.1016/j.cscm.2024.e03768.

Kayhanian, M., Li, H., Harvey, J. T., & Liang, X. (2019). Application of permeable pavements in highways for stormwater runoff management and pollution prevention: California research experiences.

International Journal of Transportation Science and Technology, *8*(4), 358–372. https://doi.org/10.1016/j.ijtst.2019.01.001.

Li, J., Liu, X., Zhao, G., Liu, Z., Cai, Y., Wang, S., Shen, C., Hu, B., & Wang, X. (2023). Piezoelectric materials and techniques for environmental pollution remediation. *Science of the Total Environment*, *869*, 161767. https://doi.org/10.1016/j.scitotenv.2023.161767.

Li, X. & Strezov, V. (2014). Modelling piezoelectric energy harvesting potential in an educational building. *Energy Conversion and Management*, *85*, 435–442. https://doi.org/10.1016/j.enconman.2014.05.096.

Liu, Q., Liu, S., Hu, G., Yang, T., Du, C., & Oeser, M. (2021). Infiltration Capacity and Structural Analysis of Permeable Pavements for Sustainable Urban: A Full-scale Case Study. *Journal of Cleaner Production*, *288*, 125111. https://doi.org/10.1016/j.jclepro.2020.125111.

Mahesh, V., Joladarashi, S., & Kulkarni, S. M. (2021). Damage mechanics and energy absorption capabilities of natural fiber reinforced elastomeric based bio composite for sacrificial structural applications. *Defence Technology*, *17*(1), 161–176. https://doi.org/10.1016/j.dt.2020.02.013.

Mayer, P. M., Moran, K. D., Miller, E. L., Brander, S. M., Harper, S., Garcia-Jaramillo, M., Carrasco-Navarro, V., Ho, K. T., Burgess, R. M., Thornton Hampton, L. M., Granek, E. F., McCauley, M., McIntyre, J. K., Kolodziej, E. P., Hu, X., Williams, A. J., Beckingham, B. A., Jackson, M. E., Sanders-Smith, R. D., & Mendez, M. (2024). Where the rubber meets the road: Emerging environmental impacts of tire wear particles and their chemical cocktails. *Science of the Total Environment*, *927*, 171153. https://doi.org/10.1016/j.scitotenv.2024.171153.

Menz, H. B., Latt, M. D., Tiedemann, A., Mun San Kwan, M., & Lord, S. R. (2004). Reliability of the GAITRite® walkway system for the quantification of temporo-spatial parameters of gait in young and older people. *Gait & Posture*, *20*(1), 20–25. https://doi.org/10.1016/S0966-6362(03)00068-7.

Mistry, R. & Mehrotra, S. (2023). Spatio-temporal Variation of the Daytime Surface Temperature in Local Climate Zones, Forming Cool Island in Bhopal. *Journal of the Indian Society of Remote Sensing*, *51*(4), 713–731. https://doi.org/10.1007/s12524-022-01658-w.

Omar, T., Mohamed, S. M., & Al-Nasr Ahmed, M. S. (2022). Applications of Smart Material to Enhance Daylighting as a tool to Improve Sustainability of Atriums. *IOP Conference Series: Earth and Environmental Science*, *1113*(1), 012022. https://doi.org/10.1088/1755-1315/1113/1/012022.

Patel, R. S., Taneja, S., Singh, J., & Sharma, S. N. (2024). Modelling of Surface Runoff using SWMM and GIS for Efficient Storm Water Management. *CURRENT SCIENCE*, *126*(4), 463.

Saadeh, S., Ralla, A., Al-Zubi, Y., Wu, R., & Harvey, J. (2019). Application of fully permeable pavements as a sustainable approach for mitigation of stormwater runoff. *International Journal of Transportation Science and Technology*, *8*(4), 338–350. https://doi.org/10.1016/j.ijtst.2019.02.001.

Sabzpoushan, S. & Woias, P. (2024). Electret-based energy harvesters: A review. *Nano Energy*, *131*, 110167. https://doi.org/10.1016/j.nanoen.2024.110167.

Saltarin-Molino, M. A., Moros-Daza, A., & Camacho-Sanchez, C. (2023). Taking steps forward: Innovative evaluation of pedestrian infrastructure through a multivariate analysis. *Case Studies on Transport Policy*, *14*, 101091. https://doi.org/10.1016/j.cstp.2023.101091.

Sharma, S. N., Kumar, A., & Dehalwar, K. (2024). The Precursors of Transit-oriented Development. *Economic & Political Weekly*, *59*(14), 16–20. https://doi.org/10.5281/zenodo.10939448.

Sharma, S. N., Lodhi, A. S., Dehalwar, K., & Jaiswal, A. (2024). Life Cycle Assessment (LCA) of Recycled & Secondary Materials in the Construction of Roads. *IOP Conference Series: Earth and Environmental Science*, *1326*(1), 012102. https://doi.org/10.1088/1755-1315/1326/1/012102.

Sharma, S. N., Prajapati, R., Jaiswal, A., & Dehalwar, K. (2024). A Comparative Study of the Applications and Prospects of Self-healing Concrete / Biocrete and Self-Sensing Concrete. *IOP Conference Series: Earth and Environmental Science*, *1326*(1), 012090. https://doi.org/10.1088/1755-1315/1326/1/012090.

Shen, Z., Liu, Y., Xu, J., Zhou, Y., Yang, L., Fang, Y., & Yuan, B. (2025). Self-powered smart fire-alarm materials: Advances and perspective. *Journal of Safety and Sustainability*, S2949926725000186. https://doi.org/10.1016/j.jsasus.2025.04.001.

Shiflett, S. A., Liang, L. L., Crum, S. M., Feyisa, G. L., Wang, J., & Jenerette, G. D. (2017). Variation in the urban vegetation, surface temperature, air temperature nexus. *Science of the Total Environment*, *579*, 495–505. https://doi.org/10.1016/j.scitotenv.2016.11.069.

Veldkamp, T. I. E., Boogaard, F. C., & Kluck, J. (2022). Unlocking the Potential of Permeable Pavements in Practice: A Large-Scale Field Study of Performance Factors of Permeable Pavements in The Netherlands. *Water*, *14*(13), 2080. https://doi.org/10.3390/w14132080.

Wang, J., Meng, Q., Tan, K., Zhang, L., & Zhang, Y. (2018). Experimental investigation on the influence of evaporative cooling of permeable pavements on outdoor thermal environment. *Building and Environment*, *140*, 184–193. https://doi.org/10.1016/j.buildenv.2018.05.033.

Wijaksono, S. & Wibowo, G. (2023). Green Community Center: Using Long-life Material for Low Maintenance and Serviceability. *IOP Conference Series: Earth and Environmental Science*, *1169*(1), 012073. https://doi.org/10.1088/1755-1315/1169/1/012073.

Wiktor, V. & Jonkers, H. M. (2011). Quantification of crack-healing in novel bacteria-based self-healing concrete. *Cement and Concrete Composites*, *33*(7), 763–770. https://doi.org/10.1016/j.cemconcomp.2011.03.012.

Zheng, H., Wang, Y., Liu, J., Wang, J., Yan, K., & Zhu, K. (2024). Recent advancements in the use of novel piezoelectric materials for piezocatalytic and piezo-photocatalytic applications. *Applied Catalysis B: Environmental*, *341*, 123335. https://doi.org/10.1016/j.apcatb.2023.123335.

Xi Wang*, Rupp Carriveau, and David S-K. Ting

Harnessing Solar Energy for Modern Heating, Ventilation, and Air Conditioning (HVAC) Systems: Recent Technological Progress

Abstract: Heating, ventilation, and air conditioning (HVAC) systems have become indispensable in modern buildings to ensure thermal comfort for occupants. Meanwhile, they are among the leading contributors to energy consumption and carbon emissions in the built environment. In light of the urgent need to reduce environmental impacts and energy costs, traditional HVAC systems must be transformed to meet the requirements of net zero energy (NZE) buildings. Among the various renewable energy options, solar energy has gained significant attention due to its abundance, reliability, and cost-effectiveness. Integrating solar technology into HVAC systems offers a promising pathway to enhance energy efficiency while reducing greenhouse gas emissions. This chapter provides a comprehensive review of recent progress in solar-assisted HVAC systems, with a focus on innovative thermo-fluid designs for solar collectors and photovoltaic panels. These designs are essential for improving energy conversion efficiency and system responsiveness to varying thermal loads. Additionally, the integration of other renewable technologies, such as heat pumps, energy storage systems, and auxiliary renewable sources, is examined to assess their synergistic impact on overall system performance. Finally, the chapter highlights the role of advanced energy management strategies, including intelligent control algorithms, in optimizing the operation of solar-based HVAC systems. These strategies are vital for achieving efficiency, reliability, and sustainability in future NZE buildings.

Keywords: Solar collector, PV panel, HVAC, AI/ML, NZE building

1 Introduction

Since the beginning of the industrial era in the eighteenth century, human activities – especially the large-scale combustion of fossil fuels – have led to an unprecedented

Acknowledgment: This work was made possible by funding from the Natural Sciences and Engineering Research Council of Canada.

*Corresponding author: Xi Wang**, Turbulence and Energy Laboratory, University of Windsor, 401 Sunset Avenue, Windsor, ON, Canada, e-mail: Wang1st@uwindsor.ca
Rupp Carriveau, David S-K. Ting, Turbulence and Energy Laboratory· University of Windsor, 401 Sunset Avenue, Windsor, ON, Canada

https://doi.org/10.1515/9783111563046-008

increase in atmospheric carbon dioxide (CO_2) concentrations. According to NASA (National Aeronautics and Space Administration), the cumulative anthropogenic carbon emissions have far surpassed the natural rise observed at the end of the last ice age, roughly 20,000 years ago. As a result, global warming has become an increasingly pressing concern. Current data show that the global average surface temperature has risen by approximately 1.2 °C compared to preindustrial levels, a change that has already had significant consequences for the planet's climate systems (International Energy Agency). It is widely recognized that excessive carbon emissions are a primary driver of global warming, which in turn is accelerating environmental degradation (International Energy Agency; Internet Geography, 2023).

The adverse impacts of climate change are becoming more apparent, with extreme weather events occurring with increasing frequency and intensity. In the summer of 2023, unprecedented heatwaves and widespread wildfires became defining features of the season (Internet Geography, 2023; Global News). Although not every extreme weather event can be directly attributed to global warming, the increasing frequency and severity of such occurrences have raised serious concerns among scientists, policymakers, and the general public.

In response to the growing environmental crisis, many of the world's leading economies have adopted policies promoting the transition to a low-carbon economy, with carbon neutrality targets set for 2050 (Global News; Asia Pacific; Lugo-Morin, 2021; Cranston and Hammond, 2010). To achieve this, the development and deployment of renewable energy technologies have become a global priority. Since 2020, global investment in renewable energy has increased by approximately 40% (International Energy Agency). By the end of 2025, renewable energy is expected to supply 35% of global electricity (International Energy Agency). Meanwhile, by 2050, renewable sources are projected to account for approximately 63% of the world's total energy consumption (International Energy Agency; Yuan et al., 2021).

In particular, the building sector is a major contributor to global energy consumption and carbon emissions. Among its components, heating, ventilation, and air conditioning (HVAC) systems are especially energy-intensive. As a critical component of modern buildings, HVAC systems play a significant role. These systems are essential for maintaining indoor air quality and thermal comfort in modern buildings by regulating temperature, humidity, and airflow. However, the combination of population growth, rising living standards, and more frequent extreme weather events is expected to increase global HVAC demand substantially. In 2020, the global HVAC market was valued at USD 175.9 billion, and it is projected to grow to USD 445.7 billion by 2033, as shown in Figure 1 (Research and Markets; Grand View Research, Accessed 16 June 2025). Despite their importance, most HVAC systems are still powered by electricity generated from fossil fuels, thereby contributing significantly to carbon emissions (World Nuclear Association). For instance, air conditioning alone accounted for 7% of global electricity consumption and 3% of CO_2 emissions in 2022 (Our World in Data). As a result, transitioning HVAC systems to operate with alternative, renewable energy sources – particularly solar – is a key strategy in achieving a low-carbon, sustainable future.

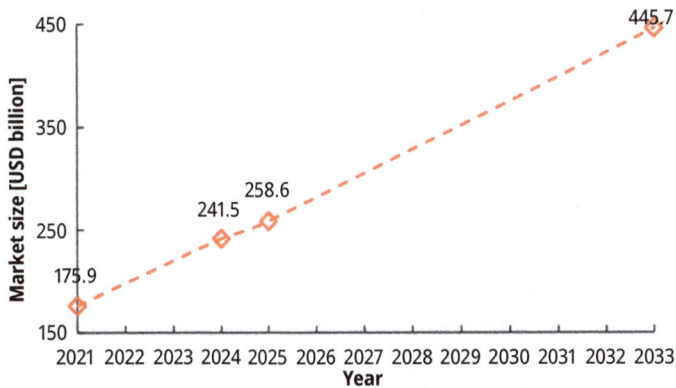

Figure 1: Global market growth of HVAC systems from 2021 to 2033 (Research and Markets; Grand View Research, Accessed 16 June 2025).

Due to its excellent cost-effectiveness and high-energy conversion efficiency, solar technology is widely regarded as one of the most promising renewable energy sources. As the demand for sustainable building operations grows, governments around the world are increasingly investing in solar technologies, particularly for residential HVAC applications. These investments aim to reduce the reliance on conventional energy sources, lower operating costs, and mitigate environmental impacts. As of 2024, the global market for solar-based air conditioning reached USD 527.28 million, and it is projected to grow to USD 658.51 million by 2033, as shown in Figure 2 (Global Growth Insights), reflecting growing interest and confidence in solar-driven HVAC systems.

In building applications, solar technologies used in HVAC systems generally fall into two main categories: solar thermal collectors and photovoltaic (PV) panels, each offering unique advantages and use cases. Solar collectors primarily harvest thermal energy to heat or cool spaces, whereas PV panels convert sunlight into electricity to power HVAC equipment such as compressors, pumps, and fans. Table 1 provides an overview of practical applications of HVAC systems powered by solar energy.

Solar thermal collectors are considered the core component in solar-assisted heating and cooling systems. They are a mature and proven technology for adjusting indoor temperatures using solar heat. For example, Tzivanidis and Bellos (2016) employed parabolic trough solar collectors to retrofit a building cooling system in Greece. Their study demonstrated that a 14-m^2 solar collector was sufficient to meet the cooling requirements of a 25-m^2 building space for 13 h/day during summer. Similarly, Al-Falahi et al. (2020) conducted a case study on a solar cooling system powered

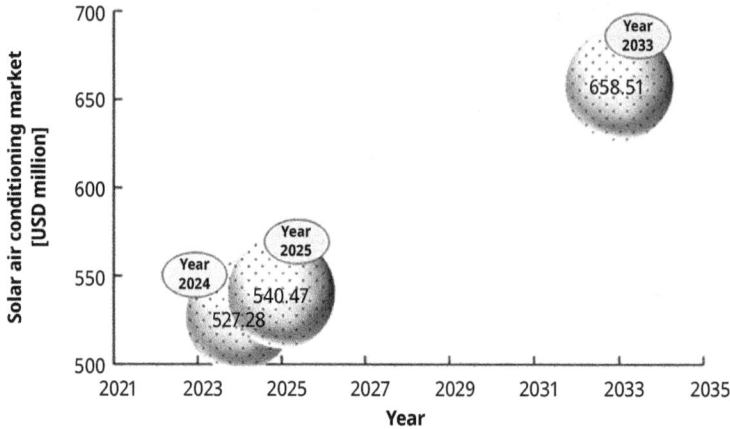

Figure 2: Solar air conditioning market (Global Growth Insights).

Table 1: Summary of practical applications of solar-based HVAC systems.

Author	Technique	Contribution	Location
Tzivanidis and Bellos (2016)	Solar thermal collectors	A 14-m^2 collector effectively cools 25 m^2 of building area	Athens
Al-Falahi et al. (2020)	Solar thermal collectors	Thermal efficiency: up to 54% (summer)	Baghdad
Suhendri et al. (2024)	Solar thermal collectors	≥1.5 kWh/day savings versus traditional systems	Madrid, Tokyo, and Isfahan
Marijuan et al. (2024)	Solar thermal collectors	21 tons of CO_2 saved annually (solar thermal HVAC)	Tartu
Opoku et al. (2018)	Solar PV	1,211 kWh/year (200 Ah, 24 V), sufficient for a 2.5-kW AC unit	Kumasi
Asif et al. (2018)	Solar PV	Their PV systems supplied 16–20% energy, cut CO_2 by 40,000 ton/year	Saudi Arabia
Albatayneh et al. (2021)	Hybrid system	3,300 MWh/year supports 700 kW cooling	Jordan
Pezo et al. ()	Hybrid system	Hybrid system cut CO_2 by 72.5% and 69.5% versus oil and gas boilers	Chile

by evacuated tube collectors (ETCs) in Baghdad, Iraq. Their results showed that the ETC achieved a peak thermal efficiency of up to 54% during summer. Suhendri et al. (2024) designed a combined solar heating and radiative cooling system for a 100-m^2 building using 9.43 m^2 of solar collectors. Their system achieved at least 1.5 kWh more daily energy savings compared to traditional solar devices. In a broader context, Marijuan et al. (2024) demonstrated the significant potential of solar collectors in supporting zero-energy buildings. In a demonstration project in Tartu, Estonia, they found that integrating solar thermal systems into HVAC operations could reduce annual CO_2 emissions by up to 21 tons.

In parallel, solar PV systems have also shown great promise for powering HVAC components electrically. These systems convert solar radiation directly into electricity, which can be used to run air conditioners, ventilation units, and other electrical loads. Opoku et al. (2018) proposed a PV-powered HVAC system for an office building in Kumasi, Ghana. Their design featured a 200-Ah, 24-V battery configuration and yielded approximately 1,211 kWh of electricity annually – sufficient to operate a 2.5-kW air conditioner. Likewise, Asif et al. (2018) assessed the performance of PV systems at the King Fahd University of Petroleum and Minerals (KFUPM) in Saudi Arabia. They compared two installation configurations – tilted and horizontal – and found that the PV systems could supply approximately 16% and 20% of the university's electricity demand, respectively. Additionally, the systems helped reduce CO_2 emissions by as much as about 40,000 tons/year.

Beyond standalone PV systems, hybrid configurations that combine PV panels with other renewable technologies are gaining traction. For instance, Albatayneh et al. (2021) proposed an air conditioning system integrated with a PV array tailored for the climatic conditions of Jordan. The system generated 3,300 MWh annually, effectively supporting a cooling system with 700 kW. Moreover, Pezo et al. (2024) introduced a hybrid system integrating heat pumps with PV panels to provide both heating and electricity. Their results indicated that, compared to conventional oil and gas boilers, the hybrid system reduced CO_2 emissions by 72.5% and 69.5%, respectively, highlighting the environmental benefits of such combined approaches.

Those practical applications mentioned before reflect the existing use of solar energy in building HVAC systems, highlighting its potential for broader adoption. According to Figure 3, the global solar energy system market (involving utility, residential, and commercial) is projected to increase significantly from $160.5 billion in 2021 to $608 billion by 2030 (Vision Research Reports). Among the various sectors, the residential building end-use sector is expected to experience notable growth during this period, driven by continued investment and policy support. Despite the fact that solar technology has been proven to be a feasible alternative to traditional HVAC systems in buildings, its intermittency and relatively low efficiency remain major challenges to widespread commercialization (Renogy et al., 2023). Over the past few decades, many engineers and researchers have devoted considerable effort to improving the energy performance and cost-effectiveness of solar-based HVAC systems by optimizing geometrical configurations and proposing innovative designs.

The primary motivation of this chapter is to provide a comprehensive review of recent technological advances in solar-based HVAC systems. To achieve this, the chapter first introduces the basic working principles and classifications of solar-assisted

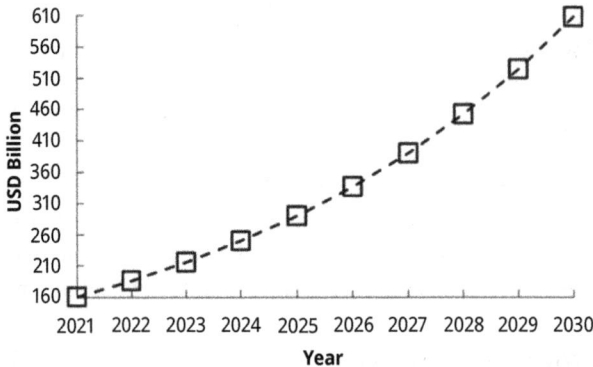

Figure 3: Global market growth of solar energy systems from 2021 to 2030 (Vision Research Reports).

HVAC systems. It then focuses on optimization strategies and novel design approaches for both solar collectors and PV panels used within HVAC applications. In addition, the integration of advanced technologies – such as hybrid energy systems and artificial intelligence (AI)-based control strategies – into solar HVAC configurations is also reviewed and discussed. Overall, this chapter aims to offer readers a clear and structured overview of the latest developments in solar energy applications for HVAC systems, highlighting both the current progress and future opportunities for achieving sustainable and energy-efficient building solutions.

2 Mechanics of Solar Technologies into a Building HVAC System

Owing to its excellent cost-effectiveness and operational reliability, solar energy technology has found wide applications in HVAC systems for both residential and commercial buildings. Based on their working principles, the solar technologies currently applied in the market can be broadly categorized into two types: solar thermal collectors and PV panels (Natural Resources Canada). In the following sections, we present a detailed discussion of the fundamental working principles of these solar technologies as they are applied in HVAC systems.

2.1 Solar Collector Technology into the HVAC System

As a specialized type of heat exchanger, the solar collector is used to convert solar radiation into internal energy within a transmission medium. Figure 4 illustrates a general schematic of an HVAC system incorporating solar collectors. The collectors serve as the core components of this system, converting solar radiation into thermal energy. This thermal energy is then transferred to a fluid – such as air or water – that flows through the collectors. The captured thermal energy can be directly utilized for heating or cooling applications. Additionally, any surplus heat can be stored in energy storage devices, enabling the HVAC system to maintain normal operation during periods of insufficient sunlight. Generally, there are two main types of solar collectors: flat plate solar collectors (FPSCs) and evacuated tube solar collectors (ETSCs) (Kalogirou, 2004).

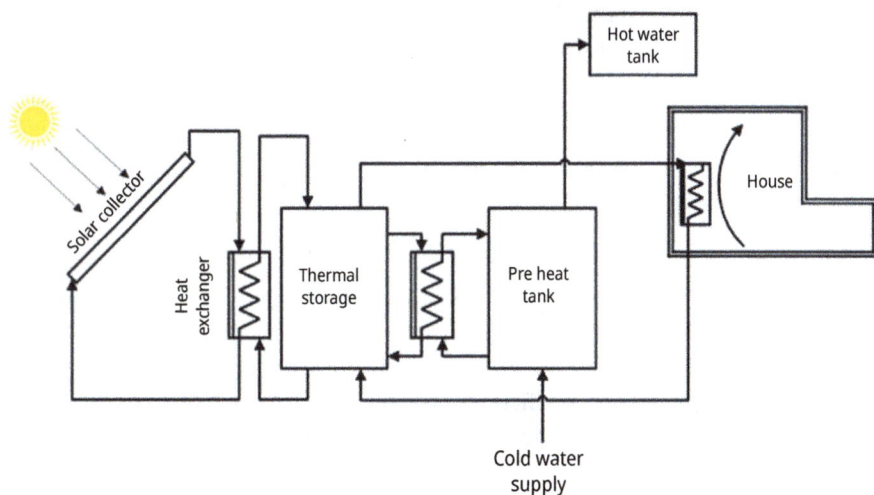

Figure 4: Schematic diagram of an HVAC system with solar collectors.

The FPSC is the most commonly used device in solar-powered HVAC systems. A typical FPSC consists of a parallel glass cover and a back plate that acts as an absorber (Figure 5), both primarily responsible for capturing and converting solar energy (Cruz-Peragon et al., 2012). Beneath the absorber, an array of copper tubes carries the working fluid and transfers the thermal energy absorbed by the plate. To minimize heat losses, thermal insulation materials are incorporated into the bottom plate and the frame. Compared to other solar technologies, the FPSC is regarded as the most cost-effective and straightforward device for solar energy collection (Miao et al., 2021). In residential applications, FPSCs are commonly used to provide hot water for users. However, their relatively low thermal efficiency remains a significant challenge for large-scale commercialization.

The basic working principle of the ETSC is to absorb solar radiation through inner tubes coated with an optical layer, which then converts the solar radiation into thermal energy (Sethi et al., 2022). This heat is subsequently transferred to the working fluid via a heat pipe. As illustrated in Figure 6, a typical ETSC consists of coated absorbers, heat pipes, and an outer glass tube (Hydro Solar Innovative Energy). Compared to

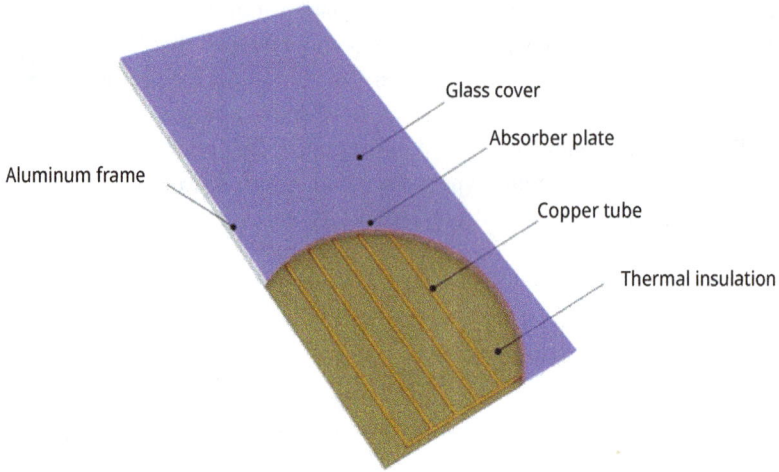

Figure 5: Basic structure of a flat plate solar collector (Cruz-Peragon et al., 2012).

FPSCs, ETSCs operate over a wider temperature range, typically between 50 °C and 200 °C. Although the market share of ETSCs has significantly increased in recent years due to their superior conversion efficiency, several challenges remain in their application for building HVAC systems, including overheating, snow accumulation, and high initial costs (Sabiha et al., 2015).

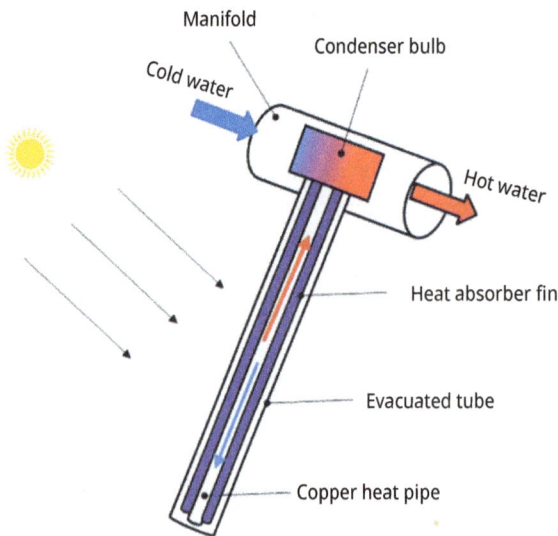

Figure 6: Basic structure of an evacuated tube solar collector (Hydro Solar Innovative Energy).

2.2 PV Technology for HVAC Applications

PV technology is widely used to generate electricity by harnessing the PV effect, providing a clean alternative to fossil fuel combustion for powering HVAC system motors (Nagaraja et al., 2025). A PV panel is a solid-state energy device that directly converts sunlight into electricity without moving parts or chemical reactions, offering long-term reliability and zero emissions (Imenes and Mills, 2004).

A typical PV cell consists of two thin semiconductor layers forming a P-N junction. When exposed to sunlight, photons with energy exceeding the material's band gap excite valence electrons, allowing them to transition to the conduction band and become free electrons (Nagaraja et al., 2025). These electrons are then driven by the internal electric field to flow through an external circuit, generating electric current, as illustrated in Figure 7.

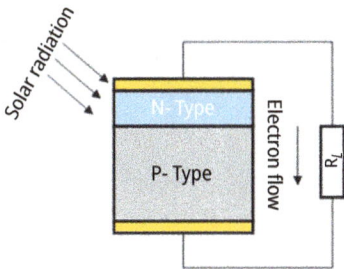

Figure 7: Schematic diagram of a PV cell (Nagaraja et al., 2025).

PV cells cannot convert all wavelengths of solar radiation into electricity, which inherently limits their efficiency. As shown in Table 2, different types of PV cells exhibit varying levels of conversion efficiency (Aurora Solar). Figure 8 illustrates that PV cells primarily utilize photons with energies exceeding the material's band gap, and maximum efficiency occurs when photon energy closely matches the band gap (Imenes and Mills, 2004). Excess photon energy that cannot be converted is released as heat, which not only goes unused but also raises the cell temperature, further reducing conversion efficiency (Imenes and Mills, 2004). Consequently, limited spectral utilization and thermal losses remain significant challenges for PV applications in building HVAC systems.

Table 2: Sample PV materials' efficiency (Aurora Solar).

PV materials	Conversion efficiency
Polycrystalline	13–16%
Monocrystalline	18–24%
Thin film	7–13%
Transparent	1–10%

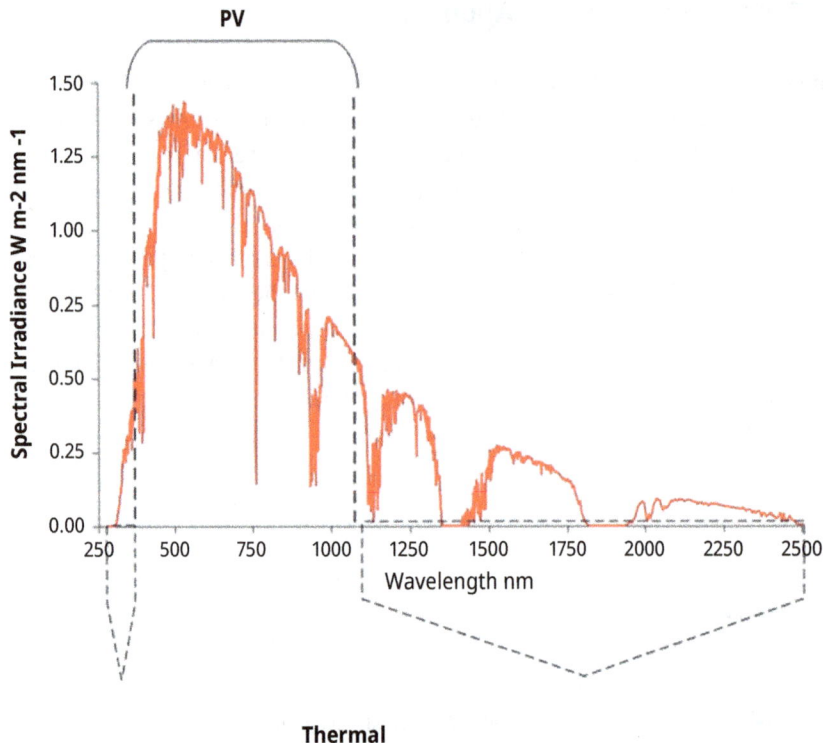

Figure 8: Solar spectrum used by PV cells (Imenes and Mills, 2004).

3 Structure Optimization and Novel Designs in the Solar-Based HVAC System

Based on the operating principles of solar-based HVAC systems, the performance of both solar thermal and PV technologies depends heavily on heat transfer processes. Since heat transfer processes directly affect the energy conversion rate and system stability, improving thermal management has become a key focus in recent research. To the purpose, various innovative designs and configurations have been developed to optimize the heat transfer characteristics within these systems. Accordingly, the following section provides a comprehensive overview of recent technological advancements and design strategies in HVAC systems that integrate solar thermal and PV components, aiming to enhance overall energy efficiency and system reliability.

3.1 Recent Progress in Solar Thermal Technology

For solar thermal devices, enhancing the heat transfer process has a direct impact on the performance of both FPSCs and ETSCs in building HVAC systems. It is well acknowledged that the flow of the working fluid plays a critical role in heat transfer efficiency. Therefore, modifying the flow channel design to increase flow area and induce turbulence is an effective strategy to improve the thermal performance of solar thermal systems.

In recent years, numerous structural optimizations and innovative designs have been proposed to enhance the thermal performance of FPSCs. For example, Verma et al. (2020) designed a spiral tube configuration for an FPSC, as illustrated in Figure 9. Experimental results show that, compared to conventional flat tubes, the spiral design results in lower pressure drops due to smoother flow paths. This helps reduce energy consumption for water pumping and improves the overall heat transfer efficiency. Under a mass flow rate of 0.026 kg/s and a solar intensity of 1,011 W/m^2, the thermal efficiency of the novel spiral-tube FPSC was found to be 21.45% higher than that of the traditional design.

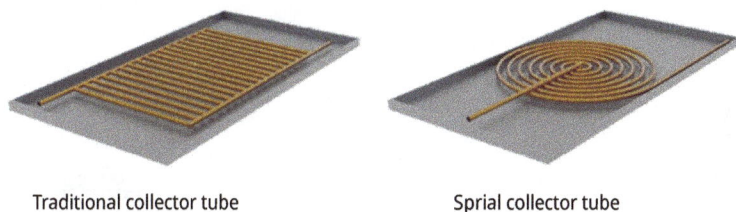

Traditional collector tube   Sprial collector tube

Figure 9: The spiral tube for an FPSC designed by Verma et al. (2020).

Similarly, by optimizing the structure of the flow channels, the effective heat transfer surface area can be increased, which is beneficial for improving the thermal efficiency of an FPSC. Elwekeel and Abdala (2023) introduced a circular collector with four different channel designs, as illustrated in Figure 10. Through simulations conducted under a heat flux of 800 W/m^2 and a mass flow rate of 0.02 kg/s, they found that the flow in Case 1 had the longest path, while the other designs generated more eddies and stagnation zones. As a result, the FPSC with channel design 1 exhibited the highest thermal efficiency among the four configurations.

Additionally, Nabi et al. (2022) aimed to enhance the thermal performance of an FPSC by increasing the turbulence intensity of the fluid within the flow channels. To achieve this, they incorporated three types of turbulence-inducing elements into the cross-sectional design of the flow channels, as illustrated in Figure 11. Simulation results showed that all three configurations effectively improved the heat transfer coefficient. Among them, Case 3 exhibited the best performance. This improvement is

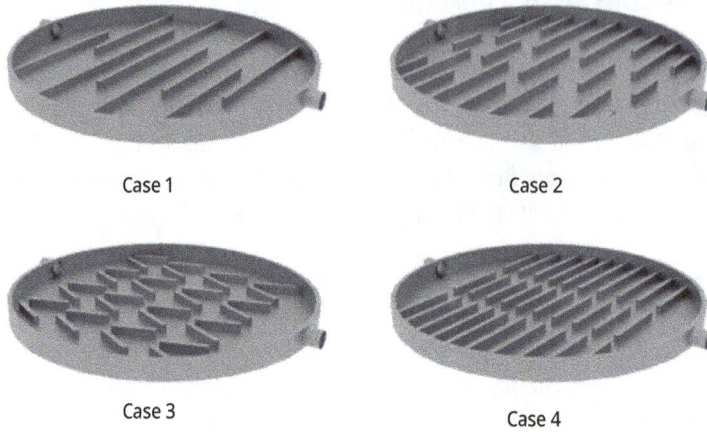

Figure 10: A circular collector with four kinds of channel designs (Elwekeel and Abdala, 2023).

mainly attributed to the ability of Case 3 to generate secondary flows, which help reduce the thickness of the thermal boundary layer and further enhance the heat transfer rate.

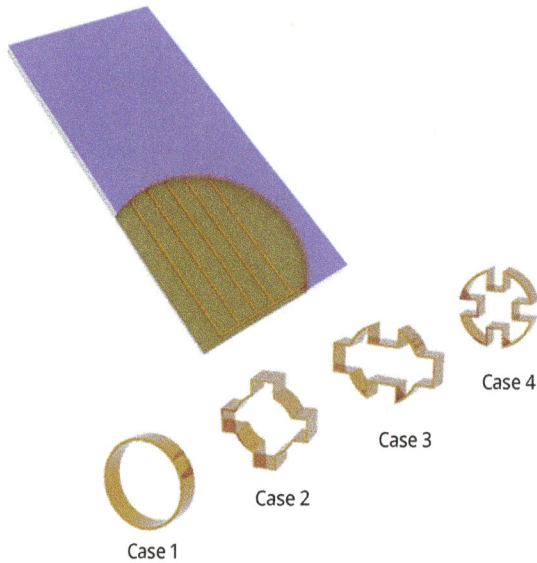

Figure 11: Cross-sectional geometry of the base pipe with multiple turbulence inducers (Nabi et al., 2022).

Similarly, enhancing heat transfer by modifying fluid flow is an effective approach to improve the performance of ETSC systems. In particular, the stagnation zone at the bottom of the evacuated tube can lead to significant energy losses. To address this issue, Agade et al. (2025) proposed the use of a perforated wavy tape, as shown in Figure 12, to enhance the performance of a solar water heater equipped with an ETSC. Experimental results demonstrate that the wavy tape effectively increases turbulence and induces swirl within the evacuated tube, thereby boosting the overall heat transfer. As a result, incorporating this type of wavy tape into the tube can increase the daily energy efficiency of the solar water heater by 27.3% compared to a traditional ETSC system.

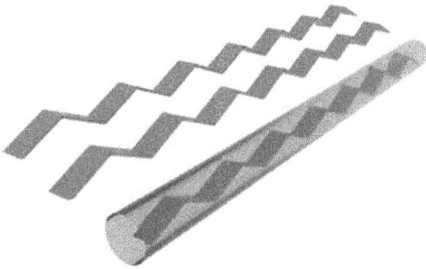

Figure 12: A wavy tape in their experimental study (Agade et al., 2025).

In addition to modifying the flow channels, several novel designs have been proposed to enhance the energy efficiency of ETSC-based HVAC systems. Teles et al. (2019) developed an ETSC heater with an eccentric structure, as illustrated in Figure 13. In this design, the absorber is positioned eccentrically within a transparent glass envelope. The resulting concentration effect caused by the eccentric alignment leads to a higher outlet temperature of the working fluid. Furthermore, the vacuum space surrounding the eccentric absorber helps reduce heat loss. As a result, this ETSC solar heater exhibits excellent thermal performance, achieving an annual efficiency of up to 61.5% under the environmental conditions of São Luís, Brazil.

Moreover, to minimize heat and water losses along the manifold of an ETSC water heater, Bouadila et al. (2023) proposed the use of phase change material (PCM) enclosed in a copper box to cover the manifold, as illustrated in Figure 14. Experimental data show that under the North African climate, their ETSC water heater is capable of providing 4.3 MJ of thermal energy per day. Additionally, considering a colder climate in Jammu, India, Pathak et al. (2024) developed a U-pipe ETSC integrated with a PCM known as PEG6000. Based on real-world testing, they confirmed that their design achieves a significantly higher maximum thermal efficiency (86.71%) compared to the traditional model (72.96%) at a flow rate of 0.50 L/min.

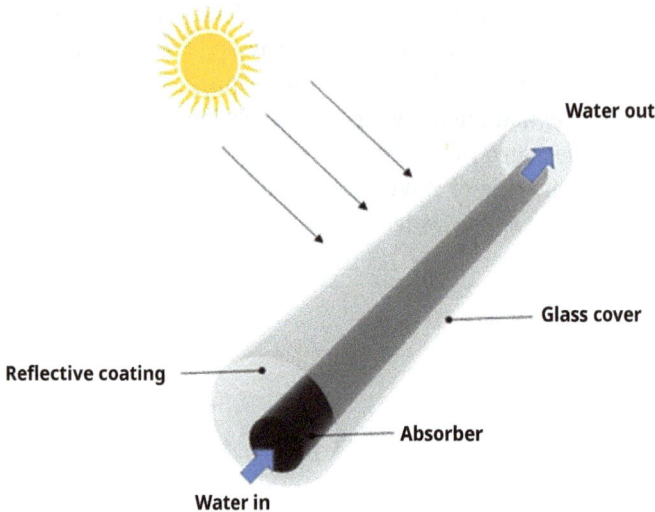

Figure 13: Eccentrically configured ETSC heater (Teles et al., 2019).

3.2 Recent Progress in Solar PV Technology

The working principle of solar PV cells indicates that they cannot convert the entire spectrum of solar radiation into electrical energy. A significant portion of the photon energy that is not absorbed by the PV cells is released as waste heat, which adversely affects the efficiency of the PV system. To enhance the overall performance of PV cells, developing advanced PV materials has been considered the most promising approach. However, due to limitations in mass production capabilities, high fabrication costs, and material availability, the widespread commercialization of such advanced materials, particularly for building-integrated applications, remains challenging. In this context, improving the thermal management of PV panels through the application of cooling technologies offers a more feasible and cost-effective solution to boost their efficiency in HVAC systems for buildings.

One passive cooling strategy is based on natural convection, which takes advantage of the buoyancy-driven movement of air, where warmer air rises and cooler air sinks (Aljuwaysir et al., 2025). This method is both energy efficient and environmentally friendly. Fossa et al. (2008) conducted experimental research to evaluate the effectiveness of natural convection in reducing the temperature of a double-skin PV façade. They examined three different PV panel layouts installed on building walls, as shown in Figure 15. The experimental results revealed that, compared to the first configuration, the other two layouts were able to reduce the panel surface temperature by approximately 10–15 °C. These findings demonstrate that rational use of natural convection can significantly enhance PV performance by mitigating overheating.

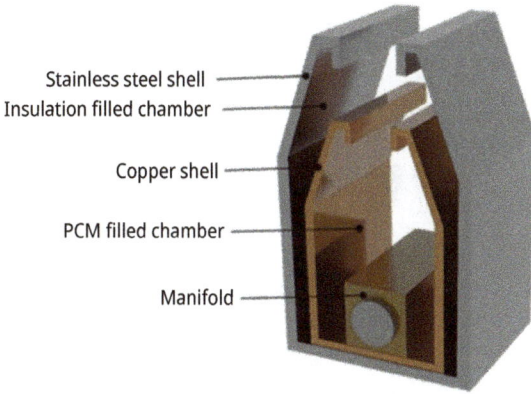

Figure 14: Schematic diagram of the heat-pipe ETSCs (Bouadila et al., 2023).

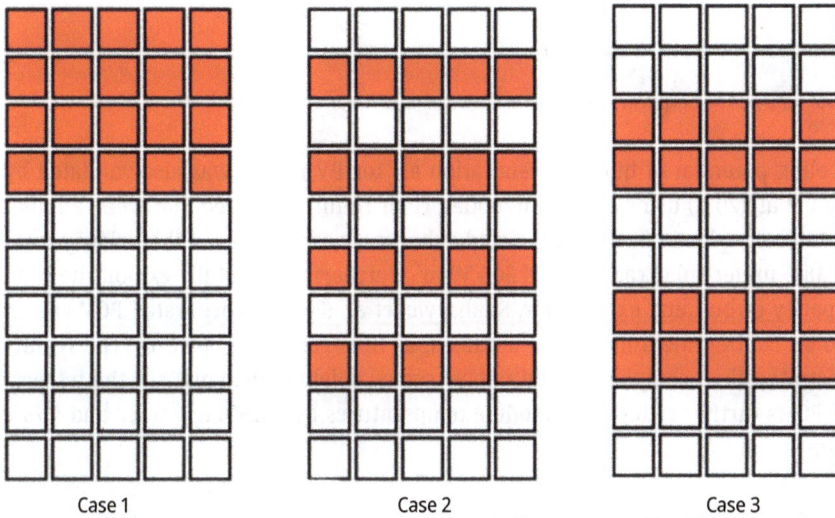

Figure 15: Three kinds of layouts of PV on the wall (Fossa et al., 2008).

In addition, ventilation and exhaust air within a building are considered to hold significant potential for cooling PV panels. Based on this principle, Shahsavar et al. (2011) developed a PV-thermal system for buildings located in Kerman Province, southern Iran, to achieve bidirectional energy utilization, as illustrated in Figure 16. Specifically, they employed the building's exhaust air to lower the temperature of PV modules, thereby enhancing their electrical efficiency. Simultaneously, the waste heat generated by the PV panels was recovered and used as a supplementary heat source to meet part of the building's thermal load through the ventilation system.

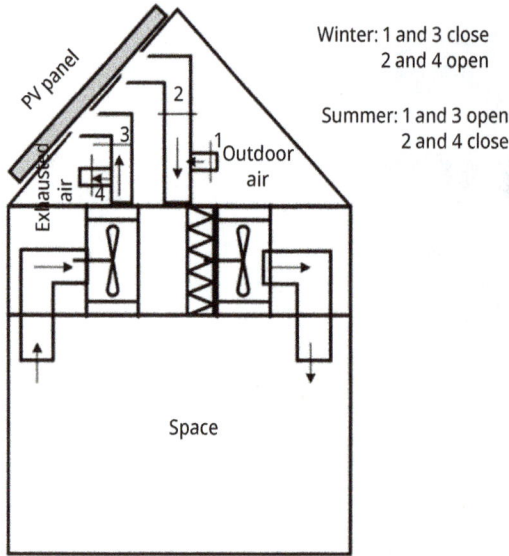

Figure 16: Schematic diagram of the PV-thermal system (Shahsavar et al., 2011).

The cooling potential of building ventilation air for PV panels was also validated by Salameh et al. (2021) using a thermal model. Their findings showed that when a building has a cooling load of 140 kW, the cold exhaust air can improve PV efficiency from 11% to 18% under solar radiation of 500 W/m². Furthermore, to fully exploit the cooling capacity of building exhaust air, Shahsavar et al. (2022) incorporated PCMs as an additional cooling medium for PV modules, as illustrated in Figure 17. The results demonstrated that, compared to PV panels cooled solely by exhaust air, the integration of PCMs further reduced PV module temperatures by 3.125% in winter and 4.78% in summer.

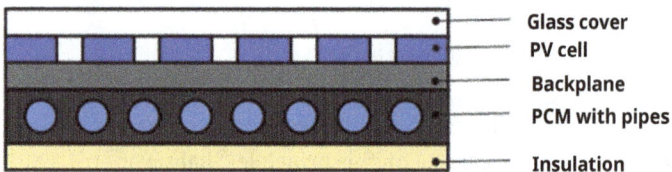

Figure 17: Cross-sectional diagrams of the PVT/PCM system (Shahsavar et al., 2022).

Moreover, a novel exhaust ventilation double-glazing PV (EVPV) curtain wall system was proposed by Tang et al. (2022). As shown in Figure 18, this system reduces the temperature of the PV curtain wall by using exhaust air as a coolant. Simultaneously, the exhaust air is recycled to preheat the dew point supply air and precool the incoming fresh air, thereby enabling the coordinated utilization of PV power generation and heat energy recovery. Based on real-world testing conducted in a restaurant in Hefei, China, the PV panel efficiency in this system reaches 7.14%, while the recovered heat amounts to 7.68 kWh/day. In summary, compared to conventional PV systems, this design achieves a daily energy savings of 63.12 kWh, representing a 19.3% improvement.

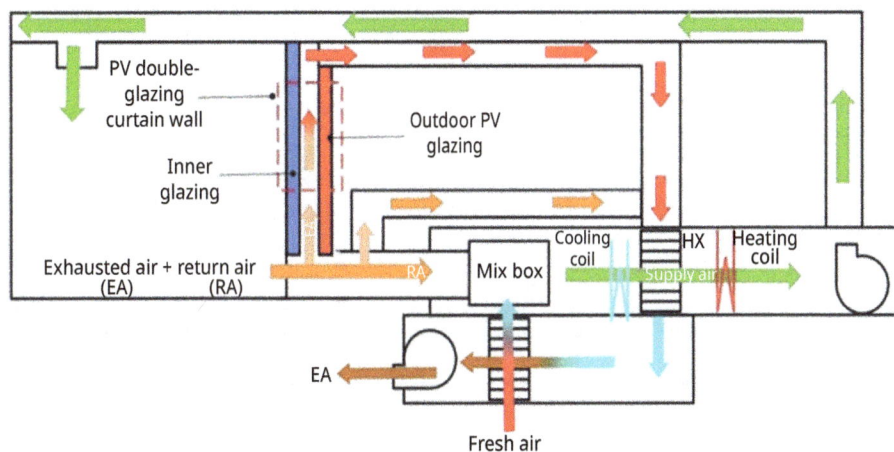

Figure 18: Working flow of the exhaust ventilation double-glazing PV curtain wall system (Tang et al., 2022).

The main contributions of the literature mentioned above are summarized in Table 3. They primarily focus on performance enhancement of solar-based systems (solar thermal collectors and PV panels) through thermofluidic system design.

Table 3: Summary of solar-based HVAC systems enhancement via a thermofluidic system design.

Author	Technique	Contribution
Verma et al. (2020)	Solar thermal collectors	The spiral-tube design boosts thermal efficiency by 21.45% over the traditional model.
Elwekeel and Abdala (2023)	Solar thermal collectors	Optimizing flow channel structure enhances heat transfer area and thermal efficiency.
Nabi et al. (2022)	Solar thermal collectors	Introducing turbulence-inducing elements enhanced heat transfer in FPSCs.
Agade et al. (2025)	Solar thermal collectors	Perforated wavy tape reduces stagnation and boosts ETSC efficiency by 27.3% through enhanced turbulence.

Table 3 (continued)

Author	Technique	Contribution
Teles et al. (2019)	Solar thermal collectors	An eccentric ETSC design enhances solar concentration and reduces heat loss, achieving up to 61.5% annual efficiency.
Bouadila et al. (2023)	Solar thermal collectors	PCM-covered manifolds reduce losses and raise ETSC efficiency up to 86.71%.
Fossa et al. (2008)	Solar PV	Optimized layouts cut PV temperatures by 10–15 °C, improving performance.
Shahsavar et al. (2011)	Solar PV	Used exhaust air to cool PV and recover heat, boosting efficiency.
Salameh et al. (2021)	Solar PV	Ventilation air cooling raised PV efficiency from 11% to 18% under 500 W/m^2 solar radiation.
Shahsavar et al. (2022)	Solar PV	PCMs enhanced exhaust air cooling, reducing PV temperatures by ~3–5%.
Tang et al. (2022)	Solar PV	EVPV cools PV panels and recovers heat, improving efficiency by 19.3%.

4 Hybrid Energy Technology and Artificial Intelligence in the Solar-Based HVAC System

In addition to optimizing the design of solar modules, integrating other energy technologies into building HVAC systems is an effective strategy to enhance overall performance, as it enables the complementary use of different technologies. Moreover, due to the intermittent nature of renewable energy sources, energy control and management have become essential components of solar-driven HVAC systems. With the advancement of intelligent algorithms – such as machine learning (ML) and AI – smart HVAC systems have gained increasing attention. Therefore, this section reviews representative hybrid HVAC systems and highlights the critical roles that AI and ML play in enabling intelligent control and optimization within these smart hybrid systems.

4.1 Hybrid Energy Technology in the Solar-Based HVAC System

The PV/T system is a commonly used hybrid HVAC solution. Its primary objective is to simultaneously generate electricity and thermal energy through PV panels and solar collectors, with both energy forms used to meet the building's heating and cooling demands. Figure 19 presents a basic schematic diagram of a typical PV/T system. Generally, the overall efficiency of a PV/T system can reach approximately 70%, comprising 15–20% electrical efficiency and about 50% thermal efficiency (Ramos et al., 2017). Furthermore, by integrating heat pumps or energy storage devices, the performance of PV/T systems in building HVAC applications can be further improved.

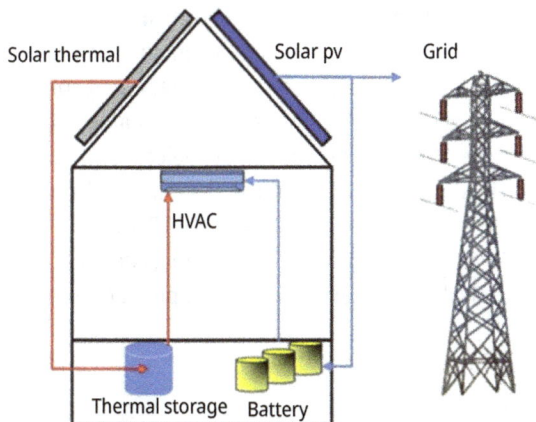

Figure 19: Schematic diagram of a PV/T system (Zanetti et al., 2020).

Rmos et al. (2017) employed the TRNSYS simulation tool to evaluate the performance of a photovoltaic/thermal (PV/T) system integrated with a heat pump in residential buildings across four European cities: Seville, Rome, Madrid, and Bucharest. The simulation results revealed that this hybrid system could satisfy approximately 60% of the total thermal demand and nearly 100% of the cooling demand in all four regions. This indicates that the integration of PV/T with a heat pump can provide a stable and reliable energy supply for various climatic conditions. Furthermore, when compared with conventional HVAC systems powered by PV modules alone, the hybrid system demonstrates improved cost-effectiveness. Specifically, the levelized cost of energy for the hybrid system was found to be 30–40% lower than that of PV-only systems, highlighting its economic advantage for long-term building operation.

Similarly, Barone et al. (2020) conducted a simulation study using two different types of PV/T collectors – one with water cooling and the other with air cooling – to assess their performance in residential buildings located in three distinct European climate zones. The findings showed that both PV/T systems offered remarkable energy savings and economic benefits. In particular, the systems achieved annual energy savings of up to 4,236 kWh, and the estimated payback periods ranged between 3 and 6 years, depending on the specific design and climate condition. These results suggest that PV/T systems are not only technically feasible but also economically viable solutions for energy-efficient buildings.

Moreover, integrating solar collectors with other renewable energy technologies, such as heat pumps, is increasingly regarded as a promising pathway toward achieving net-zero or zero-carbon buildings. For instance, Chae et al. (2023) proposed a PV/T system combined with an air-source heat pump (ASHP), as illustrated in Figure 20, to deliver comprehensive HVAC services. ASHPs extract heat from the ambient air; however, in warm mixed climates, their heating capacity may be significantly reduced due to unfavorable temperature conditions. To address this limitation, the hybrid system utilizes thermal energy collected by the PV/T modules as an auxiliary heat source, thereby enhancing the ASHP's performance. This not only ensures a stable indoor thermal environment but also reduces the system's reliance on grid electricity during peak demand hours. According to their simulation, the hybrid system could reduce total energy consumption by up to 43% compared to traditional HVAC systems, and the payback period was estimated to be around 9.4 years. Such findings demonstrate the potential of hybrid PV/T and ASHP systems to serve as sustainable and efficient HVAC solutions for future buildings.

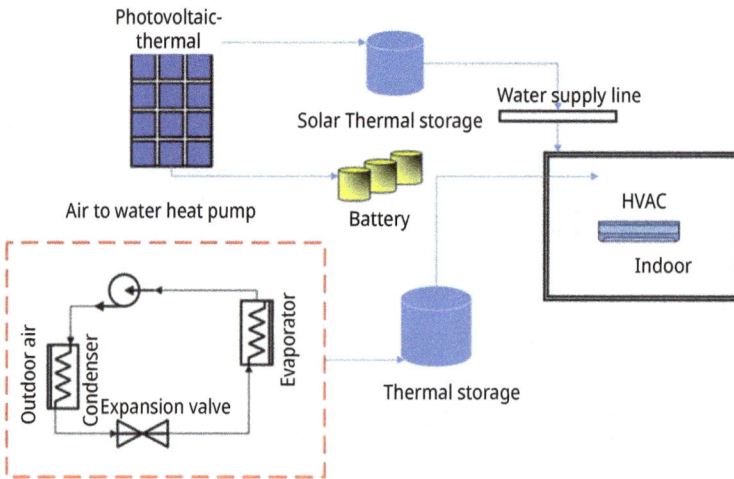

Figure 20: Basic working flow diagram of the PV/T-heat pump system (Chae et al., 2023).

Furthermore, several recent studies have explored the application of hybrid HVAC systems powered by combinations such as solar-wind and solar-geothermal energy. For example, Deymi-Dashtebayaz et al. (2022) developed a hybrid solar-wind energy system (shown in Figure 21) designed to meet the heating, cooling, and electricity demands of a building located in St. Petersburg, Russia. The system primarily comprises a wind turbine, solar thermal loop, absorption chiller, and an energy storage unit. According to their simulation results, the solar thermal component can supply approximately 61% of the building's annual heating demand. Meanwhile, the wind turbine, in conjunction with the energy storage system, is capable of covering nearly 99% of the building's electricity load. This integrated approach not only improves energy self-sufficiency but also significantly reduces environmental impact. In particular, the hybrid system has the potential to cut carbon emissions by approximately 14,000 kg/year, demonstrating its environmental and economic benefits.

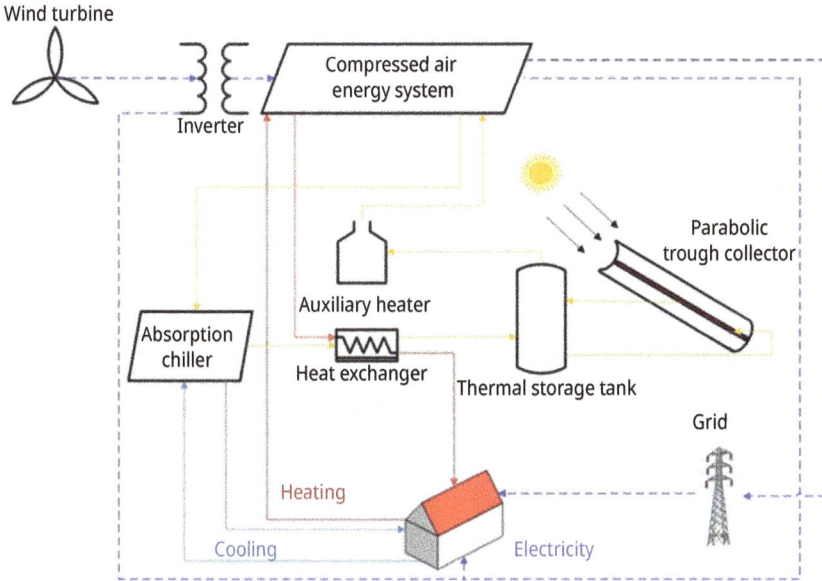

Figure 21: Basic working flow diagram of the solar-wind HVAC system (Deymi-Dashtebayaz et al., 2022).

Wang and You (2023) proposed an integrated HVAC system that combines solar and geothermal energy, as illustrated in Figure 22. This hybrid system comprises two main subsystems: energy piles and a PV/T unit. The energy piles utilize geothermal energy from the soil to operate a heat pump, thereby meeting the building's heating and cooling demands. The PV/T subsystem is responsible for generating electricity for the building. To prevent overheating of the PV panels, a thermal storage device is incorporated into the system to facilitate both cooling and thermal energy recovery. Based on their simulation results, the PV/T subsystem achieves 23% higher efficiency compared to conventional PV modules. Additionally, the coefficient of performance (COP) of the energy piles remains stable at 3.21 over a 10-year period. The system demonstrates strong energy-saving potential, with total energy savings estimated at 46.47 MWh.

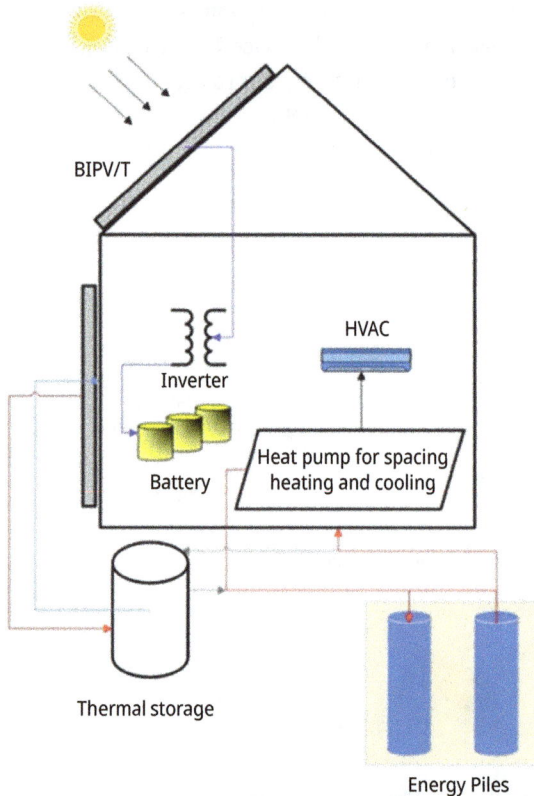

Figure 22: Basic working flow diagram of the solar-geothermal HVAC system (Wang and You, 2023).

4.2 Energy Management and Control in a Solar-Based HVAC System

It is well acknowledged that, to address the challenges of intermittency and low efficiency, solar-based HVAC systems are often integrated with multiple auxiliary devices such as batteries, heat pumps, and thermal energy storage tanks. However, the addition of these components makes the energy control and management of solar-based HVAC systems significantly more complex than that of conventional systems. In this context, a well-designed energy control model plays a critical role in enhancing overall system performance. Typically, energy control models can be classified into two main types: physics-driven models and data-driven models (Ma et al., 2025).

A physics-driven model refers to an energy control strategy based on thermodynamic principles and economic modeling. Under this approach, the various subsystems within a solar-based HVAC system can operate in coordination according to single or multiobjective criteria. For example, Zanetti et al. (2020) developed an optimal energy management strategy using nonlinear programming for a hybrid PV/T–ASHP HVAC system. Figure 23 illustrates the basic flow diagram of their proposed control strategy. When minimizing energy costs as the objective, their model achieved up to 20% cost savings. In addition, compared to conventional control methods, the proposed approach improved PV self-consumption by approximately 30%.

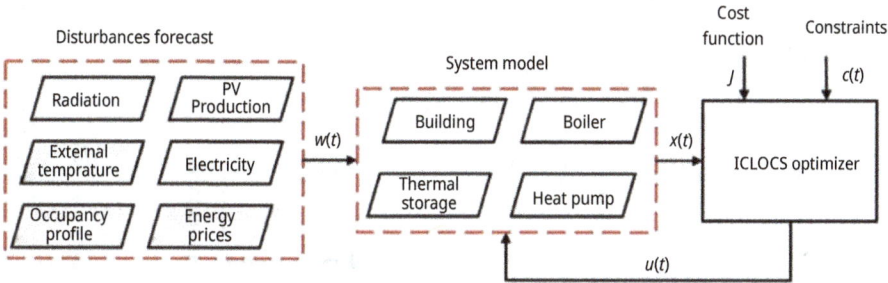

Figure 23: Diagram of the optimal control process (Zanetti et al., 2020).

Differing from the physics-driven model, the data-driven model is built upon historical datasets, enabling it to capture complex patterns and relationships that may not be easily modeled using physical laws alone. With the advancement of sensor technologies and intelligent algorithms, data-driven models have become increasingly applicable for predicting the operational states of solar-based HVAC systems. For instance, Alden et al. (2021) developed a novel ML model by integrating deep learning techniques with datasets on solar HVAC energy usage and weather conditions. Specifically, they employed a long short-term memory encoder-decoder model to forecast annual energy consumption of a solar-based HVAC system using future weather predictions. Validation tests showed that their method achieved hourly and daily root mean square errors of 29.4% and 11.1%, respectively, which align with accepted academic standards and the ASHRAE building model calibration criteria.

In another study, AlShammari (2025) developed an artificial neural network (ANN) model to predict the energy efficiency of a solar-powered absorption cooling system. A solar-powered absorption cooling system is regarded as a sustainable technology for meeting cooling demands in arid regions. However, its performance is highly sensitive to climatic and operational variations, which presents significant challenges for optimization. As ANNs do not depend on explicit physical equations, data-driven ANN models are well-suited to accurately and efficiently predict system performance under fluctuating operating conditions. The basic framework of the ANN model used in AlShammari's study (2025) is illustrated in Figure 24. Typically, an ANN consists of three components: an input layer, one or more hidden layers, and an output layer. In this study, the input layer includes key variables that influence system performance such as solar radiation, ambient temperature, humidity, component efficiencies, and flow rates. The output layer corresponds to the COP, which serves as a key indicator of the cooling system's efficiency.

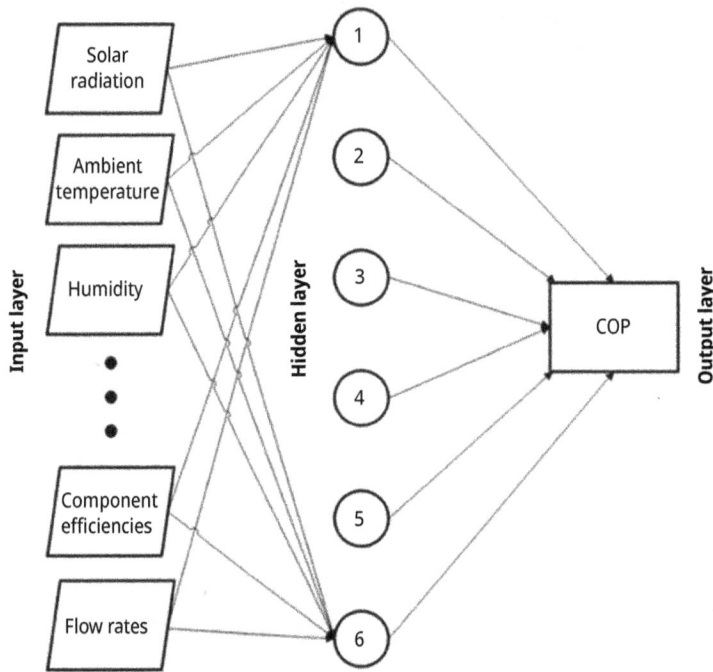

Figure 24: Schematic diagram of the ANN model (AlShammari, 2025).

Furthermore, to improve prediction accuracy, their ANN model was enhanced using nature-inspired optimization algorithms such as chicken swarm optimization (CSO), moth flame optimization algorithm (MFOA), and whale optimization algorithm (WOA) (AlShammari, 2025). The results demonstrated that the prediction accuracies of ANN-CSO, ANN-MFOA, and ANN-WOA were improved by 11.46%, 10.36%, and 10.54%, respectively, compared to the standard ANN model.

The reliability and predictive accuracy of data-driven models highlight their promising role in the energy control and management of solar-based HVAC systems. For example, Senjaliya and Tejani (2020) applied ML algorithms to forecast energy demands in buildings, facilitating real-time optimization of energy distribution between solar thermal systems and heat pumps. Their model achieved up to 35% more energy savings compared to conventional strategies. Furthermore, Morovat et al. (2024), using data from a school building in Québec, Canada, developed a data-driven control strategy aimed at optimizing HVAC energy consumption. Their results revealed that when integrated with PV/T systems, the school's HVAC setup was capable of reducing peak energy demand by up to 100% during high-demand periods compared to a fully electric baseline system.

5 Conclusions

Among numerous renewable energy technologies, solar energy is widely regarded as the most promising for commercialization due to its excellent energy conversion efficiency and cost-effectiveness. Specifically, solar technology has been extensively applied in buildings – ranging from individual dwellings to modern office complexes – to provide both thermal comfort and electrical power for occupants. Musall et al. (2010) investigated the adoption rates of energy technologies in more than 280 international net zero energy (NZE) buildings. Figure 25 illustrates the adoption rates of solar technologies across different building types. Here, the adoption rate refers to the percentage of buildings that have incorporated these technologies into their energy systems. Among them, PV rooftop systems and solar thermal domestic hot water (DHW) systems are the most commonly used solar technologies in building projects. However, to realize the goal of NZE buildings, it is essential to further enhance the performance of solar-based HVAC systems.

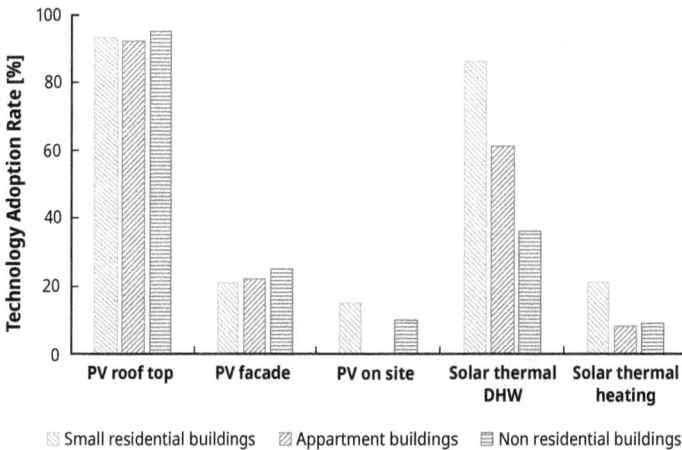

Figure 25: Solar technology adoption rate if different types of buildings (Musall et al., 2010).

This chapter has primarily focused on reviewing recent advances in solar-based HVAC systems. Firstly, for both solar thermal and PV systems, thermo-fluid dynamics play a critical role in determining energy conversion efficiency. Consequently, many novel designs and optimization strategies have recently been proposed to improve the performance of solar collectors and PV panels. Moreover, to achieve NZE buildings, the integration of solar energy with other renewable technologies – such as heat pumps, wind turbines, and geothermal systems – has attracted increasing attention from researchers aiming to develop hybrid HVAC systems. Furthermore, advanced algorithms like ML and AI have demonstrated great potential to accurately predict energy usage in solar-based HVAC systems based on future weather data. Meanwhile, energy control and management strategies founded on these advanced algorithms provide effective means to optimize energy distribution within solar-based HVAC systems, thereby enabling substantial energy savings.

Nevertheless, the development of solar-based HVAC systems is ongoing, and further advancements are needed to optimize their performance. Future research should address inherent challenges associated with solar energy, such as intermittency and relatively low efficiency, by pursuing technological breakthroughs in novel energy materials, energy storage technologies, and structural optimizations. In particular, applying data-driven models to analyze thermodynamic processes in solar systems holds great promise. Compared to traditional approaches like simulations and experimental studies, data-driven models offer more efficient and flexible tools for structural optimization of solar modules. Additionally, it is crucial to continuously refine AI- and ML-based energy management models by fine-tuning parameters to balance prediction accuracy with computational efficiency. Simultaneously, the introduction of multicriteria optimization strategies in these models will enable the design of optimal frameworks that balance solar power with other integrated energy sources, allowing building HVAC systems to simultaneously achieve thermal comfort, economic savings, and environmental sustainability.

References

Agade, P., Agrawal, R., & Dubey, N. (2025). Augmenting the performance of evacuated tube solar water heater using perforated wavy tape. *Solar Energy and Sustainable Development Journal, 14*, 258–278.

Albatayneh, A., Jaradat, M., Al-Omary, M., & Zaquot, M. (2021). Evaluation of coupling PV and air conditioning vs. solar cooling systems – Case study from Jordan. *Applied Sciences, 11*(2), 11020511.

Alden, R. E., Gong, H., Jones, E. S., Ababei, C., & Ionel, D. M. (2021). Artificial intelligence method for the forecast and separation of total and HVAC loads with application to energy management of smart and NZE homes. *IEEE Access, 9*, 160497–160509.

Al-Falahi, A., Alobaid, F., & Epple, B. (2020). A new design of an integrated solar absorption cooling system driven by an evacuated tube collector: A case study for Baghdad, Iraq. *Applied Sciences, 10*(10), 10103622.

Aljuwaysir, S., Osman, K., Abidin, U., Ahmad, M., Mohamed, E., & Farooq, M. (2025). Review of cooling techniques for improving solar photovoltaic panel efficiency. *Journal of Advanced Research in Fluid Mechanics and Thermal Sciences, 125*(1), 193–219.

AlShammari, N. K. (2025). Metaheuristic energy efficiency optimization of solar-powered absorption cooling systems under operating climactic conditions integrated with explainable AI. *Case Studies in Thermal Engineering, 69*, 106016.

Asia Pacific, [Online], Typhoon makes landfall in China as toll rises in the Philippines, Typhoon Doksuri Makes Landfall in China as Toll Rises in the Philippines –The New York Times. Accessed 16 June 2025.

Asif, M., Hassanain, M., Nahiduzzaman, K. M., & Sawalha, H. (2018). Techno-economic assessment of application of solar PV in building sector – A case study from Saudi Arabia. *Smart and Sustainable Built Environment*, 8, 34–52.

Aurora Solar, [Online], A guide to solar panel efficiency, A guide to solar panel efficiency | Aurora Solar. Accessed 16 June 2025.

Barone, G., Buonomano, A., Forzano, C., Giuzio, G. F., & Palombo, A. (2020). Passive and active performance assessment of building integrated hybrid solar photovoltaic/thermal collector prototypes: Energy, comfort, and economic analyses. *Energy*, 209, 118435.

Bouadila, S., Rehman, T., Baig, M. A. A., Skouri, S., & Baddadi, S. (2023). Energy, Exergy and Economic (3E) analysis of evacuated tube heat pipe solar collector to promote storage energy under North African climate. *Sustainable Energy Technologies and Assessments*, 55, 102959.

Chae, S., Bae, S., & Nam, Y. (2023). Economic and environmental analysis of the optimum design for the integrated system with air source heat pump and PVT. *Case Studies in Thermal Engineering*, 48, 103142.

Cranston, G. R. & Hammond, G. P. (2010). North and south: Regional footprints on the transition pathway towards a low carbon, global economy. *Applied Energy*, 87(9), 2945–2951.

Cruz-Peragon, F., Palomar, J. M., Casanova, P. J., Dorado, M. P., & Manzano-Agugliaro, F. (2012). Characterization of solar flat plate collectors. *Renewable and Sustainable Energy Reviews*, 16(3), 1709–1720.

de P. R. Teles, M., Ismail, K. A. R., & Arabkoohsar, A. (2019). A new version of a low concentration evacuated tube solar collector: Optical and thermal investigation. *Solar Energy*, 180, 324–339.

Deymi-Dashtebayaz, M., Baranov, I. V., Nikitin, A., Davoodi, V., Sulin, A., Norani, M., & Nikitina, V. (2022). An investigation of a hybrid wind-solar integrated energy system with heat and power energy storage system in a near-zero energy building-A dynamic study. *Energy Conversion and Management*, 269, 116085.

Elwekeel, F. N. M. & Abdala, A. M. M. (2023). Numerical and experimental investigation of the performance of a new circular flat plate collector. *Renewable Energy*, 209, 581–590.

Fossa, M., Ménézo, C., & Leonardi, E. (2008). Experimental natural convection on vertical surfaces for building integrated photovoltaic (BIPV) applications. *Experimental Thermal and Fluid Science*, 32(4), 980–990.

Global Growth Insights, [Online], Solar air conditioning market size, share (window solar air conditioner, cassette solar air conditioner), by applications covered (residential, commercial), regional insights and forecast to 2033,. Solar Air Conditioning Market Size & Growth Trends 2033 Accessed 16 June 2025.

Global News, [Online], B.C. declares state of emergency amid 'worst wildfire season in our province's history', B.C. declares state of emergency amid 'worst wildfire season in our province's history' | Globalnews.ca. Accessed 16 June 2025.

Grand View Research, [Online], HVAC systems market size, share & trends analysis report by equipment (heating, cooling, ventilation), by application (residential, commercial, industrial), by distribution channel (online, retail stores, wholesale stores), by geography, and segment forecasts, 2025 – 2033, HVAC Systems Market Size & Share | Industry Report, 2033. Accessed 16 June 2025.

Hassan, Q., Algburi, S., Sameen, A. Z., Salman, H. M., & Jaszczur, M. (2023). A review of hybrid renewable energy systems: Solar and wind-powered solutions: Challenges, opportunities, and policy implications. *Results in Engineering*, 20, 101621.

Hydro Solar Innovative Energy, [Online], How do vacuum tubes solar collectors work?, How do Vacuum Tubes Collectors Work?. Accessed 16 June 2025.

Imenes, A. G. & Mills, D. R. (2004). Spectral beam splitting technology for increased conversion efficiency in solar concentrating systems: A review. *Solar Energy Materials and Solar Cells*, 84(1), 19–69.

International Energy Agency, [Online], IEA: More than a third of the world's electricity will come from renewables in 2025,Renewable energy will produce 35% of global electricity by 2025: IEA |World Economic Forum. Accessed 16 June 2025.

International Energy Agency, [Online], Rapid progress of key clean energy technologies shows the new energy economy is emerging faster than many think, Rapid progress of key clean energy technologies shows the new energy economy is emerging faster than many think – News – IEA. Accessed 16 June 2025.

International Energy Agency, [Online], World energy outlook 2023, Executive summary – World Energy Outlook 2023 – Analysis – IEA. Accessed 16 June 2025.

Internet Geography, [Online, 2023 – A summer of extreme weather, 2023 – A Summer of Extreme Weather – Internet Geography. Accessed 16 June 2025.

Kalogirou, S. A. (2004). Solar thermal collectors and applications. *Progress in Energy and Combustion Science, 30*(3), 231–295.

Lugo-Morin, D. R. (2021). Global future: Low-carbon economy or high-carbon economy?. *World, 2*(2), 175–193.

Ma, Z., Jiang, G., Hu, Y., & Chen, J. (2025). A review of physics-informed machine learning for building energy modeling. *Applied Energy, 381*, 125169.

Marijuán, A. G., Gómez, N. V., Elguezabal, P., Álava, I., & Martínez, A., (2024). ', *9th International Multidisciplinary Conference on Computer and Energy Science*, p. 07.

Miao, R., Hu, X., Yu, Y., Zhang, Y., Wood, M., & Olson, G. (2021). Experimental study of a newly developed dual-purpose solar thermal collector for heat and cold collection. *Energy and Buildings, 252*, 111370.

Morovat, N., Athienitis, A., & Candanedo, J. (2024). Design of a model predictive control methodology for integration of retrofitted air-based PV/T system in school buildings. *Journal of Building Performance Simulation*, 1–19.

Musall, E., Weiss, T., Voss, K., Lenoir, A., Donn, M., Cory, S., & Garde, F., (2010). 'Net zero energy solar buildings: An overview and analysis on worldwide building projects', *EuroSun conference*, pp. 7–8.

Nabi, H., Pourfallah, M., Gholinia, M., & Jahanian, O. (2022). Increasing heat transfer in flat plate solar collectors using various forms of turbulence-inducing elements and CNTs-CuO hybrid nanofluids. *Case Studies in Thermal Engineering, 33*, 101909.

Nagaraja, M. R., Biswas, W. K., & Selvan, C. P. (2025). Advancements and challenges in solar photovoltaic technologies: Enhancing technical performance for sustainable clean energy – A review. *Solar Energy Advances, 5*, 100084.

National Aeronautics and Space Administration, [Online], Carbon dioxide, Carbon Dioxide | Vital Signs – Climate Change: Vital Signs of the Planet. Accessed 16 June 2025.

Natural Resources Canada, [Online], Solar energy, Solar energy – Natural Resources Canada. Accessed 16 June 2025.

Opoku, R., Mensah-Darkwa, K., & Muntaka, A. S. (2018). Techno-economic analysis of a hybrid solar PV-grid powered air-conditioner for daytime office use in hot humid climates – A case study in Kumasi city, Ghana. *Solar Energy, 165*, 65–74.

Our World in Data, [Online], Air conditioning causes around 3% of greenhouse gas emissions. How will this change in the future?, Air conditioning causes around 3% of greenhouse gas emissions. How will this change in the future? – Our World in Data. Accessed 16 June 2025.

Pathak, S. K., Tyagi, V. V., Chopra, K., & Pandey, A. K. (2024). Solar thermal potential of phase change material-based U-pipe ETSCs for different climatic zones: Evaluating energy matrices and economic viability. *Sustainable Materials and Technologies, 40*, e00857.

Pezo, M., Cuevas, C., Wagemann, E., & Cendoya, A. (2024). NET zero energy building technologies – Reversible heat pump/organic Rankine cycle coupled with solar collectors and combined heat pump/photovoltaics – Case study of a Chilean mid-rise residential building. *Applied Thermal Engineering*, 252, 123683.

Ramos, A., Chatzopoulou, M. A., Guarracino, I., Freeman, J., & Markides, C. N. (2017). Hybrid photovoltaic-thermal solar systems for combined heating, cooling and power provision in the urban environment. *Energy Conversion and Management, 150*, 838–850.

Renogy, [Online], How solar panel efficiency and cost changed over time., How Solar Panel Efficiency and Cost Changed Over Time – Renogy United States Accessed 16 June 2025.

Research and Markets, [Online], Global HVAC systems market report 2022 – Pressing need to cut energy consumption and operational costs impels energy efficient systems, Global HVAC Systems Market Report 2022 – Pressing Need to. Accessed 16 June 2025.

Sabiha, M. A., Saidur, R., Mekhilef, S., & Mahian, O. (2015). Progress and latest developments of evacuated tube solar collectors. *Renewable and Sustainable Energy Reviews, 51*, 1038–1054.

Salameh, W., Castelain, C., Faraj, J., Murr, R., Hage, H. E., & Khaled, M. (2021). Improving the efficiency of photovoltaic panels using air exhausted from HVAC systems: Thermal modelling and parametric analysis. *Case Studies in Thermal Engineering, 25*, 100940.

Senjaliya, N. & Tejani, A. (2020). Artificial intelligence-powered autonomous energy management system for hybrid heat pump and solar thermal integration in residential building. *International Journal of Advanced Research in Engineering & Technology, 11*, 1025–1037.

Sethi, M., Tripathi, R. K., Pattnaik, B., Kumar, S., Khargotra, R., Chand, S., & Thakur, A. (2022). Recent developments in design of evacuated tube solar collectors integrated with thermal energy storage: A review. *Materials Today: Proceedings, 52*, 1689–1696.

Shahsavar, A., Askari, I. B., & Dovom, A. R. M. (2022). Energy saving in buildings by using the exhaust air and phase change material for cooling of photovoltaic panels. *Journal of Building Engineering, 53*, 104520.

Shahsavar, A., Salmanzadeh, M., Ameri, M., & Talebizadeh, P. (2011). Energy saving in buildings by using the exhaust and ventilation air for cooling of photovoltaic panels. *Energy and Buildings, 43*(9), 2219–2226.

Suhendri, S., Hu, M., Dan, Y., Su, Y., Zhao, B., & Riffat, S. (2024). Building energy-saving potential of a dual-functional solar heating and radiative cooling system. *Energy and Buildings, 303*, 113764.

Tang, Y., Ji, J., Wang, C., Xie, H., & Ke, W. (2022). Performance prediction of a novel double-glazing PV curtain wall system combined with an air handling unit using exhaust cooling and heat recovery technology. *Energy Conversion and Management, 265*, 115774.

Tzivanidis, C. & Bellos, E. (2016). The use of parabolic trough collectors for solar cooling – A case study for Athen's climate. *Case Studies in Thermal Engineering, 8*, 403–413.

Verma, S. K., Sharma, K., Gupta, N. K., Soni, P., & Upadhyay, N. (2020). Performance comparison of innovative spiral shaped solar collector design with conventional flat plate solar collector. *Energy, 194*, 116853.

Vision Research Reports, [Online], Solar energy systems market (by product: solar panels, batteries, inverters; by source: new installation, MRO; by end-use: residential, commercial) – global industry analysis, size, share, growth, trends, revenue, regional outlook and forecast 2022–2030, Solar Energy Systems Market Size, Share | Report 2022–2030. Accessed 16 June 2025.

Wang, F. & You, T. (2023). Synergetic performance improvement of a novel building integrated photovoltaic/thermal-energy pile system for co-utilization of solar and shallow-geothermal energy. *Energy Conversion and Management, 288*, 117116.

World Nuclear Association, [Online], Carbon Dioxide Emissions from Electricity., Carbon Dioxide Emissions From Electricity – World Nuclear Association Accessed 16 June 2025.

Yuan, X., Chen, Z., Liang, Y., Pan, Y., Jokisalo, J., & Kosonen, R. (2021). Heating energy-saving potentials in HVAC system of swimming halls: A review. *Building and Environment, 205*, 108189.

Zanetti, E., Aprile, M., Kum, D., Scoccia, R., & Motta, M. (2020). Energy saving potentials of a photovoltaic assisted heat pump for hybrid building heating system via optimal control. *Journal of Building Engineering, 27*, 100854.

Beth-Anne Schuelke-Leech
Three Competing Visions for the Future of Engineering Education

Abstract: From the time engineering was first recognized as a distinct profession, there have been calls for change and reforms to engineering education and practice. Many of these calls are rooted in questions about the roles that engineers should perform in society and in organizations, and correspondingly, how to educate engineers to fulfill these roles. Currently, there are three distinct visions for engineering reform. The first is to make engineering education more responsive to industry needs, here called *industry-servicing engineering*. The second one responds to emerging and disruptive technologies, here denoted as *technology-responsive engineering*. The last reform looks to integrate greater social-consciousness, environmental sustainability, social justice, and equity into the engineering curriculum, here termed *socially responsive engineering*. Each of these visions has related skills and knowledge. This chapter presents an overview of the impetus for change in engineering and three visions of engineering that are currently being proposed. This chapter is intended to prompt discussion and reflection about these competing visions.

1 Introduction

Calls for reforms to engineering education have been ongoing since engineering became more formalized as a separate profession in the nineteenth century (cf. Eddy, 1897; Ginter, 1955; Walker, 1971; Woods, 1983; ABET, 1995; Institution of Engineers Australia, 1996; Peterson, 1996; National Academy of Engineering, 2004; CAE, 2005; Goldberg and Somerville, 2014; Crosthwaite, 2021). The history of engineering education reform can be considered a struggle over the nature of engineering, what should be taught, and the role of the engineer in society. There are many different justifications for engineering education reform and substantial variety in the prescriptions for reform.

In the early modern era, many engineers learned their trade through apprenticeships and hands-on practice. In the nineteenth century, increasing developments and applications of technologies in military and civilian organizations led to greater demands for technical experts. Formal engineering programs emerged, taught at universities and polytechnic institutions. Some of the early calls for reform focused on the academic rigor of the engineering curriculum and the need to ground engineering on mathematics and science courses (cf. Eddy, 1897). Accreditation of engineering programs in the United States began in the 1930s and this resulted in a significant tighten-

Beth-Anne Schuelke-Leechs, University of Windsor, e-mail: basl@uwindsor.ca

https://doi.org/10.1515/9783111563046-009

ing of curriculum leeway (Grinter, 2014). There was a great desire to leverage science, technology, engineering, and mathematics (STEM) to ensure economic growth and prosperity after World War II (WWII) (Schuelke-Leech and Leech, 2018). When the Soviet satellite Sputnik became the first man-made object to orbit the Earth in 1957, the focus on STEM education intensified (Grinter, 2014). The emphasis on greater technical theory created tensions with those that wanted engineering education to ensure that students were trained with industry needs in mind. Grinter (1955) called for a division of engineering into two tracks: one for those who aspired to go on to graduate school, which required advanced science and mathematics, and a second track for professional practice, which would place greater emphasis on engineering design and practical analysis. Much to Grinter's (2014) chagrin, no agreement on the separate tracks could be reached and the tensions between academic rigor and professional practice continued.

In the 1990s, several engineering organizations put out reports calling for changes in engineering culture and curriculum. The Institution of Engineers Australia (1996) criticized engineering faculties for being slow to reform and recognize the changing needs of industry and society in engineering. The Accreditation Board for Engineering and Technology, Inc., or ABET (1995), presented a new vision for engineering education, focused on graduate abilities and performance outcomes of graduates, rather than the education process. Since many engineers were (and are) employed in industry, it was logical to focusing engineering education on meeting industry needs. Engineers have been focused on improving efficiencies and doing analysis, design, drafting, and managing projects and they frequently viewed management as a natural career progression (Ritti, 1971). Thus, reform calls focused on meeting industry needs (cf. Geiger and Kapoor, 1995; Onar et al., 2018) and improving professional skills (cf. Duderstadt, 2008; Mitchell et al., 2021) are common.

Improving engineering graduate abilities in design, innovation, and problem-solving has been another area of reform proposals. The National Academy of Engineering (2004) in the United States presented an alternative vision of what an engineer should be, with a focus on design abilities, creativity, ingenuity, leadership, and high ethical standards. The following year, the National Academy of Engineering (2005) explained that the engineer of 2020 would not only have a strong grounding in mathematics and science, but also have a solid foundation in humanities, social sciences, and economics.

Yet another focus of reform calls has been on engineers' responsibilities to society and sustainability. Significant attempts to integrate social responsibility occurred in the 1930s (Layton, 1986) and the 1960s (Wisnioski, 2012). More recently, there have been attempts to integrate social justice, equity, diversity, and inclusiveness (cf. Cech, 2013b; Leydens and Lucena, 2017; Baillie, 2020; Sorby et al., 2021), sustainability (cf. Von Blottnitz et al., 2015), and addressing large societal problems (cf. National Academy of Engineering, 2008; Duderstadt, 2010a; UNESCO, 2010; Engineering Deans Canada, 2020; Shahidul, 2020; Shayesteh, 2025) into the curriculum. Thus, engineering for sustainability becomes another pathway competing for attention, commitment, and champions. Integrating sustainability and social justice into engineering competes with pressures for greater emphasis on the needs of industry and responding to emerging technologies.

Though there are many proposals and assertions that engineering education needs to change, there is little agreement about what needs to change or how to do it. This chapter presents an overview of the current impetus for change and three major visions for reform, specifically industry-servicing, technology-responsive, and socially responsive engineering education. The first section looks at engineering identity and current education. The second section discusses commonly presented reasons for change and issues of change management. Then, the three competing visions are each considered, including considerations for their implementation. Finally, the implications of these issues are discussed. Though there is a long-standing recognition that lectures and passive learning are not the best method for delivering all curriculum and that problem-based and experiential learning give students better opportunities to engage with material (cf. Woods, 1975, 1994, 2013), calls for reforms in how the engineering curriculum is delivered is beyond the scope of this chapter. Instead, this chapter focuses on the content of engineering programs and the calls for reform in the curriculum itself, rather than methods of delivering that content.

2 The Engineering Identity

In looking at the need for change, we need to consider the nature of engineering education, including its culture and norms, and to ask what this education conveys to our students and graduates. Engineering has a distinctive culture (Pilotte, 2013). This culture comprises the shared values, beliefs, and practices (Schein, 1988; Kotter and Heskett, 1992; Kunda, 2009), which are developed through education and professional networks. Professional socialization and education form powerful, and often unconscious, expectations and perspectives about the way that the world works and how one is to behave (Tajfel, 1974; Becher, 1989; Huber and Shaw, 1992). Identifying with any group is a powerful influence on one's behavior (Abrams, 1996). People work hard to conform to the expectations, standards, and values of a group that they claim membership in (Turner, 1981). This membership becomes their identity and social capital (Bordieu, 1986; Lin, 2001). Acceptance in the group becomes essential to one's identity and this need for acceptance can create exceptional pressure for individuals to maintain this membership. People will often censor themselves rather than disagree with other members or be viewed as dissenting from group consensus (Janis, 1971).

Engineering students are quickly socialized into seeing themselves as part of an elite intellectual group, and they retain this identity regardless of whether they go on to other education or careers (Becker and Carper, 1956). Identifying with any group is a powerful influence on one's behavior (Abrams, 1996). People work hard to conform to the expectations, standards, and values of a group that they claim membership in (Turner, 1981). This membership becomes their identity and social capital (Bordieu, 1986; Lin, 2001). Acceptance in the group becomes essential to one's identity and this need for acceptance can create exceptional pressure for individuals to maintain this membership. People will often censor themselves rather than disagree with other members or be viewed as dissenting from group consensus (Janis, 1971). Thus, changing the engineering culture means addressing the expectations of engineering practice and behavior. It also takes understanding what attracts people to engineering and the engineering approach to solving problems.

Engineers were overly represented in the leadership of the Russian Communist Party, the Chinese Communist Party, and the German National Socialist Party (the Nazis) (Gambetta and Hertog, 2017). This begs the question as to why engineers are attracted to these types of organizations. In their research, Gambetta and Hertog (2017) asked the question about why engineers were overly represented in right-wing extremist groups. They found that engineers have significant in-group bias, a need for social order, a desire to perfect and purify their environment, and a preference for unambiguous intellectual environments (Gambetta and Hertog, 2017). More simply, engineers prefer to look at the world in black-and-white terms, where their task is to physically and technically improve the world. They want certainty, where the rules are clear and unequivocal. This desire for social order and control is reflected by the over-representation of engineers in the leadership of autocratic organizations.

Gambetta and Hertog (2017) also found that engineers are disproportionately present in groups where liberal arts graduates and women are absent. This is not to say that any individual engineer does not like women or the liberal arts. Rather, it is that environments that seem to be attractive to engineers are often not really appealing to individuals with liberal arts backgrounds or to women.

In North America, engineers are overwhelmingly Caucasian men. For the past 20 years, the proportion of men to women in engineering has remained essentially constant at about 80% men to 20% women, up from about 16% in the early 1990s.[1] The majority of engineering students identify as Caucasian, with 41.8% identifying as something other than Caucasian.[2] Those identifying as women comprise 35.5% of the students majoring in the physical sciences, 44.1% of the students majoring in the life sciences, 29.6% of computer science and mathematics students, but only 14.2% of engineering graduates and 21% of engineering students (U.S. Census, 2014). In Canada, women are 22% of the

---

[1] Of the 3.34 million engineering graduate aged 25–64 years old, 85.8% are men and 14.2% are women (U.S. Census, 2014) (see Table 1). Approximately, 21% of engineering students are identified as women.
[2] About 63.3% of engineering graduates in the United States identify as Caucasian, 22.0% as Asian, 8.2% as Hispanic, and just 4.6% as Black or African American. The demographics of engineering students show slightly more diversity than the profession itself, as shown in Table 1.

undergraduate engineering degree holders (Kerr et al., 2025). Table 1 shows the demographic breakdown of engineering graduates in the United States and Canada.

Table 1: Demographic breakdown of engineering graduates (U.S. Census, 2014; Kerr et al., 2025).

	Engineering graduates	U.S. engineering graduates (%)	Canadian engineering graduates (%)
Total	3,340,430	100.0	100.0
Men	2,867,400	85.8	78.0
Women	473,025	14.2	22.0
Caucasian	2,113,930	63.3	44.0
Black and African American	154,555	4.6	
Asian	735,785	22.0	
Hispanic	273,870	8.2	

Engineering is more ethnically diverse than the sciences. While engineering is 63.3% Caucasian, 71.2% of students in the physical sciences and 75.6% of students in the life sciences are Caucasian. Like engineering, 63.3% of students in computer science and mathematics are Caucasian (U.S. Census, 2014). Nonetheless, recruitment and retention of women, and racialized minorities continue to be a challenge in engineering (cf. Beasley and Fischer, 2012; Silbey, 2016a, 2016b).

Given that men make up 85.8% of the profession in the United States, it is not surprising that these numbers are reflected in employment, where about one-third of male engineering graduates are working as engineers, a little over 18% are working as managers, and just under half are not working as engineers or managers. The data present a slightly different picture for women and minorities. For these groups, approximately 60% of engineering graduates leave engineering, whereas for men, it is under 48%, and for Caucasians, it is about 43%. This means that the engineering profession is even more skewed towards white and male than engineering graduates. Likewise, the data show that Caucasians and men disproportionately move into management. In the US, 14.3% of women are managers (U.S. Census, 2014); in Canada, it is about 15% of the managers in STEM fields who are women (Randstad, 2024). Though research supports the proposition that women leave engineering because of lack of equitable and fair compensation, poor and inflexible working conditions, and few opportunities for advancement and recognition (Fouad et al., 2017), there is little support for the notion that women leave because of family obligations (Hunt, 2015; Fouad et al., 2017). Instead, discrimination from managers and coworkers, and a lack of supportive networks and mentors are significant factors in women's decisions to leave the profession (Hunt, 2015).

Table 2 shows the employment for engineering graduates. As the data indicate, many engineering graduates leave the profession after graduation. Of all working engineering graduates aged 25–64 years, approximately 33% are still working within engineering and identify themselves as engineers, with an additional 18% having moved into management. The remaining 49.2% have left engineering. Thus, much of the skills and knowledge developed by engineering graduates are not being employed within the engineering profession.

Table 2: Employment for U.S. engineering graduates aged 25–64 years old (U.S. Census, 2014).

	Engineering graduates (%)	Working as engineers (%)	Managers (%)	Not engineer or manager (%)
Total	100.0	32.8	18.0	49.2
Men	100.0	33.7	18.6	47.7
Women	100.0	27.3	14.3	58.4
Caucasian	100.0	37.0	20.4	42.6
Black or African American	100.0	26.0	16.4	57.6
Asian	100.0	25.3	12.1	62.6
Hispanic	100.0	25.0	16.6	58.4

3 Engineering Methodologies

Understanding what engineers do and the methodologies that they use for solving engineering problems are really inseparable from understanding what engineers know, why engineers behave as they do, and what we can (and should) expect from them (Vincenti, 1990). Unlike science, which has thousands of books and articles presenting and analyzing the scientific method, literature that analyzes and critiques the engineering methodologies and mindset is quite sparse.[3] Instead, engineering is often presented as "applied science," and studies of engineers have been subsumed by studies of technologies themselves. Though there is a strong scholarly community around the philosophy and study of technology, just as there is one around the philosophy of science, there is only a small (and relatively recent) scholarly community looking at the philosophy of engineering.

Above anything else, engineers are problem-solvers, albeit particular types of problems.[4] Engineers focus on technical problems: problems that involve an understanding of the physical world, which are solved by some application of technology.

---

3 This is not to imply that there are not plenty of resources purporting to present various aspects of the engineering method to students. Rather, there is little literature critiquing what engineering is exactly and what makes an engineer, an engineer.

4 For a discussion of the types of problems that engineering students typically encounter, see Schuelke-Leech (2020). For a discussion of what graduate engineers classify as "real" engineering, see Anderson et al. (2009).

As an important part of this problem-solving process, engineers design and invent new technologies and find novel ways to apply their knowledge. A lot of what engineers do involves trial-and-error, experimenting until something works as desired. Engineers start with some problem or question or objective. Typically, these are defined by other people, though not always. The goal is to solve the problem. Often this begins with trying to determine what exactly is required and what is known about the current situation.

Engineers have developed many heuristics (i.e., common methods and rules of thumb) that guide engineering problem-solving and design. They provide guidance in the solution search and idea generation. In one study of engineering design, Daly et al. (2012) found that engineering students used 62 design heuristics during idea generation. In his explanation of inventive problem-solving (known as TRIZ), Gerhard Altshuller identified 40 inventive principles and 76 standard solutions. He developed these after reviewing over 40,000 Russian patents. He concluded that there were very few truly inventive or original ideas. Rather, most patent applications were really variations in applications to a common set of heuristics. These heuristics often provide the foundation for new ideas and designs. But engineers often have to refine and develop their ideas in an iterative or trial-and-error process.

Regardless of the specific application, people that identify as engineers claim methods and ways of solving problems that make them who they are (Martin et al., 2025). That is, they claim there is something specific to the domain of engineering that makes people, practicing in this specific way, *engineers*. Thus, understanding what engineers do in their work, and determining what knowledge, methods, and methodologies are unique to engineering, is not as straightforward as it initially seems.

What counts as engineering? Many engineers and engineering students distinguish "real" or "authentic" engineering from other activities. Authentic engineering is technical (Anderson et al., 2009). Frequently, it is focused on analyzing physical problems, and then applying design techniques to create new solutions, products, or services. This problem-solving process is intended to lead to practical, implementable solutions to a particular problem, as opposed to theoretical knowledge about the problem.

There are different descriptions of engineering problem-solving, its goals, and process. Hayes (1981) outlines four steps of engineering problem-solving: (1) finding a problem, (2) representing the problem in order to understand the nature of the problem, (3) planning a solution, and (4) evaluating the solution. Harris (2023) expands these steps to include establishing specific goals, and implementing the solution. However, these higher level stages or steps often involve many substages, as represented in Figure 1.

Problem-solving begins with identifying a need or problem to be addressed. For many engineers, this step is relatively straightforward because the problem is often defined for them. Many of the problems that engineers deal with come as relatively

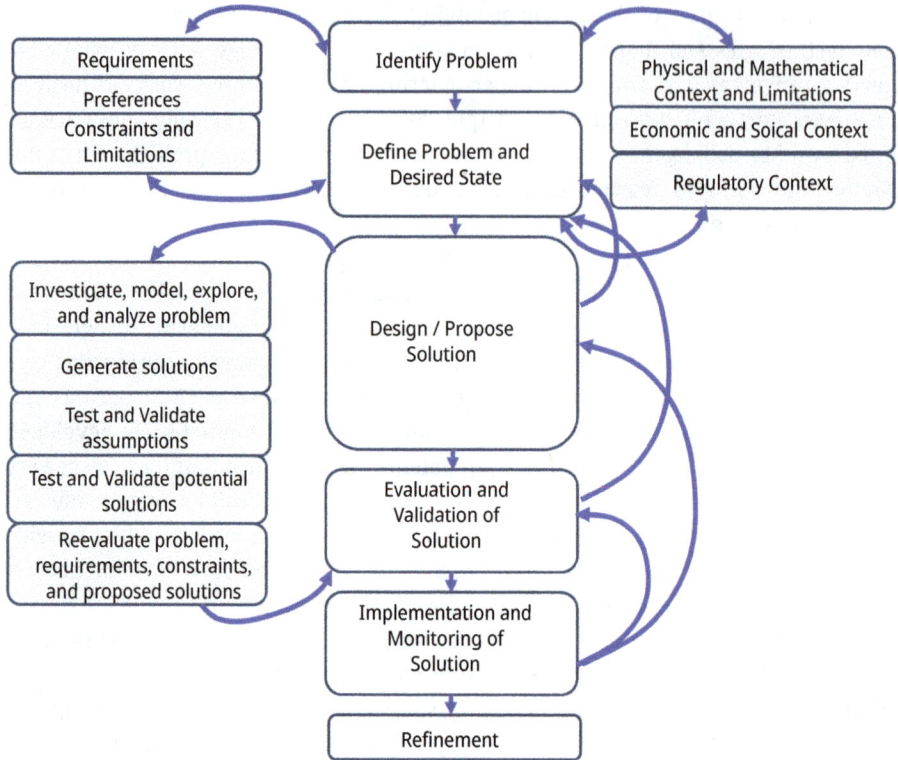

Figure 1: Engineering problem-solving method.

specific, well-defined, and constrained problems. Since most engineers work in organizations, the overall scope of the problems is set by their managers. Problems start off as high-level requirements or objectives, which are successively refined and constrained through levels in an organization so as to make them manageable. Thus, a broad, ill-defined problem is refined into specific requirements and constraints, until the scope becomes manageable for an individual engineer. This hierarchical nature of the engineering process means that engineers get problems that are smaller, manageable subproblems defined by successive levels in an organization (Vincenti, 1990).

The joy that engineers derive from solving technical problems is difficult to describe to others. Some people will describe how they get lost in reading a book, watching a movie, playing a video game, failing to notice the time they are absorbed in the activity. Technical problem-solving can have the same effect on some engineers. There is a satisfaction in struggling through a problem that requires piecing together parts of an intellectual puzzle. It satisfies a desire to understand the world and to make something work (Allred, 2012). However, these problems are defined in terms of the technical feasibility of a solution.

4 Engineering Education

Though many engineers have been fascinated with how things work since they were children, the process of being trained as an engineer also influences the way that engineers think and work. The educational foundation of engineering is mathematics, specifically algebra and calculus. Algebra has been the backbone of high school curriculum for many years. Algebra has a definite process for formulating physical questions as mathematical problems and then finding a solution (Rhine et al., 2018). Part of the strength of algebraic problem representation is that it allows for consistent structuring and manipulation of the problem. It is a very specific, disciplined way of representing a problem. It also provides a language through which others can understand the formulation of the problem under consideration. Though there are other ways to work through a quantitative problem, using algebra provides a common foundation for problem-solving. Thus, the foundation of engineering is a disciplined and specific method for analysis. There is little place for creativity in this form of problem representation and analysis.

Engineering students are overwhelmingly taught using well-structured physical problems with predefined solutions (Wood, 1983; Schuelke-Leech, 2020). The problems are limited in scope and the students are provided with the prescribed method for solving these problems. Thus, engineers learn to solve problems that always have solutions that are unambiguous and mostly academically straightforward. For instance, a typical homework or exam question would look like this:

> The hot combustion gases of a furnace are separated from the ambient air and its surroundings, which are at 25°C, by a brick wall 0.15 m thick. The brick has a thermal conductivity of 1.2 W/m·K and a surface emissivity of 0.8. Under steady-state conditions an outer surface temperature of 100°C is measured. Free convection heat transfer to the air adjoining this surface is characterized by a connection coefficient h=20 W/m²·K. What is the brick inner surface temperature? (Incropera and Dewitt, 1990, p. 20)

Over 95% of the courses that engineering students take use this kind of well-structured problem (Schuelke-Leech, 2020). This does not mean that the problems are not difficult for the students to solve; they frequently are. However, there is a limited amount of intellectual ambiguity and uncertainty. In most of the problems that engineering students encounter, there is a correct solution. They rarely encounter problems that do not have a straightforward solution. When engineering students do encounter problems that are not well-defined and unambiguous, they are often taken aback. Their first efforts are to constrain the problem in such a way as to make it well-defined. Then they use common solution methods and heuristics they have learned to see whether they can find a solution.

It is hard to know whether engineers are born as engineers or whether they are trained to be the way that they are. In truth, it is probably a combination of both of these, and each person is a little different. However, engineering education enforces particular mental models (Schmidt and Müller, 2023). Engineering education is extraordinarily structured. For the first 2 years, engineering students focus on foundational science and mathematics courses. Though there is significant evidence that most engineers do not use this theoretical knowledge in their careers, it is generally believed that this foundation provides for the development of logical and systemic thought.[5] There is often little context placed around these subjects. Instead, these subjects are treated as isolated topics, with little synthesis or connection to the other topics taught. Thus, engineering students are taught that the world outside of equations, physical relationships, and modeling is unimportant to engineering. It is hardly surprising that engineering students quickly begin to disregard social and environmental contexts. In addition, students come to believe in the superiority of rigorous mathematics and scientific principles as the foundation to solving all meaningful problems. They learn to dismiss any discussion of context or implications of technologies and engineering work as irrelevant to engineering. Empathy and caring are seen as unimportant to engineering, by both students and engineering faculty members (Strobel et al., 2013).

Looking at engineering education, it is possible to draw some conclusions about what engineering students learn. Students learn that the world is generally physical and unambiguous, with correct solutions to problems. The engineering curriculum is perceived by many students as an intellectual obstacle course (Florman, 1994; Godfrey et al., 2010). Engineering students feel under constant pressure and stress (Jensen and Cross, 2019; Jensen et al., 2023). Students have virtually no time for reflection on what they are learning or what it means for others. On the contrary, students are taught not to question what they are asked to do or learn. They have very few electives (Ellis, 2021). Thus, they have few opportunities to explore their own interests, ideas, or questions. This may help to explain why engineers are generally less assertive and visionary than other people (Williamson et al., 2013). It may also help to explain why engineers are so committed to social order (Gambetta and Hertog, 2017).

In engineering education, there is little value placed on nonquantitative things, whether this is communications, writing, ethics, liberal arts, or feelings. Though writing and communication are presented as important skills, they are not taught as integral to engineering, since the vast majority of evaluations are through numeric, calculation-based assignments and exams. Despite lip service about the importance of oral and written communication skills, these are often taught in separate classes and rarely incorporated into the technical curriculum. Likewise, ethics are irrelevant and external to technical discussions and decisions. The ethics that engineering students are exposed to are often lumped together with legal and regulatory compliance and professional obligations, so that students rarely grapple with the social and ethical

---

5 Arguably, any academic discipline will claim that they are teaching their students to think in a disciplined and systematic way (Zussman, 1985).

context of the technical material they encounter. "Complementary" liberal arts studies are typically taken from nonengineering instructors and are derided as being easy and unconnected to the (important) technical subjects they study. Thus, the subjects that cover the ambiguity and complexity of society and human processes are considered tangential to engineering. Empathy, compassion, and an understanding of the human condition are unrelated to real engineering practice and technological problem-solving. In fact, these are typically viewed as irrelevant to engineering education and practice. Nontechnical subjects are also viewed as intellectually simple and often derided as being unimportant, even ridiculous. In practice, engineers are often confronted with complex, ill-defined problems. However, it can be difficult for engineers to integrate broader social and ethical contexts with their work.

Though there is some rhetoric by leaders in engineering faculties and professional associations on the importance of the social context of engineering decisions and technologies, most students would not really get this from their education. In fact, while graduate engineers acknowledge the nontechnical requirements of their work, they are reluctant to include these in their definition of "authentic engineering" (Anderson et al., 2010). Little in their engineering training would lead them to believe that these activities should be considered "engineering."

This paradigm makes it easy for engineering students to dismiss work that is not solving equations or designing new products as nonengineering work, and thus, as not really within the scope of their concern. Over the course of their education, engineering students actually become less concerned about public welfare (Cech, 2013a), as well as less innovative and creative (Sola et al., 2017).

The other thing that is conveyed clearly through engineering education is the importance of obedience and compliance to authority. Engineering students are not taught to question what they are asked to do, to see ambiguity over questions and issues, or be creative for its own sake. In their few design courses, they get minor (if any) exposure to complex problems (Schuelke-Leech, 2020), which require synthesizing components from many of their courses or integrating interdisciplinary perspectives. Thus, students are not taught to deal with ambiguity or to have to negotiate a conflict. Rather, they are taught to comply with authority within the organizations where they work. Engineers are generally less assertive and visionary than other people (Williamson et al., 2013). Engineers have great affinity and loyalty to their organizations (Perrucci, 1971). They frequently view management as a natural, progressive step in their careers (Allen, 1988) and they learn that being successful means being good, obedient workers and team players (Kelman and Hamilton, 1989; Milliken et al., 2003).

Since engineers prefer environments with little intellectual ambiguity (Gambetta and Hertog, 2017), they can frequently be overly harsh and judgmental of others. They are consistently more conservative than other professional groups (Snow, 1954; Gambetta and Hertog, 2017). They often see things outside of their engineering paradigm as irrelevant and unnecessary.

Though social context and ethical implications are mentioned as components of the engineering profession, these are not central to the daily curriculum, and engineering students can be forgiven for thinking that the nontechnical aspects of work are outside the scope of genuine engineering. When asked, graduate engineers often discuss the nontechnical aspects of their work as not really being engineering (Anderson et al., 2010). Instead, professional and interpersonal skills are called "soft skills," implying that they are not as "hard" as real engineering.

In modern industrial societies, engineers are the ones that figure out how to create technologies and solve technical problems. They are the ones that create new technologies and get them to work better. They make machines and processes more efficient. They have been responsible for our modern supply chain system, for manufacturing that creates inexpensive products for mass consumption, for the energy grid that supplies power in support of the modern lifestyle. However, engineers generally do not set their own technical goals. Instead, these are provided by others within the organization that employs them. In addition to defining the goals, these organizations provide the resources that engineers need to develop and realize technologies.

When problems are ill-structured or too broadly defined, engineers are confronted with feedbacks and interactions that make it impossible to determine the priorities and trade-offs needed for a solution (Jonassen, 1997). Though practicing engineers are confronted with ill-structured problems, they work to refine the problem until it is possible to come up with a practical solution that addresses as much of the need as possible within the constraints defined (Koen, 2003). Thus, when engineers are faced with ill-structured problems, they will transform these problems until they can be addressed (Rice, 1994; Ogilvie, 2007). Engineering students are generally not confronted with wicked problems (Schuelke-Leech, 2020). Wicked problems, as originally defined by Rittel and Webber (1973), are problems that are ill-defined, with no clear boundaries, stopping points, and solutions. They are unique with no enumerable solution set. They are governed by feedbacks and interactions that are difficult to define and control. Thus, it can be very difficult, if not impossible, to refine a wicked problem to the point that it can be solved using traditional engineering conceptualizations and methodologies.

The transformation of engineering into a mass occupation created conflicts over the definition of engineering, engineers' responsibility to society, and how they should be trained (Meiskins, 1996). There were also questions about whether an engineering education should focus on practical training and primarily on current business needs or whether there should be greater emphasis on the academic subjects of science and mathematics. That is, the evolution of engineering education and the tension between theoretical and practical subjects have existed since the beginning of engineering as a formal field of academic training.

There are a lot of reasons why each country made decisions over their engineering curriculum, but it is impossible to separate the academic curriculum from the social class of the engineers in each country or the ideological, political, and economic goals of the elites. Who has access to higher education, how it is funded, who is employed there, and what is taught are all reflective of the society in which the education takes place. For instance, in the United Kingdom, engineering was considered within the purview of the lower classes, requiring hands-on technical training rather than advanced (i.e., university) degrees. Business managers came from the ranks of the elites, who studied classics at the premier institutions in the UK, such as Oxford and Cambridge Universities (Crawford, 1996). Thus, the engineering curriculum in the UK focused on apprenticeships and on-the-job training (Glover and Kelly, 1987), and less on the theoretical foundations of mathematics and sciences. It was not until after WWII that university graduates in engineering started to become more common in industry. However, UK industry still did not support licensure for engineers, which would have shifted some of the control over production and resources to engineers as professional experts (Crawford, 1996). Thus, engineers in the UK do not have the same status and autonomy as other professionals in the UK or that of engineers in other countries enjoy.

Engineers in the United States and Canada followed a different path. In the late nineteenth and early twentieth centuries, the rise of formal engineering programs in the United States, combined with rising demand for engineering skills and expertise, led to the rise of engineers as a distinct profession (Meiskins, 1996). Businesses liked the knowledge and expertise that engineers brought into their organizations, but they were concerned with the financial premium and autonomy that engineers demanded (Meiksins and Watson, 1989). However, engineers demonstrated that they were not competing for control of the organization, and engineers were integrated into businesses as a unique class of employees, separate from front-line workers, but not managers either. Engineers worked hard for professional recognition and autonomy. They had some independence and autonomy in their work, but they were generally supportive of achieving organizational goals and maintaining existing power structures. American engineers rely heavily on the integration of academic subjects, such as mathematics and science into their formal training (Meiskins, 1996), as do French engineers, where training in academic (and theoretical) subjects has existed in engineering since the eighteenth century (Crawford, 1996).

Canadian engineering followed the model of the United States rather than the United Kingdom. Engineers were active participants in the building of the country in the nineteenth century through such projects as the transcontinental railroad, the Welland Canal, and the Rideau Canal (Ball, 1987). Early preparation programs were designed to train technicians and mechanics (Harris, 1976). When formal programs of engineering were proposed, American university programs at elite universities, such as Harvard, Yale, and MIT were the models and standards aspired to (Harris, 1976). Thus, like American universities, curriculum in Canadian engineering programs relied heavily on academic subjects. In 1896, the first professional engineering association in Canada was formed in the province of Manitoba (Engineers Canada, 2020). Canadian provincial legislation for the licensing of engineers is fairly restrictive and professional associations in each province restrict who gets to claim the title of engineer. Thus, engineers in Canada are recognized professionals and enjoy a fair level of autonomy and prestige.

Whether engineering was taught with a heavy reliance on theoretical subjects or a focus on hands-on training seems to be related to the prestige and status of the engineering profession in various countries. Before the 1950s, academic subjects were viewed as being the purview of the privileged, whereas practical training was viewed as belonging to the working classes. Thus, the foundations of the theoretical subjects are actually closely related to how engineering is viewed, both by engineers themselves and by their societies at large.

It should hardly be surprising with how and what we teach engineering students that graduates are accused of being ill-prepared for industrial practice, unable to effectively communicate, and oblivious to the larger context of engineering and technologies.

5 Impetus for Change

With so many calls for change, it would seem unnecessary to ask whether change is actually necessary. However, it is a question worth asking. It is also worth asking about the exact nature of any proposed change.

There are several proposed reasons for change:

Business leaders, academics, and policymakers consistently talk about a shortage of engineers.

The shortage of engineers is so commonly discussed and asserted that it is often accepted without question. For instance, Engineers Canada (2012) projected that there would be shortfall of 95,000 engineers in Canada. There are also anecdotal stories of companies having difficulties filling specific jobs. However, it is unclear that there is really a shortage of engineers. There is actually significant evidence that the labor market is not suffering from a significant shortage of engineers (Rand Corporation, 2004; Lowell and Salzman, 2007; Teitelbaum, 2014). There is no evidence that engineering salaries have significantly outpaced the increase in other professional and skilled labor salaries, which would be a clear indication of a shortage in the market. Rather, engineering salaries have remained fairly consistent and have been growing at a more modest rate than many other professions, such as consultants and managers (Brown, 2009). In addition, there are indications that there is a substantial surplus of STEM doctoral graduates (Weissman, 2013) and a surplus of STEM workers for the jobs available (Ospe, 2015). This would seem to indicate that there is not a market shortage of engineers.

The idea of a shortage is often based on predictions about current engineering employment, projected retirement, projected job growth, and graduation rates. There are advantages to promoting the idea of a shortage. It encourages governments to keep investing in STEM education and it allows companies to push for immigration policies that allow for the importation of STEM talent from abroad. There has been a globalization of engineering workers and it has become common for engineers to be recruited internationally. International students make up an increasing percentage of engineering students, and the numbers of graduates in STEM fields are growing in many countries, such as China and India, even as they decline in Western democracies (Freeman, 2005). Though this may suggest that firms are unable to get engineering talent locally, it may also suggest that companies can get economic benefit from sourcing this talent internationally. That is, they can keep salary costs lower by recruiting globally.

As indicated here, the evidence is ambiguous about a shortage of engineers; it does not necessarily support the conclusion that there is a lack of engineering labor available, or that engineering education needs to change to correct this specific situation. Industry has long complained that engineering programs are not sufficiently supporting their needs and that there is a gap between graduating engineers' capabilities and industry's desires.

Rather than a shortage, therefore, it might be better to consider the gap a "skills mismatch" (Joppen, 2012). As engineers retire, there is a gap between those leaving and the younger engineers that replace them (Brightwing Talent Experts, 2012; Engineers Canada, 2012). This is hardly surprising. Experience and understanding gained over many years of practice cannot be easily replaced by young, less experienced engineers. It is also difficult to develop these skills through formal programs. There is evidence that there is a declining willingness by companies to invest in further developing the skills and knowledge of employees (Employee Benefits, 2019). This means that there can be complaints about a lack of appropriate engineering skills, even though there may not be a lack of engineers themselves. This mismatch provides an opportunity to offer short-term credentials and programs in these higher level skills, such as management, leadership, and innovation.

Innovation is essential for national competitiveness. More engineers are needed to support domestic technological innovation and production. Thus, business and society need to have more people studying engineering in order to meet industry's needs for technical talent.

In the early 1950s, the economist Robert Solow (1956, 1957) demonstrated that economic growth since the eighteenth century has been driven by technological change and innovation. Building on this foundation, economists and innovation scholars have demonstrated the importance of innovation to the economy and business. The 1950s and 1960s were a time of significant economic prosperity. Since that time, there has been a decline in growth rates and a stagnation of wages for many workers (Schuelke-Leech and Leech, 2018). Policymakers have promoted the idea that innovation is the foundation of our economy and scientific and engineering experts, as well as entrepreneurs and innovators, are needed to build a strong, prosperous economy. As part of this narrative, more engineers are needed as the foundation to the innovation economy.

There is certainly justification for the idea that engineers are an important part of the innovation economy. Engineers develop technical products and improvements that make new commercial products, services, and applications possible. Engineers are equally important to domestic manufacturing and production.

There are significant technical and social problems in the world. As experts in (technical) problem-solving, the skills and knowledge of engineers are essential for addressing these problems.

Undeniably, significant problems confront the world. Many of these problems involve technological elements. Engineers are viewed as superb technical problem-solvers and it is natural to want to leverage that ability to address many of the problems that now challenge humanity.

The United States National Academy of Engineering (2008) announced its grand challenges for the twenty-first century, with the goal of inspiring engineers and engineering educators to address sustainability and social problems. Likewise, the Royal Engineering Society of the United Kingdom has promoted a vision of the future in which engineers develop ethical and sustainable technologies to make the world better (Royal Academy of Engineering, 2019). The Engineering Deans of Canada have embarked on their own grand engineering challenges to inspire and encourage engineers to improve society (Wells, 2019). There are efforts around the world to incorporate the United Nations Sustainable Development Goals (SDGs) into engineering curricula and to make engineers the problem-solvers for all of society's problems, whether these are genuinely technological problems or not. However, without a fundamental shift in how engineers are trained and socialized into the profession, and what they are taught to value, exposure to these concepts do not necessarily result in changes to engineering thinking or practice.

In the early twentieth century, both Frederick Taylor and Thorstein Veblen had visions in which engineers would lead revolutions. Both viewed engineers as being able to separate themselves from the elite power structure, so that they were able to act with autonomy. Taylor's vision was based on efficiency and rationality, in which engineers would rationally optimize society because of their technical expertise and objectivity. Veblen (1921) envisioned engineers as the originators of a social revolution, where production would be organized and controlled by engineers and workers. Veblen's vision never materialized, fundamentally because he misunderstood and discounted the affinity that engineers have with business and their reluctance to challenge existing organizational hierarchies (Layton, 1962). Taylor's vision arguably did revolutionize industrial production throughout the world, but largely because the theory of scientific management was used to promote the goals of elites. Indeed scientific management was absorbed and practiced in authoritarian regimes from across the ideological spectrum, such as Communist Russia and Nazi Germany (Olson, 2015). That is, scientific management was not a threat to the existing business elites or political order, whereas Veblen's vision was. This suggests that a vision of engineering change must align with the current values and expectations of the engineering community.

Edwin Layton's (1986) book on *The Revolt of the Engineers* details the struggle of engineers in the 1920s and 1930s to take on greater professionalism and social responsibility. The engineering profession grappled with two visions of engineering – one where the engineer was an autonomous professional with the service to society being paramount and the other where service and loyalty to business was emphasized. As Layton details, the latter vision won out. Matt Wisnoiski (2012) outlines essentially the same struggle in the 1960s and 1970s in his book *Engineers for Change: Competing Visions of Technology in 1960's America*. Both books detail the forces that pushed for reform for socially responsive engineering and those that pushed back towards the industry-servicing model. Ultimately, the drive for socially responsive engineering did little to actually change either the practice of engineering or the education of engineers.

Engineering oversight and accreditation are founded on the principle that the paramount duty of the engineer is to protect the public good. In practice, this ideal is a goal, rather than a guiding principle for the daily practice of engineers or of engineering education. There is a recognition of the role that engineers play in creating, deploying, and managing technologies. However, there is virtually no real meaningful guidance for engineers on how to protect society within the institutional structures and expectations of their employment. In North America, there are relatively weak whistleblower protections. Engineers and workers are frequently punished economically and professionally ostracized when they speak out publicly against employers. On the contrary, those that comply, or even actively engage, in corporate malfeasance are rarely penalized. Quite often, they are rewarded for loyalty to the organization.

The engineering profession's emphasis on technical solutionism, and its male-dominated culture, continue to support and perpetuate an unjust economic and social system. Social and environmental justice demand that the engineering profession needs to change in order to be more inclusive and equitable, as well as to take a wider perspective of the factors that make engineering a benefit to all society.

While it is unclear that more engineers are needed, it is clear that there is an ongoing need for engineers and technically trained experts. At the same time, there is evidence that engineering is not necessarily an attractive educational program and occupation, particularly for women and minorities (cf. Marra et al., 2012; Geisinger and Raman, 2013; Sallai et al., 2023).

There are often discussions about the need to change perceptions of engineering (cf. National Science Board, 2007; Royal Academy of Engineering, 2007b; Myers, 2016) or calls for programs to encourage women and underrepresented minorities to go into the profession (cf. Milgram, 2011; Engineers Canada, 2019). However, there is little substantive reflection on the engineering culture itself or how it influences how people perceive engineering. The engineering profession is systemically discriminatory against women and underrepresented minorities and biased in favor of Caucasian men, as is engineering education. Recruitment programs have been largely unsuccessful over the long-term because they do not deal with the systemic discrimination and bias. Efforts focused on changing the demographic composition of educational programs, without addressing the engineering culture may not succeed. Engineering culture is such a strong force in reproducing the standards, norms, values, practices, and expectations of the profession (Carberry and Baker, 2018). Instead, engineering reform must meaningfully address the structure and expectations among engineering professionals and practice.

6 Three Competing Visions of Reform

Currently, there are three distinct and competing visions for engineering reform, as shown in Figure 2. The first is to make engineering education more responsive to industry needs, here called *industry-servicing engineering*. The second one responds to emerging and disruptive technologies, here referred to as *technology-responsive engineering*. The last reform looks to integrate greater social-consciousness, environmental sustainability, social justice, and equity into the engineering curriculum, here termed *socially responsive engineering* or *engineering for humanity*. Each of these visions has corresponding desired skills and knowledge. In the next section, each of these visions and what would be required to actually realize the proposed changes are discussed.

6.1 Industry-Servicing Engineering

Throughout the history of engineering as a separate profession, there have been struggles over where engineers fit in business and how their training should be related to the needs of business. Most engineers view management as an appropriate progressive step in their careers (Ritti, 1971). Engineers generally develop great affinity for, and loyalty to, the organization that they work for (Ritti, 1968, 1971; Kennedy et al., 1997). Most engineering graduates work for large corporations (U.S. Census, 2014) and most engineering undergraduates aspire to work for these organizations (Sheppard et al., 2009). Thus, there is a natural affinity between industry and engineers.

Businesses have consistently complained that engineering graduates still require training and that they lack many of the skills necessary to be successful within industry.

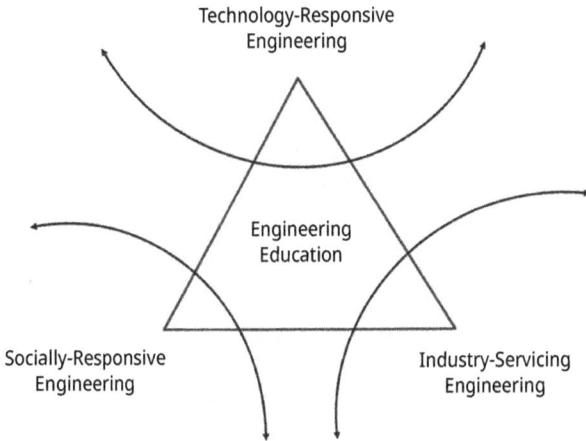

Figure 2: Three competing visions of engineering reform.

In his 1955 report on engineering education, Linton Grinter initially recommended that engineering education should be divided into two streams: the first stream would be grounded in basic science and mathematics with a goal of preparing students for graduate school and the second stream would emphasize engineering design and analysis and prepare students for professional engineering practice (Grinter, 1955). Grinter recognized that the focus on science and mathematics was a necessary foundation for research, but that it did not serve graduates well for industrial practice. Grinter's report was not well received, and the final version of his report dropped the "two tier" recommendation and instead advocated for greater emphasis on science and mathematics (Seely, 1999). Thus, there has been increased stress placed on learning the theoretical "basics" of engineering.

At the same time, there have been calls for improved communication skills and interpersonal skills, as well as for other complementary skills, such as innovation, creativity, design skills, design thinking, systems thinking, and entrepreneurial thinking (cf. National Academy of Engineering, 2004; Royal Academy of Engineering, 2007a; Duderstadt, 2008; Grasso and Martinelli, 2010; Asme, 2011; Brunhaver et al., 2017). The demands for these complementary skills in the engineering curriculum have existed since the nineteenth century. In his 1897 presidential address to the Society for the Promotion of Engineering Education (later renamed the American Society of Engineering Education), Henry Eddy noted that "[Engineers'] preparation in the use of language for writing and speaking has been too meager" (Eddy, 1897, p. 503).

Many of the changes in program requirements have really come through accreditation requirements. In response to requirements for communication skills, complementary studies, and "soft" skills, universities have added courses and graduate degrees that incorporate these "other" skills. However, these courses are often taught tangentially to the core curriculum. Frequently, they are taught by other faculties or clinical and adjunct faculty, giving the impression that these courses are not integral to engineering and that the skills and knowledge taught in these courses are not really "engineering." In addition, the material covered in these courses requires a new way of thinking and cannot typically be absorbed through one class or a few assignments (Pulko and Parikh, 2003). Without reinforcement throughout the engineering curriculum, students are not likely to really acquire these skills.

What businesses want or need from employees is not always consistent or clear. For instance, while organizations say that they want innovative and creative employees, this must be within the boundaries and structure of the organization. They do not want dissenters or people who question the goals and mission of the organization. Organizations simply could not function if engineers were questioning everything that they were asked to do. This leaves universities to try and interpret what skills and knowledge are needed, and how best to fit them into their core curriculum.

In finance, there is a concept of fiduciary responsibility, in which a *principal* employs an *agent* to act on its behalf. The agent has a fiduciary (financial) responsibility to act for the benefit of the principal, rather than for its own benefit. The ethical standards that an agent might employ in personal decisions are expected to be subjugated to the values of the principal. For instance, a financial professional may believe that the use of petroleum products is immoral and, thus, no investments should be made in petroleum companies. However, if the principal does not hold this same ethical standard and believes that petroleum companies are good investments, the agent would have to make investment decisions based on the principal's values, allowing for petroleum investments, even if the agent does not personally agree this is an ethical investment. The agent is, thus, expected to separate the ethical values that they exercise in their own personal choices from those that are employed when acting on behalf of another.

Given this, it can be useful to think of the ethics that people have as being divided into different categories, dependent on the context. Those of us who work in organizations can often have an ethical standard within that context that is different from the one that we have in our personal lives. These, in turn, can be different from the ethics that we hold while engaged in other activities, such as volunteering for a charitable organization or engaging in professional or industrial associations. That is, we experience *ethical polyfurcation*, in which our ethics are not just contextual, but rather are actually separated depending on the activity that we are doing (Schuelke-Leech et al., 2018).

Engineers have the same relationship and responsibility for any organization that they work for. They are generally hired to act as technical agents, *technical fiduciaries* if you will, and their responsibilities are then to act for the benefit of their organization. So, engineers must balance their responsibilities to their organization, their profession, themselves, and society in general. This can be a difficult balancing act.

In a study examining business values and ethics, Lynn Paine (2002) explored the differences in what people think is right when they were acting in the role of private citizen versus when they were acting on behalf of a company as a member of its board of directors. In the study, Paine presented a case to groups of executives, managers, and business students at the Warton School of Business, University of Pennsylvania. The case involved an imaginary pharmaceutical company that became aware that one of its leading drugs was causing approximately 14–22 needless deaths each year. The company was also contemplating their actions in light of the knowledge that regulators in their domestic market were looking to ban the drug. The mock board of director members were asked to decide what course of action that they should take. More than 80% of these Boards decided to continue marketing and distributing the drug domestically and abroad, and to take aggressive legal and political actions to prevent regulators from banning the drug. None of the Boards recommended recalling the drug. When individuals were presented with the same case and asked for their personal opinion of the decision to continue marketing and distributing the drug, 97% of the respondents felt that the decision was unethical and socially irresponsible. In other words, 80% of people acting as agents of the company would make a decision that 97% of individuals acting on their own behalf deemed unethical and socially irresponsible.

Engineers simply cannot question every instruction or task that they are given. It is impractical to evaluate whether each activity that they undertake is in the *public interest*. They would be mentally paralyzed if they had to question everything they were asked to do. More importantly, the institutions that they work for would not tolerate the time and energy required for this questioning. And what would engineers do if they decided that what they were being asked to do was not in the best interests of the public? Would they be expected to quit, giving up the financial security of their employment for some ambiguous societal good (even if the idea of "societal good" could be appropriately and universally defined)?

Companies say they want "job ready" engineers. Though some of the skills and knowledge that companies demand is specific to company applications, there are opportunities for higher education to respond to this need. Engineers will have an ongoing need for continual learning and institutions of higher education can provide these skills and knowledge, particularly if they are willing to offer micro-credentials and short-term courses where enrollment is easy and delivery flexible and responsive to the learners' needs. However, many of the skills that businesses say that they want are not ones that engineering faculties excel at providing. Thus, bridging the skills mismatch will take thoughtful reconceptualizing how these skills fit into the current curriculum and how they can be reinforced throughout the engineering program.

6.2 Technology-Responsive Engineering

The second proposed vision for engineering education is focused on responding to emerging technologies and technological opportunities, primarily coming from information and computer technologies (ICTs).

Emerging and disruptive technologies are driving changes to the opportunities and needs of society and industry. ICTs and the proliferation of sensors and connected systems are the foundation to smart cities and communities, Industry 4.0, artificial intelligence, smart and autonomous systems, and data analytics. They also present a new set of problems that need to be thoughtfully addressed. There is a general understanding that data, information, and cyber-systems are beginning to dominate many industries, but the full implications of ICTs for education, business, and governance are not yet known. Autonomous and intelligent systems shift the agency in the system from humans to cyber systems and algorithms, which may not be understood by the designers of the system. These new technologies are creating opportunities and challenges that many current businesses have not envisioned, and certainly, do not know how to deal with. Thus, businesses do not necessarily know what skills they will need going forward. Though the goal may still lie in servicing industry, when new industries and opportunities are being created, it is not possible to ask industry what they are going to need as the opportunities unfold. That is, the technologies themselves become the drivers of the opportunities, employment, and foundation for new skills and knowledge requirements.

Additionally, the international environment is changing and there is a corresponding power shift in industries across the globe. Many countries are making significant investments in research, development, and the commercialization of these technologies. The technological leadership of Western democracies, primarily the United States, is being challenged now. China has become a powerhouse for artificial intelligence and it is looking to be the leader in the global provision and use of these technologies (Lee, 2018). China is also leading innovation in advanced energy systems and other technologies (Huang and Sharif, 2016). Therefore, the business environment is changing and what employees need to be able to do is also changing.

Unfortunately, it is very difficult for companies and policymakers to determine what is needed to respond to the changing environment and challenges. They are confronted with disruptive technologies, changing demographics affecting employees and consumers, increasing global competition, and systems that they do not necessarily understand. Defining the skills and knowledge required to respond to these needs takes more than just asking businesses what they need or want; their answers are going to be based on their needs for today (or yesterday). They are rarely going to be answers for what they will need 5 or 10 or 20 years in the future, which is what universities need for designing their programs and shaping their curricula.

The shifting conditions of employment – from long-term stable corporate employment to much more precarious and shorter-term employment – means that the burden of skills and knowledge development and economic security are falling to the individual. We know that engineers are going to need greater abilities to deal with data and ICTs. They will also need to be able to recognize new and emerging opportunities and to be able to champion their ideas within their organizations, including persuading managers and executives to back those ideas and provide resources for them. That is, they are going to need to learn to be more innovative and entrepreneurial. They are also going to need to be able to understand and explain their decisions, since for many of these technologies the engineers are the only ones that have a detailed understanding of how the technologies and systems actually works. They are also going to need to be much more personally responsible for their employment and advancement. These skills and abilities are currently not integrated into engineering education.

Many leaders have called for a transformation of education to deal with these new technologies. Duderstadt (2010b), former Dean of Engineering at the University of Michigan, explained that we needed transformation, rather than reform, in engineering education in order to deal with new challenges, technologies, and problems.

In 1996, a report by the Institution of Engineers Australia stated:

> . . . the engineering curriculum has been slow to respond and while there has been some reform over the past 15 years, the educational model we use is still not much different from that of 30 years ago, and while the pace of change in the world has increased significantly, the pace of change in engineering education has been far too slow. (Institution of Engineers Australia, 1996)

Unfortunately, many engineering programs continue to hold on to a curriculum and style of teaching that is strongly reminiscent of past methods and needs. In reporting a large-scale study on engineering education, Sheri Sheppard and her colleagues put it this way:

> In the midst of a profound, worldwide transformation in the engineering profession, undergraduate engineering education in the United States is holding on to an approach to problem solving and knowledge acquisition that is consistent with the practice that the profession has left behind. (Sheppard et al., 2009, xxi)

Technologies are changing society and the needs around us. We cannot continue to hold on to the educational model that worked for mass production systems and long-term stable employment. Students need a much greater understanding of how computer and autonomous systems work and how to design and use these systems. They also need abilities to recognize and develop commercial opportunities for new products and services. This takes teaching students about innovation and entrepreneurial thinking throughout their education. It takes a shift in mindset from expecting a problem to be provided and the scope of responsibility being restricted to simply addressing the technical aspects of the problem. The technologies themselves are changing business and engineering. New skills and knowledge are needed by virtue of the nature of the technologies themselves. How well these fit in with the industry-servicing model of engineering education is not clear yet. While it is tempting to believe that they are synonymous, they are not. Often the calls for industry-servicing engineering are coming from large corporations that are looking for more effective employees. Technology-responsive engineering is more focused on the changing skills and knowledge in response to emerging technologies and societal changes, which current industry may or may not have yet incorporated into their operations.

6.3 Socially Responsive Engineering – Engineering for Humanity

Like the calls for greater responsiveness to industry, the engineering profession has been confronted with calls to be more socially responsive and to address humanity's major problems. The push to address these problems, and for engineers to be the leaders in solving society's problems, is just the latest version of calls for engineers to be socially responsive and accountable to society-at-large, not simply to their employers. Engineering faculty are more likely than their peers in other research disciplines to believe that funding decisions should be based on the application of the research, rather than for purely intellectual curiosity (Schuelke-Leech, 2011). Calls for engineering programs to incorporate society's "grand challenges" and sustainability and to address societal injustices have become more common (cf. National Academy of Engineering, 2008; Duderstadt, 2010a; UNESCO, 2010; Leydens and Lucena, 2017; Baillie, 2020; Engineering Deans Canada, 2020; Shahidul, 2020; Shayesteh, 2025).

It is not clear if students are looking for more socially responsive education. Research has shown that women are much more likely to be interested in university programs that consider and address social and environmental issues (Silbey, 2016a, 2016b). Additionally, students are looking for meaningful employment and opportunities to contribute (Yang and Guy, 2006; Ng et al., 2010). Millennials believe that they are responsible for making the world a better place and contributing to society (Mcglone et al., 2011). Many feel a calling for serving society (Henstra and Mcgowan, 2016). Attracting these students to engineering is not about a PR campaign or exposing them to "engineering." It takes recognizing that they have a fundamentally different motivation for their careers. They are looking to solve different types of problems – not just traditional industry problems of efficiency and optimization or developing and improving commercial products.

For meaningful socially responsive engineering, the curriculum will need to emphasize greater social-consciousness, nuanced and critical thinking, the ability to question authority, ethics and integrity (with an accompanying ability to champion social and ethical positions in organizations), and genuine professional autonomy and independence. As previously discussed, this is *not* how engineering education is currently designed. Engineering students are far more likely to see the world in unambiguous ways (i.e., in black-and-white terms) than are students in the liberal arts or the humanities (Paulsen and Wells, 1998). Getting engineering students to see the world differently means helping them to see the complexity of the world and the factors affecting engineering systems. That is, it means helping them to see the shades of gray in the world. What is unclear is how introducing greater ambiguity in the scope of problems and the complexity of factors influencing solutions will affect engineering graduates. Right now, they are taught through a process of being presented with theory and then testing this in laboratories. Their assessments are tightly coupled with the theories that they have been taught. Through this process, students learn about the certainties of theory and application. They learn that there are solutions that can be found to problems. Once they have this solid grounding in scientific and mathematical theories, they go on to put additional constraints around the problems (such as financial and time constraints), but the goal is still finding a solution to a problem based on understanding the nature of the (tangible and physical) problem and the objectives and constraints for a solution.

Engineering for humanity requires students and graduates to gain an understanding of the complexity of problems and their context. It requires that them to incorporate alternative perspectives and social constraints into their problem-solving framework. It also requires them to question and push back on authority and to question the problems that they are given. It is not just a matter of layering on another course on UN SDGs or exposing them to sustainability (or grand) challenges. If all that is expected to change is that students graduate with an enhanced idea of the kinds of problems that exist in the world, without necessarily expecting engineers themselves to change, then no meaningful change is needed in the goals of engineering education. On the contrary, if genuine transformation is desired in the way that engineers think and practice, this will require much more thoughtful and deliberate action.

Socially responsive engineering is really about changing the engineering mindset to reflect the responsibilities that we have to society and the public. It is more than just about protecting public safety. It is about recognizing that engineers must be reflective about what they are being asked to do with their skills and knowledge. It is about integrating values and societal criteria and constraints into their work, without explicitly being given permission to do so by their employers. It requires a substantial shift in mindset from being an agent of a company to being an autonomous actor with responsibilities beyond those required by an employment contract.

There have been some attempts in engineering to lead transformative change. Olin College of Engineering was founded in 1997 as an elite engineering institution designed to transform engineering and its education. In their book *A Whole New Engineer: The Coming Revolutions in Engineering Education*, David Goldberg and Mark Sommerville (2014) describe how the new College wanted to build a different culture in engineering: one focused on design thinking, solving real-world problems, and integrating innovation and creativity. Olin College is an elite teaching college and its faculty are hired to align with its curriculum. It is not concerned with research or hiring scholars. It is also highly selective and caterers to a small number of students. Its total enrollment is 402 undergraduate students, of which 48% are female (US News & World, 2025).

There have been other attempts to integrate nontechnical knowledge, such as entrepreneurship, innovation, creativity, arts, ethics, systems thinking, and design thinking, into the engineering curriculum. For instance, Peter Senge's *The Fifth Discipline* sparked the incorporation of systems and holistic thinking into organizations, including engineering faculties. *Holistic engineering*, spearheaded at the University of Vermont by Grasso and Martinelli (2010), sought to integrate nontechnical perspectives and knowledge – such as social justice, strategy, communications, and psychology – into solving technical problems. There have been some attempts to integrate engineering and social justice. Catalano and Baillie (2006) and Baillie and Kadetz (2024), for example, have written about how to incorporate social justice and peace studies into engineering, as well as how to focus engineering more on sustainable and community development. Leydens and Lucena (2017) have been explicit about trying to create a socially just engineering profession. Kevin Pissano (2009) at the Ohio State University has been educating students on humanitarian engineering for decades. *Value sensitive design* at the University of Washington School of Information was developed by Batya Friedman and David Henry (2019) to help students to integrate an understanding of ethics and values into new technologies. More recently, there have been calls to more fully incorporate equity, social justice, and inclusivity initiatives into engineering. All of these attempts are important steps in showing engineering faculty and students that there can be a different way of doing engineering and a different set of priorities and problems that we engage with. While these calls have been taken up by many institutions, engineering culture is very difficult to change and remains entrenched in past practices and norms.

7 Change Management

Reform in any large organization or system is difficult (Hunsucker and Loos, 1989; Kanter, 1992; Khaw et al., 2023). Resistance to change is common. Any meaningful change is difficult and requires overcoming the resistance that people have to change (Weick and Quinn, 1999; Kaufman, 2017). There are vested, elite, and societal interests, which can compete and conflict. Changing a profession requires a clear, consistent vision of the desired change and effective leadership (Kotter, 1996). It also requires an imperative to change (Oakland and Tanner, 2007).

Engineering programs are already full of courses and simply layering new courses and ideas onto existing offerings is not necessarily going to lead to meaningful change. There are multiple vested interests in the current engineering educational system. To start with, the professors who teach engineering are embedded in an academic system that rewards research productivity and funding. Teaching is a secondary concern. Tenure and promotion are tied to research, not to teaching or ensuring that the engineering curriculum meets the needs of graduates or their employers. Few professors have spent any significant time as employees in industry. Few engineering faculty members have any interaction with industry and the majority have no interaction at all (Schuelke-Leech, 2011). Even when faculty members belong to interdisciplinary research centers, these tend to be dominated by their own academic specialty (Schuelke-Leech, 2011). Thus, engineering faculty members who teach engineering students rarely have any meaningful understanding of actual engineering practice, and they have little incentive for engaging with anyone outside of their field of research interest. Instead, faculty members are much more likely to be focused on graduate student training, which is essential for research publications and securing grant funding. This necessarily biases them to a curriculum that promotes the skills and knowledge needed for advanced research. In addition, faculty are rarely interested in the overall engineering curriculum or engineering education beyond their specific area of research. Curriculum committees will discuss particular courses, but they rarely discuss holistic changes to the engineering programs or propose visions of educational reform.

Likewise, university administrators generally have little reason to push for substantive change. Public university revenues are a combination of government funding based on domestic enrollment, tuition and fees from students, revenues from ancillary services (e.g., dorms, food service, and parking fees), research contracts and grants, restricted and unrestricted donations, and investment revenues. Unless programs are declining in enrollment or are so small so as to make it difficult to continue, there is significant momentum to keep the status quo. There is always an impetus to grow and expand, but fiscal austerity, downsizing, and substantial program changes are disliked and difficult. Overcoming faculty resistance requires sustained efforts. Deans and senior administrators must decide where to use their political capital, time, and resources. Long-term systemic change requires prioritization and compromises among competing stakeholders (Kezar, 2009). It also requires an understanding of the factors that lead to successful change. Few university managers have any training in management, strategy, or organizational behavior. Most university administrators have advanced into administration because they were successful faculty members and administration was an avenue for advancement. They learned to manage their organizations and people while doing it, rather than because they have any expertise at it ex ante. This means that they are relatively inexperienced at large-scale organizational change or transformation. Successful change requires significant efforts (Kezar, 2018).

These visions of reform are really about where we are going as a society and what role engineers are going to play in that society. Unquestionably, engineers need to be better communicators. They should also be able to solve technical problems effectively and to make design decisions that optimize designs relative to functions, features, and constraints. However, as De George (1981) pointed out in his article on engineering ethics and the Ford Pinto case, engineers can do cost-benefit and risk analyses, but they cannot decide if these are acceptable and this is the purview of management.

Each reform vision has its own advocates. These visions of the future of engineering education are not consistent and they are not wholly compatible with each other or our current education and practice of engineering. Nonetheless, we rarely talk about competing visions of reform, and most advocates assume that "reform" means changing engineering for the better, typically in whatever way they see as being "better." However, there is little genuine investigation of what reform will mean. There is also little appetite for wholly redesigning or restructuring the engineering curriculum, and the profession more broadly.

So, reform typically comes down to making incremental changes to engineering practice and the curriculum for engineering students. There are recommendations to change or add a few courses or to add a new program that will service a particular industrial segment, technology, or calls for justice. The core of engineering remains the same, and other knowledge and skills are added on – problem-based learning, design, communications, people skills, design thinking, systems thinking, innovation, entrepreneurship, ethics, management, and sustainability thinking – are treated as adjunct to the core of engineering (i.e., the "real" engineering courses). That is, these other courses are not treated as integral to the development of the technical engineer by faculty and engineering students alike. They do not recognize them as foundational to engineering.

Partly, this is a function of the way that universities operate. Faculty members are technical experts that have training in a specific area of a discipline. Their success comes from publishing and teaching in their area of expertise. As any faculty member knows, changing a curriculum can be a difficult process. Faculty members will advocate for their technical areas (and courses) of expertise, while they can discount or ignore other knowledge areas, particularly if these appear to be nontechnical.

As long as professional programs and nontechnical courses are treated as adjunct to the core of engineering, no meaningful reform is really going to happen. Most institutions will not take on the task of substantive reform because the risks are far too great. Accreditation becomes the overarching requirement of program structures. While we are prepared to tweak the edges and offer some additional programs and courses, these remain adjunct to the core of our engineering undergraduate education.

One of the things that leaders need to decide as they are advocating for reform is what vision of engineering they are pursuing. While there are certainly skills and knowledge that are foundational to all, such as an ability to communicate and deal with people, engineering students cannot be asked to take on greater and greater course loads as we struggle with what we really want them to be and do.

Leading change requires a clear vision of *what* must change and *how* to ensure the change fits within accreditation requirements and faculty resources. Currently, there is no widespread vision and leadership for change that will bring about substantive changes to engineering education. Reforming the curriculum requires both a holistic view of engineering education and a vision of reform.

What we need to determine is how much we want engineering students and graduates to change. Do we really expect them to change from being compliant and obedient employees to being able to critically analyze and question what they are being asked to do as engineers? Do we expect engineers to be leaders throughout society? This will require a revolution in the role that we expect engineers to play in society, and correspondingly, how engineers are trained to meet these new expectations. The desire to apply engineering skills and knowledge to the technical and socio-technical problems of society is understandable. However, making meaningful change will not be accomplished by merely adding in a couple of courses aimed at introducing either the UN sustainability goals or environmental challenges.

An analysis of the contemporary engineering curricula in North America shows some consistent ideas that are being taught and reinforced throughout various programs. Specifically:

- The current engineering curriculum is heavily weighted towards mathematics and science, particularly in the first few years. There is little explanation of the context to the subjects, regardless of whether one is looking for an industrial or social context. Engineering students receive the message that the world outside of equations and technical boundaries is completely unimportant. Thus, it should be unsurprising that engineering graduates develop an arrogance over the superiority of math and science and tend to dismiss any discussions over context or implications of technologies and engineering work.
- Writing and communication are also unimportant, since the vast majority of the evaluations are through numeric, calculation-based assignments and exams. While there is discussion of the importance of oral and written communication skills, these are often taught in separate classes and rarely incorporated into the technical curriculum.
- Ethics are irrelevant and external to technical discussions and decisions. The ethics that engineering students are exposed to are often lumped together with legal and regulatory compliance and professional obligations, so that, students rarely grapple with the social and ethical context of the technical material they encounter.
- Likewise, "complementary" liberal arts studies are typically taken from nonengineering instructors and are derided as being easy and unconnected to the (important) technical subjects they study. Thus, the subjects that cover the ambiguity and complexity of society and human processes are considered tangential to engineering. Empathy, compassion, and an understanding of the human condition are unrelated to real engineering practice and technological problem-solving.

–   Efficiency and optimization are typically the primary factors in a decision. While there may be other important factors are important, are often considered as secondary.

If an institution truly wants to transform their engineering program and its culture, then there are some meaningful steps that can be taken:

–   The values that are desired of the graduate must be incorporated into the curriculum from the beginning. If we want engineering students (and thus graduates) to have an interdisciplinary, systems perspective of problems, they must be presented with these types of knowledge and problems as soon as they begin their studies. It may even be better to slow down the curriculum in core mathematics and science and to spread these out to allow for the presentation of design and systems thinking early.
–   Students must be given an opportunity to experiment and question, without adverse consequences. This is a hard balance. Courses that have little academic costs or penalties are likely to be ignored by students as they focus their time on (as they perceive them) more rigorous and demanding courses. Thus, the curriculum that exposes students to ethics, systems, design thinking, innovation, and creativity, must be engaging and rigorous in the process, but still allow for genuine experimentation and failure.

Olin College demonstrates this alternative ways of thinking and assessments by a commitment to (Goldberg and Somerville, 2014):

–   Challenging students to experiment, explore, and take risks
–   Distinguishing between situations that are predictable and stable with those where risk, uncertainty, failure, and learning are inevitable
–   Reframing failure and mistakes as opportunities to learn and reflect
–   Focusing on process rather than outcomes, particularly with respect to traditional measures of student achievement and grades
–   Motivating and inspiring students with efforts to change the world

–   Wicked problems, as previously defined, should be integrated into the curriculum, and students should be given a chance to explore the problem and explore potential solutions. (Olin College presents a good model of how this can be done.) However, it is essential to realize that not all engineering students are interested in this type of problem-solving. Thus, a diversity of problems must be presented to students to allow them to solve the types of problems that they are truly interested in. For students interested in more complex social problems, much more mentoring is going to be required to help them to move beyond traditional engineering problem-solving.
–   Helping students to demonstrate inclusivity and tolerance, including having places where students can explore their ideas and questions. In a culture that has no

sincere tolerance for exploration, students will learn to mimic tolerance, but they will not genuinely incorporate inclusivity and openness. Thus, the culture must allow for reflection, change, and growth in a nonaccusatory and riskless way. Everyone must feel truly valued, which requires having faculty members open to listening and learning themselves and to recognizing the difficulties that significant transformation presents to everyone.

8 Conclusions

Engineers have a profound impact on our society. They develop many of the products and services that we use daily. Their knowledge and practice provide a mechanism for society to develop and address many problems. What makes engineers so great at understanding and solving technical problems may also be what makes them entrenched in a culture that is exclusionary and resistant to change. Simply calling for reform does not provide a pathway for achieving it. Additionally, there must be agreement about the goal and the strategy for achieving it. The future of engineering education, and of the profession itself, is dependent on the path that we choose. Making positive changes in engineering cultures and institutions is incredibly difficult. If engineering for sustainability is the goal, then there must be consensus and an imperative for change.

The discussion of engineering education reform exists within the context of the extreme volatility and disruption that is confronting humanity: the development and deployment of autonomous and intelligent systems; climate change; and economic, political, and social stresses, both domestic and global. These disruptions are going to fundamentally change the way that society functions and organizes. To survive and prosper, we need to think about how we are educating for this future. Simply continuing to educate our students with a focus on current practices and consumption patterns is a disservice to them. We will not be teaching them how to contribute to the future health and success of humanity. Instead, we need to focus on helping them to develop their abilities to frame problems differently, to come up with creative and innovative solutions, and to work for all of humanity. This focus includes integrating more open and inclusive problem definitions, innovation, entrepreneurship, creativity, and social awareness into engineering education.

Engineering is fundamental to solving the problems facing humanity. Moreover, engineers must understand their role in both solving these problems, but also in how they have been created and how we can engineer differently. That is truly the challenge of engineering education. If and how we reform engineering education, and whether reformation is possible, is dependent on how we define the reformation that we want to make happen and on its implementation. The processes that we develop for transformation are paramount to its success. Not everyone is going to agree.

There is going to be resistance to change, particularly by those that enjoy the benefits of the current system.

Engineering oversight and accreditation are founded on the principle that the paramount duty of the engineer is to protect public safety. In practice, this ideal is a goal, rather than a guiding principle for the daily practice of engineers or of engineering education. There is a recognition of the role that engineers play in creating, deploying, and managing technologies. However, there is virtually no real meaningful guidance for engineers on how to protect society within the institutional structures and expectations of their employment. In North America, there are relatively weak whistleblower protections. Engineers and workers are frequently punished economically and professionally ostracized when they speak out publicly against employers. On the contrary, those that comply, or even actively engage, in corporate malfeasance are rarely penalized. Quite often, they are rewarded for loyalty to the organization.

The struggle over the place of professionals in many countries reflected the fears of the elites over control and power. Professionals in general have an ability to operate outside the existing power structure. Their independent knowledge and abilities make it impossible for the elites to completely control the outcome of professional decisions and actions. To elites, there is risk and uncertainty in allowing professional autonomy. As importantly, professionals have an ability to command loyalty and respect from others that is outside the normal power structure. This can make professionals a real threat to the elites.

Elites have recognized this threat. Their actions and how engineers have responded create the core fabric of the history and culture of engineering practice. The actions of engineers in different political and social regimes demonstrate the norms of the profession. In general, engineers have not been social reformers. Typically, they have vigorously supported the dominant social structure. Engineers have done great things in society, such as designing and building sanitization systems, transportation and communication networks, and infrastructure to harness and distribute energy. At the same time, they have also been responsible for products and services that have produced significant societal harms. They have designed chemical and biological weapons and industrialized death camps. They have overseen projects that have been environmental disasters. They have designed addictive products and services, as well as ones that invade our privacy, track our movements, and pollute our environment.

The world is confronted with many challenges. Two in particular are entwined with engineering and the current paradigm of technological development and problem-solving. The first is the climate change and the need to adapt to a warming world. The second is the emergence of autonomous and intelligent systems, where technologies are being developed and deployed, which are not necessarily designed to serve humanity or to preserve and protect the values that we cherish.

Technologies are not neutral, but they are not wholly good or bad generally. Rather most technologies are simultaneously potentially good and bad. That is, they can be used for good purposes, but they can also have negative side effects or be used for bad purposes. Facebook, for instance, can be used to share information about charitable events or it can be used to coordinate illegal activities. From this perspective, Facebook is neutral. However, there is also what the technology is designed to do. Many social media platforms are designed to be addictive; to get users to spend more and more time on the site, absorbing the content and being receptive to advertisements, making the platform money and providing user data that can be mined and exploited.

Technologies are unquestionably going to continue to be part of our society and to provide people with new services that leverage the data generated by the billions of devices and sensors proliferating around us all. Technologies are often discussed as an essential component to addressing societal and environmental problems, whether this is through the development and adoption of "smarter" products and services or as a means to help individuals, public officials, and businesses to understand people and society better. At the same time, it is essential to understand how the developers of these technologies – namely, the engineers – solve problems and how they understand their role in developing these technologies.

The proposals for reform are really about the goals of engineering practice and how to best train students to do this work. However, the three proposed reforms are actually pulling in different directions. Engineering that is industry-servicing necessarily defers to the goals and needs of private corporations that set their own demands. Engineers defer to these demands in the service of achieving greater profits though greater efficiencies, meeting and extending customer needs, and incorporating technologies. Technology-responsive engineering, on the contrary, focuses on how technologies are evolving and looks at how these technologies can be further developed and utilized. This reform necessarily requires teaching students about entrepreneurship and innovation, as well as the technical and commercial aspects of technology development. Engineering skills are used in the service to developing and responding to technologies and how they can be used in the market place. While engineers may be developing the technologies, the focus is on commercial opportunities defined in the marketplace or by needs of entities, such as the military. Socially responsive engineering has entirely different focus. It is about engineering with society and society's needs in the forefront of engineering education and practice. Engineering students must be taught not just how to solving technical problems, but also the nature and value of the problems that they are being asked to solve. This change in focus requires a significant transformation in the nature of engineering.

As I have tried to layout in this chapter, engineering education has been focused on teaching students to be effective technical problem-solvers and designers, through an understanding of the physical world (via STEM courses) and design methodologies. Problem-solving has been a deductive process to constrain given problems to the point that they can be effectively manipulated, refined, and resolved. A socially responsive engineering requires a reframing of how engineering educators view the goals of engineering. Engineering becomes more about how larger societal goals can be achieved, rather than whether particular technologies are commercializable or in direct response to market forces. This would be a revolution in engineering education.

Engineers heretofore have been responsible for solving the problems that have generally been set by others, especially corporate managers. Moving to a model where engineers must now assess the societal impact of problems, including their ethical, environmental, and social qualities requires a skillset that engineering educators have not yet embedded into engineering education. Nor have attempts at reforms aimed at large scale change in engineering been successful. Instead, reforms have generally been more incremental.

Thus, any large-scale change to socially responsive engineering must involve widespread consensus of the imperative to change and a pathway to achieve it successfully. It requires both the policy and a path for implementation. I hope that this chapter provides some background and understanding of the challenges that we face and the options that are open to those looking to reform engineering education to respond to the sustainability challenges that are facing us. Reform that focuses on sustainability and socially responsive engineering looks very different from reform that is focused on greater responsiveness to industry needs or to the current direction of technological development. It is uncertain if these reforms are at all compatible with each other since they require different paradigms of engineering practice and problem-solving. They result in different problem requirements and constraints, as well as different potential solution spaces. This, in turn, requires different educational goals and curriculum. The purpose of this chapter has been to provide a foundation for considering the types of reforms that are proposed and the issues associated with major reforms. Future work must consider how consensus can be reached about what kind of engineering that we want and how we can lay out a pathway to achieve it.

References

ABET, (1995), The Vision for Change: A Summary Report of the ABET/NSF/Industry Workshops, Accreditation Board for Engineering and Technology, Inc., Retrieved on August 4, 2025 from http://bioinfo.uib.es/~joe/semdoc/PlansEstudis/ABET_Criteria_PTE/Vision.pdf.

Abrams, D. (1996). Social Identify, Self as Structure and Self as Process. In Robinson, W. P. (ed). *Social Groups and Identities: Developing the Legacy of Henri Tajfel* (pp. 143–167). Boston MA: Butterworth-Heinemann.

Allen, T. J. (1988). Distinguishing Engineers from Scientists. In Katz, R. (ed). *Managing Professionals in Innovation Organizations* (pp. 3–18). Cambridge, MA: Ballinger Publishing.

Allred, R. (2012). *I Am an Engineer: A Memoir.* Creative Arts & Sciences House: Indian Harbor Beach, FL:.

Anderson, K., Courter, S., McGlamery, T., Nathans-Kelly, T., & Nicometo, C. (2009). *Understanding the current work and values of professional engineers: Implications for engineering education.* Paper presented at the 2009 Annual Conference &Exposition.

Anderson, K. J. B., Courter, S. S., McGlamery, T., Nathans-Kelly, T. M., & Nicometo, C. G. (2010). Understanding engineering work and identity: A cross-case analysis of engineers within six firms. *Engineering Studies, 2*(3), 153–174.

ASME, (2011), *Vision 2030: Creating the Future of Mechanical Engineering Education, Phase I Final Report*: American Society for Mechanical Engineers.

Baillie, C. (2020). Engineering and Social justice. In Michelfelder, D. P. & Doorn, N. (eds). *The Routledge Handbook of the Philosophy of Engineering* (pp. 674–686). New York, NY: Routledge.

Baillie, C. & Kadetz, P. I. (2024). Introduction to the Book. In Baillie, C. & Kadetz, P. I. (eds). *Reimagining Engineering Education: Health. Justice. Sustainability* (pp. 1–13). Singapore: Springer Nature Singapore.

Ball, N. R. (1987). Mind, Heart, Vision: Professional Engineering in Canada, 1887 to 1987. Ottawa, ONT: National Museum of Science and Technology, National Museums of Canada, in cooperation with the Engineering Centennial Board.

Beasley, M. A. & Fischer, M. J. (2012). Why they leave: The impact of stereotype threat on the attrition of women and minorities from science, math and engineering majors. *Social Psychology of Education, 15*(4), 427–448.

Becher, T. (1989). *Academic Tribes and Territories, First Edition*. Philadelphia, PA: Open University Press.

Becker, H. S. & Carper, J. (1956). The Elements of Identification with an Occupation. *American Sociological Review, 21*(3), 341–348.

Bordieu, P. (1986). The forms of social capital. In Richardson, J. G. (ed). *Handbook of Theory and Research for the Sociology of Education*. New York, NY: Greenwood.

Brightwing Talent Experts, (2012), Research on the Engineering Talent Shortage Says We Should Start Worrying, February 12, 2019, Retrieved on March 26, 2021 from https://gobrightwing.com/2019/02/12/research-engineering-talent-shortage/#!/.

Brown, A. S., (2009), What Engineering Shortage?, Retrieved on March 17, 2021 from file:///C:/Users/BETH-A~1/AppData/Local/Temp/What-engineering-shortage_summer2009_Brown.pdf.

Brunhaver, S. R., Korte, R. F., Barley, S. R., & Sheppard, S. D. (2017). Bridging the Gaps Between Engineering Education and Practice. In Freeman, R. B. & Salzman, H. (eds). *Engineering in a Global Economy, Conference Held September 26–27, 2011*. Chicago, IL: Chicago University Press.

CAE (2005), Task Force on the Future of Engineering: A Framework for Discussion, Canadian Academy of Engineering, December 2005, Retrieved on February 28, 2025 from https://www.cae-acg.ca/wp-content/uploads/2014/01/2005_Major%20Directions.pdf.

Carberry, A. R. & Baker, D. R. (2018). The impact of culture on engineering and engineering education. In *Cognition, Metacognition, and Culture in STEM Education: Learning, Teaching and Assessment*. Springer, 217–239.

Catalano, G. & Baillie, C. (2006). *Engineering, social justice and peace: A revolution of the heart*. Paper presented at the 2006 Annual Conference &Exposition.

Cech, E. A. (2013a). Culture of Disengagement in Engineering Education?. *Science, Technology, & Human Values, 39*(1), 42–72.

Cech, E. A. (2013b). The (mis) framing of social justice: Why ideologies of depoliticization and meritocracy hinder engineers' ability to think about social injustices. In *Engineering Education for Social Justice: Critical Explorations and Opportunities*. Springer, 67–84.

Crawford, S. (1996). The Making of the French Engineer. In Meiksins, P. & Smith, C. (eds). *Engineering Labour: Technical Workers in Comparative Perspective* (pp. 98–131). New York, NY: Verso.

Crosthwaite, C., (2021), Engineering futures 2035 engineering education programs, priorities & pedagogies. *Australian Council of Engineering Deans, Report*, Vol. No.

Daly, S. R., Yilmaz, S., Christian, J. L., Seifert, C. M., & Gonzalez, R. (2012). Design heuristics in engineering concept generation. *Journal of Engineering Education, 101*(4), 601–629.

De George, R. T. (1981). Ethical Responsibilities of Engineers in Large Organizations: The Pinto Case. *Business & Professional Ethics Journal, 1*(1), 1–14.

Duderstadt, J. J. (2008). *Engineering for A Changing World: A Roadmap to the Future of Engineering Practice, Research, and Education*. Ann Arbor, MI: University of Michigan.

Duderstadt, J. J. (2010a). Engineering for a Changing World. In Grasso, D. & Burkins, M. B. (eds). *Holistic Engineering Education: Beyond Technology* (pp. 17–36). New York, NY: Springer.

Duderstadt, J. J. (2010b). New University Paradigms for Technological Innovation. In Weber, L. & Duderstadt, J. J. (eds). *University Research for Innovation* (pp. 237–250). London: Economica.

Eddy, H. T. (1897). Address by the President Before the Society for the Promotion of Engineering Education. *Science*, *6*(144), 502–508.

Ellis, J. D. (2021). *Lessons learned while (maybe) educating optomechanical engineers*. Paper presented at the Optomechanics and Optical Alignment.

Employee Benefits, (2019), Why is employer training on the decline? November 4th, 2019, Retrieved on March 26, 2021 from https://employeebenefits.co.uk/why-is-employer-training-on-the-decline/.

Engineering Deans Canada, (2020), Call to Action – Canadian Engineering Grand Challenges Published, Retrieved on March 26, 2021 from https://engineeringdeans.ca/en/call-to-action-canadian-engineering-grand-challenges-published/.

Engineers Canada, (2012), The Engineering Labour Market in Canada: Projections to 2020, Final Report October 2012, Retrieved on March 26, 2021 from https://engineerscanada.ca/sites/default/files/w_Engineering_Labour_Market_in_Canada_oct_2012.pdf.

Engineers Canada, (2019), Recruitment, Retrieved on March 21, 2021 from https://engineerscanada.ca/diversity/women-in-engineering/recruitment.

Engineers Canada, (2020), About Engineers Canada, Retrieved on January 22, 2021 from https://engineerscanada.ca/about/about-engineers-canada.

Florman, S. C. (1994). *The Existential Pleasures of Engineering*. New York, NY: St. Martin's Griffin.

Fouad, N. A., Chang, W.-H., Wan, M., & Singh, R. (2017). Women's reasons for leaving the engineering field. *Frontiers in Psychology*, *8*(875).

Freeman, R. B., (2005), Does Globalization of the Scientific/Engineering Workforce Threaten U.S. Economic Leadership? National Bureau of Economic Research, Working Paper 11457, July 2005, Retrieved on March 17, 2021 from https://www.nber.org/papers/w11457.

Friedman, B. & Hendry, D. G. (2019). *Value Sensitive Design: Shaping Technology with Moral Imagination*. Cambridge, MA: MIT Press.

Gambetta, D. & Hertog, S. (2017). *Engineers of Jihad: The Curious Connection between Violent Extremism and Education*. Princeton, NJ: Princeton University Press.

Geiger, R. D. & Kapoor, R. R. (1995). Reengineering the Engineering Education. *Technology Management*, *1995*(1), 182–186.

Geisinger, B. & Raman, D. R. (2013). Why they leave: Understanding student attrition from engineering majors. *International Journal of Engineering Education*, *29*(4), 914.

Ginter, L. E. (1955). Report on the Evaluation of Engineering Education. *Journal of Engineering Education*, *1955*(September), 25–60.

Glover, I. & Kelly, M. P. (1987). *Engineers in Britain: A Sociological Study of the Engineering Dimension*. London, UK: Allen & Unwin.

Godfrey, E., Aubrey, T., & King, R. (2010). Who leaves and who stays? Retention and attrition in engineering education. *Engineering Education*, *5*(2), 26–40.

Goldberg, D. E. & Somerville, M. (2014). *A Whole New Engineer: The Coming Revolution in Engineering Education*. Douglas,MI: ThreeJoy Associates, Inc.

Grasso, D. & Martinelli, D. (2010). Holistic Engineering. In Grasso, D. & Burkins, M. B. (eds). *Holistic Engineering Education: Beyond Technology* (pp. 11–16). New York, NY: Springer.

Grinter, L. E. (1955). Report on the Evaluation of Engineering Education. *Journal of Engineering Education*, *1955*(September), 25–60.

Grinter, L. E. (2014). Engineering and Engineering Technology Education. *Journal of Engineering Technology*, *31*(2), 8–11.

Harris, R. (2023). *Creative Problem Solving: A Step-by-step Approach*. Taylor & Francis.

Harris, R. S. (1976). *A History of Higher Education in Canada 1663–1960*. University of Toronto Press.

Hayes, J. R. (1981). *The Complete Problem Solver*. Philadelphia, PA: The Franklin Institute.

Henstra, D. & McGowan, R. A. (2016). Millennials and public service: An exploratory analysis of graduate student career motivations and expectations. *Public Administration Quarterly, 490–516*.

Huang, C. & Sharif, N. (2016). Global technology leadership: The case of China. *Science and Public Policy*, *43*(1), 62–73.

Huber, L. & Shaw, G. (1992). Towards a New Studium Generale: Some Conclusions. *European Journal of Education, 27*(3), 285–301.

Hunsucker, J. L. & Loos, D. (1989). Transition management – An analysis of strategic considerations for effective implementation. *Engineering Management International, 5*(3), 167–178.

Hunt, J. (2015). Why do Women Leave Science and Engineering?. *ILR Review, 69*(1), 199–226.

Incropera, F. P. & DeWitt, D. P. (1990). *Introduction to Heat Transfer Second Edition*. New York,NY: John Wiley & Sons.

Institution of Engineers Australia. (1996). *Changing the Culture: Engineering Education into the Future, Review of Engineering Education Steering Committee*. Barton,A.C.T: Institution of Engineers Australia.

Janis, I. (1971). *Groupthink*. Boston, MA: Houghton Mifflin.

Jensen, K. & Cross, K. J. (2019). *Student perceptions of engineering stress culture*. Paper presented at the 2019 ASEE Annual Conference &Exposition.

Jensen, K. J., Mirabelli, J. F., Kunze, A. J., Romanchek, T. E., & Cross, K. J. (2023). Undergraduate student perceptions of stress and mental health in engineering culture. *International Journal of STEM Education, 10*(1), 30.

Jonassen, D. H. (1997). Instructional Design Model for Well-Structured and Ill-Structured Problem-Solving Learning Outcomes. *Educational Technology Research & Development, 45*(1), 65–95.

Joppen, L., (2012), hortage of engineers starting to impact industry, Stainless Steel World Web Article, May 16, 2020, Retrieved on March 26, 2021 from https://www.stainless-steel-world.net/webarticles/2020/05/16/shortage-of-engineers-starting-to-impact-industry.html.

Kanter, R. M. (1992). *Challenge of Organizational Change: How Companies Experience It and Leaders Guide It*. New York, NY: Simon and Schuster.

Kaufman, H. (2017). *The Limits of Organizational Change*. New York, NY: Routledge.

Kelman, H. C. & Hamilton, V. L. (1989). *Crimes of Obedience: Towards a Social Psychology of Authority and Responsibility*. New Haven, CT: Yale University Press.

Kennedy, J. M., Pinelli, T. E., Barclay, R. O., & Bishop, A. P. (1997). Distinguishing Engineers and Scientists – The Case for an Engineering Knowledge Community. In Pinelli, T. E., Barclay, R. O., Kennedy, J. M., & Bishop, A. P. (eds). *Knowledge Diffusion in the U. S. Aerospace Industry, Part A* (pp. 177–214). Greenwich, C: Ablex Publishing Corporation.

Kerr, V., Chan, A., Moore, E., & Weissling, L. (2025). *Towards an Understanding of the Contemporary Engineering Profession: Occupational Outcomes of Engineering Graduates in Canada by Gender, Race, and Location of Study, March 2025*. (Vol. University of Toronto, Ontario Society of Professional Engineers). Toronto, Ontario.

Kezar, A. (2009). Change in higher education: Not enough, or too much?. *Change: The Magazine of Higher Learning, 41*(6), 18–23.

Kezar, A. (2018). *How Colleges Change: Understanding, Leading, and Enacting Change*. New York, NY: Routledge.

Khaw, K. W., Alnoor, A., Al-Abrrow, H., Tiberius, V., Ganesan, Y. et al. (2023). Reactions towards organizational change: A systematic literature review. *Current Psychology, 42*(22), 19137–19160.

Koen, B. V. (2003). *Discussion of the Method: Conducting the Engineer's Approach to Problem Solving.* New York, NY: Oxford University Press.

Kotter, J. P. (1996). *Leading Change.* Boston, MA: Harvard Business Review Press.

Kotter, J. P. & Heskett, J. L. (1992). *Corporate Culture and Performance.* New York, NY: The Free Press.

Kunda, G. (2009). *Engineering Culture: Control and Commitment in a High-tech Corporation.* Philadelphia, PA: Temple University Press.

Layton, E. T. Jr. (1962). Veblen and the Engineers. *American Quarterly, 14*(1), 64–72.

Layton, E. T. Jr. (1986). *The Revolt of the Engineers: Social Responsibility and the American Engineering Profession.* Baltimore, MD: The Johns Hopkins University Press.

Lee, K.-F. (2018). *AI Superpowers: China, Silicon Valley, and the New World Order.* Boston, MA: Houghton Mifflin Harcourt.

Leydens, J. A. & Lucena, J. C. (2017). *Engineering Justice: Transforming Engineering Education and Practice.* John Wiley & Sons.

Lin, N. (2001). *Social Capital: A Theory of Social Structure and Action.* New York, NY: Cambridge University Press.

Lowell, B. L. & Salzman, H., (2007), Into the Eye of the Storm: Assessing the Evidence on Science and Engineering Education, Quality, and Workforce Demand, The Urban Institute, October 2007, Retrieved on March 17, 2021 from https://www.urban.org/sites/default/files/publication/46796/411562-Into-the-Eye-of-the-Storm.PDF.

Marra, R. M., Rodgers, K. A., Shen, D., & Bogue, B. (2012). Leaving engineering: A multi-year single institution study. *Journal of Engineering Education, 101*(1), 6–27.

Martin, S., Eckert, C., Pirtle, Z. G., Poznic, M., Schuelke-Leech, B.-A. et al. (2025). Methods as a form of engineering knowledge. *Design Science, 11.*

McGlone, T., Spain, J. W., & McGlone, V. (2011). Corporate Social Responsibility and the Millennials. *Journal of Education for Business, 86*(4), 195–200.

Meiksins, P. F. & Watson, J. M. (1989). Professional Autonomy and Organizational Constraint: The Case of Engineers. *The Sociological Quarterly, 30*(4), 561–585.

Meiskins, P. F. (1996). Engineers in the United States: A House Divided. In Meiksins, P. & Smith, C. (eds). *Engineering Labour: Technical Workers in Comparative Perspective* (pp. 61–97). New York, NY: Verso.

Milgram, D., (2011), How to Recruit Women and Girls to the Science, Technology, Engineering, and Math (STEM) Classroom, Retrieved on March 21, 2021 from https://www.iteea.org/File.aspx?id=137394&v=340d4cae.

Milliken, F. J., Morrison, E. W., & Hewlin, P. F. (2003). An exploratory study of employee silence: Issues that employees don't communicate upward and why. *Journal of Management Studies, 40*(6), 1453–1476.

Mitchell, J. E., Nyamapfene, A., Roach, K., & Tilley, E. (2021). Faculty wide curriculum reform: The integrated engineering programme. *European Journal of Engineering Education, 46*(1), 48–66.

Myers, M., (2016), Changing the perception of engineering in the community, Retrieved on March 21, 2021 from https://www.engineersaustralia.org.au/News/changing-perception-engineering-community.

National Academy of Engineering. (2004). *The Engineer of 2020: Visions of Engineering in the New Century.* Washington, DC: The National Academies Press.

National Academy of Engineering. (2005). *Educating the Engineer of 2020: Adapting Engineering Education to the New Century, National Academy of Engineering.* Washington, DC: The National Academies Press.

National Academy of Engineering, (2008), National Academy of Engineering Grand Challenges for Engineering, Retrieved on September 12, 2020 from http://www.engineeringchallenges.org/challenges.aspx.

National Science Board. (2007). *Moving Forward to Improve Engineering Education, November 2007.* Washington DC: National Science Board, National Science Foundation.

Ng, E. S. W., Schweitzer, L., & Lyons, S. T. (2010). New Generation, Great Expectations: A Field Study of the Millennial Generation. *Journal of Business and Psychology, 25*(2), 281–292.

Oakland, J. S. & Tanner, S. (2007). Successful Change Management. *Total Quality Management & Business Excellence*, *18*(1–2), 1–19.

Ogilvie, C. A. (2007). Moving Students from Simple to Complex Problem Solving. In Jonassen, D. H. (ed). *Learning to Solve Complex Scientific Problems* (pp. 159–186). New York, NY: Lawrence Erlbaum Associates, Taylor & Francis Group.

Olson, R. G. (2015). *Scientism and Technocracy in the Twentieth Century: The Legacy of Scientific Management*. Lenham, MD: Lexington Books.

Onar, S. C., Ustundag, A., Kadaifci, Ç., & Oztaysi, B. (2018). "The changing role of engineering education in industry 4.0 Era". In *Industry 4.0: Managing the Digital Transformation*. Springer, 137–151.

OSPE, (2015), Crisis in Ontario's Engineering Labour Market: Underemployment Among Ontario's Engineering-Degree Holders, Retrieved on March 7, 2021 from https://www.ospe.on.ca/public/documents/advocacy/2015-crisis-in-engineering-labour-market.pdf.

Paine, L. S. (2002). *Value Shift: Why Companies Must Merge Social and Financial Imperatives to Achieve Superior Performance*. New York, NY: McGraw Hill Professional.

Passino, K. M. (2009). Educating the humanitarian engineer. *Science and Engineering Ethics*, *15*, 577–600.

Paulsen, M. B. & Wells, C. T. (1998). Domain differences in the epistemological beliefs of college students. *Research in Higher Education*, *39*(4), 365–384.

Perrucci, R. (1971). Engineering: Professional servant of power. *American Behavioral Scientist*, *14*(4), 492–506.

Peterson, G. D. (1996). Engineering critera 2001: The ABET vision for change. *JoM*, *48*(9), 12–14.

Pilotte, M. K., (2013), *Engineering: Defining and Differentiating Its Unique Culture*.

Pulko, S. H. & Parikh, S. (2003). Teaching 'Soft' Skills to Engineers. *The International Journal of Electrical Engineering & Education*, *40*(4), 243–254.

Rand Corporation, (2004), The U.S. Scientific and Technical Workforce: Improving Data for Decisionmaking, Edited by Kelly, T., Butz, W., Carroll, S. J., Adamson, D. M., & Bloom, G., Retrieved on March 17, 2021 from https://www.rand.org/pubs/conf_proceedings/CF194.html.

Randstad, (2024), Smart and talented women: there's a place for you in engineering, February 29th, 2029, Retrieved on May 5, 2025 from https://www.randstad.ca/job-seeker/career-resources/career-development/smart-talented-women-place-engineering/.

Rhine, S., Harrington, R., & Starr, C. (2018). *How Students Think When Doing Algebra*. Charlotte, NC: Information Age Publishing.

Rice, P. (1994). *An Engineer Imagines*. London, UK: Batsford.

Rittel, H. W. J. & Webber, M. M. (1973). Dilemmas in a general theory of planning. *Policy Sciences*, *4*(2), 155–169.

Ritti, R. R. (1968). Work Goals of Scientists and Engineers. *Industrial Relations*, *7*(2), 118–131.

Ritti, R. R. (1971). *The Engineer in the Industrial Corporation*. New York, NY: Columbia University Press.

Royal Academy of Engineering. (2007a). *Educating Engineers for the 21st Century*. London, UK: Royal Academy of Engineering.

Royal Academy of Engineering, (2007b), Public Attitudes to and Perceptions of Engineering and Engineers 2007: A study commissioned by The Royal Academy of Engineering and the Engineering and Technology Board, Retrieved on March 21, 2021 from https://www.raeng.org.uk/publications/other/public-attitude-perceptions-engineering-engineers.

Royal Academy of Engineering, (2019), Global Grand Challenges Summit 2019, Retrieved on September 12, 2020 from https://www.raeng.org.uk/global/international-partnerships/international-policy/welcome.

Sallai, G. M., Bahnson, M., Shanachilubwa, K., & Berdanier, C. G. P. (2023). Persistence at what cost? How graduate engineering students consider the costs of persistence within attrition considerations. *Journal of Engineering Education*, *112*(3), 613–633.

Schein, E. H. (1988). *Organizational culture* (Working paper 2088–88). Retrieved from Cambridge, MA: http://www.jstor.org/stable/2393715

Schmidt, J. & Müller, R. (2023). Discipline differences in mental models: How mechanical engineers and automation engineers evaluate machine processes. *Human Factors and Ergonomics in Manufacturing & Service Industries, 33*(6), 521–536.

Schuelke-Leech, B.-A., (2011), *Strangers in a Strange Land: Industry and Academic Researchers*, PhD *Dissertation*, Public Administration and Policy, University of Georgia (PhD).

Schuelke-Leech, B.-A. (2020). *The Place of Wicked Problems in Engineering Problem Solving: A Proposed Taxonomy*. Paper presented at the 2020 IEEE International Symposium on Technology and Society, November 12–15, 2020.

Schuelke-Leech, B.-A. & Leech, T. C. (2018). Innovation in the American Era of Industrial Pre-eminence: The Interaction of Policy, Finance, and Human Capital. *Journal of Policy History, 30*(4) October, 727–753.

Schuelke-Leech, B.-A., Leech, T. C., Barry, B., & Jordan-Mattingly, S. R. (2018). *Ethical Dilemmas for Engineers in the Development of Autonomous Systems*. Paper presented at the IEEE International Symposium on Technology and Society, November 13th-14th, Washington DC. http://www.sciencedirect.com/science/article/pii/S1364032114001488.

Seely, B. E. (1999). The other re-engineering of engineering education, 1900–1965. *Journal of Engineering Education, 88*(3), 285–294.

Shahidul, M. I. (2020). Engineering education for achieving sustainable development goals by 2030: Revealing the paths for challenging climate change and Covid 19. *Sustainable Development, 5*(2), 403–410.

Shayesteh, H. (2025). *Integrating the United Nations Sustainable Development Goals into Engineering Education: A Practical Framework for Developing Future Leaders in Sustainability*. Paper presented at the 2025 IEEE Global Engineering Education Conference (EDUCON).

Sheppard, S. D., Macatangay, K., Colby, A., & Sullivan, W. M. (2009). *Educating Engineers: Designing for the Future of the Field*. San Francisco, CA: Jossey-Bass.

Silbey, S. S. (2016a). Why do so many women who study engineering leave the field. *Harvard Business Review, 1–2*.

Silbey, S. S. (2016b). Why do so many women who study engineering leave the field. *Harvard Business Review, 23*(1–8).

Snow, C. P. (1954). *The New Men*. London, UK: Macmillan.

Sola, E., Hoekstra, R., Fiore, S., & McCauley, P. (2017). An Investigation of the State of Creativity and Critical Thinking in Engineering Undergraduates. *Creative Education, 8*(09), 1495.

Solow, R. M. (1956). A contribution to the theory of economic growth. *The Quarterly Journal of Economics, 70*(1), 65–94.

Solow, R. M. (1957). Technical change and the aggregate production function. *The Review of Economics and Statistics, 39*(3), 312–320.

Sorby, S., Fortenberry, N. L., & Bertoline, G. (2021). Stuck in 1955, Engineering Education Needs a Revolution. *Issues in Science and Technology, 2021*(September 13), 1–4.

Strobel, J., Hess, J., Pan, R., & Wachter Morris, C. A. (2013). Empathy and care within engineering: Qualitative perspectives from engineering faculty and practicing engineers. *Engineering Studies, 5*(2), 137–159.

Tajfel, H. (1974). Social identity and intergroup behaviour. *Social Science Information, 13*(2), 65–93.

Teitelbaum, M. S., (2014), The Myth of the Science and Engineering Shortage, The Atlantic, March 19, 2014, Retrieved on March 19, 2021 from https://www.theatlantic.com/education/archive/2014/03/the-myth-of-the-science-and-engineering-shortage/284359/.

Turner, J. C. (1981). Towards a cognitive redefinition of the social group. In Tajfel, H. (ed). *Social Identity and Intergroup Relations* (pp. 15–40). New York, NY: Cambridge University Press.

U.S. Census, (2014), Where do College Graduates Work? A Special Focus on Science, Technology, Engineering, and Math, Retrieved on January 31, 2024 from https://www.census.gov/dataviz/visualiza tions/stem/stem-html/.

UNESCO. (2010). *Engineering: Issues, Challenges, and Opportunities for Development.* Paris, France: United Nations Educational, Scientific and Cultural Organization.

US News & World, (2025), Franklin W. Olin College of Engineering, Retrieved on May 20, 2025 from https://www.usnews.com/best-colleges/franklin-w-olin-college-of-engineering-39463.

Veblen, T. (1921). *The Engineers and the Price System.* New York, NY: B. W. Huebsch.

Vincenti, W. G. (1990). *What Engineers Know and How They Know It: Analytical Studies from Aeronautical History.* Baltimore, MD: Johns Hopkins University Press.

von Blottnitz, H., Case, J. M., & Fraser, D. M. (2015). Sustainable development at the core of undergraduate engineering curriculum reform: A new introductory course in chemical engineering. *Journal of Cleaner Production, 106,* 300–307.

Walker, E. A. (1971). The major problems facing engineering education. *Proceedings of the IEEE, 59*(6), 823–828.

Weick, K. E. & Quinn, R. E. (1999). Organizational change and development. *Annual Review of Psychology, 50*(1), 361–386.

Weissman, J., (2013), The Ph.D Bust: America's Awful Market for Young Scientists – in 7 Charts: Perhaps it's time to start talking about a STEM surplus?, The Atlantic, February 20, 2013, Retrieved on March 7, 2021 from https://www.theatlantic.com/business/archive/2013/02/the-phd-bust-americas-awful-market-for-young-scientists-in-7-charts/273339/.

Wells, M., (2019), Realizing the potential of Canadian engineers to change the world, December 31, 2019, Retrieved on September 28, 2020 from https://www.thestar.com/opinion/contributors/2019/12/31/re alizing-the-potential-of-canadian-engineers-to-change-the-world.html.

Williamson, J. M., Lounsbury, J. W., & Han, L. D. (2013). Key personality traits of engineers for innovation and technology development. *Journal of Engineering and Technology Management, 30*(2), 157–168.

Wisnioski, M. (2012). *Engineers for Change: Competing Visions of Technology in 1960s America.* Cambridge, MA: The MIT Press.

Wood, P. K. (1983). Inquiring Systems and Problem Structure: Implications for Cognitive Development. *Human Development, 26*(249–265).

Woods, D. R. (1975). Teaching Problem Solving Skills. *Engineering Education, 66*(3), 238–243.

Woods, D. R. (1983). Introducing explicit training in problem solving into our courses. *Higher Education Research and Development, 2*(1), 79–102.

Woods, D. R. (1994). *Problem-based Learning: How to Gain the Most from PBL.* Waterdown, Ont: Woods.

Woods, D. R. (2013). Problem-oriented learning, problem-based learning, problem-based synthesis, process oriented guided inquiry learning, peer-led team learning, model-eliciting activities, and project-based learning: What is best for you?. *Industrial & Engineering Chemistry Research, 53*(13), 5337–5354.

Yang, S.-B. & Guy, M. E. (2006). GenXers versus boomers: Work motivators and management implications. *Public Performance & Management Review, 29*(3), 267–284.

Zussman, R. (1985). *Mechanics of the Middle Class: Work and Politics among American Engineers.* Berkeley, CA: University of California Press.

Index

https://doi.org/10.1515/9783111563046-010

www.ingramcontent.com/pod-product-compliance
Lightning Source LLC
Chambersburg PA
CBHW080926220326
41598CB00034B/5688